SEALAB

America's Forgotten Quest
to Live and Work on the Ocean Floor

BEN HELLWARTH

SIMON & SCHUSTER
New York   London   Toronto   Sydney   New Delhi

SIMON & SCHUSTER
1230 Avenue of the Americas
New York, NY 10020

First Simon & Schuster hardcover edition January 2012

SIMON & SCHUSTER and colophon are
registered trademarks of Simon & Schuster, Inc.

For information about special discounts for bulk purchases,
please contact Simon & Schuster Special Sales at
1–866–506–1949 or business@simonandschuster.com.

The Simon & Schuster Speakers Bureau can bring authors
to your live event. For more information or to book an event,
contact the Simon & Schuster Speakers Bureau at
1–866–248–3049 or visit our website at www.simonspeakers.com.

Design by Esther Paradelo

Manufactured in the United States of America

1   3   5   7   9   10   8   6   4   2

Library of Congress Cataloging-in-Publication Data
Hellwarth, Ben.
Sealab : America's forgotten quest to live and work on the
ocean floor / Ben Hellwarth.—1st Simon & Schuster hardcover ed.
p.   cm.
Includes bibliographical references and index.
1. Project Sealab. 2. Manned undersea research stations—History—20th century.
3. Deep diving—History—20th century. 4. Divers—United States—Biography.
5. United States. Navy—Biography. I. Title.
GC66.H45 2012
551.460973—dc22   2011015725
ISBN 978–0–7432–4745–0
ISBN 978–1–4391–8042–6 (ebook)

Insert Photo Credits: U.S. Navy photographs: 1, 3–9, 11–18, 20–23, 28;
Collection of George Bond Jr.: 2, 19; Robert Sténuit photo: 10;
The Link Collections, Binghamton University: 24; Alain Tocco/Comex: 25;
Collection of Drew Michel: 26, 27.

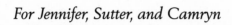

For Jennifer, Sutter, and Camryn

CONTENTS

SEALAB

1

A DEEP ESCAPE

Three hundred feet down, deeper than most divers were equipped to go, the World War II–era submarine *Archerfish* prepared to release two passengers into a dim sea with little more than the breath they held in their lungs. A spate of foul weather and strong currents had complicated some trial runs, but now, early on the first day of October 1959, the sun came up with just a few wispy clouds on the horizon. In calm water, the sub set out from the U.S. Navy base at Key West, heading southwest into the Gulf of Mexico. The mission's unusual aim was to prove that a trapped submariner could reach the distant surface on his own—and live to tell about it. No escape quite like this had ever been made, and not everyone was convinced it was a good idea to try. But one of the passengers, Commander George Bond, a Navy doctor, had argued for attempting to make the escape. As was typical of him, he had also argued that he should be the one to do it.

George Bond was forty-three, though his soft features and jowls made him look older. He was a couple of inches over six feet, barrel-chested and thickly built, with close-cropped salt-and-pepper hair and owlish dark eyebrows. Bond had joined the Navy just a few years before but he already had a reputation for being a maverick with a fondness for showboating. He preferred not to be photographed without his pipe. Daring, gregarious, kind-hearted, even his detractors found him difficult not to like. He often lapsed into the disarming Appalachian brogue of the clients he had served not so long ago as a country doctor. His resonant baritone, noticeably seasoned somewhere south of the Mason-Dixon line, had the soothing, unhurried intonation of a storyteller sent from heaven. It was a voice that had served him well as a lay preacher back home in the Blue Ridge Mountains.

Dr. Bond had recently become head of the Medical Research Laboratory at the U.S. Naval Submarine Base at New London, Connecticut. Most American submariners got their specialized schooling at the base, which was named for New London but was actually in neighboring Groton, on the eastern bank of the Thames River. Scientists employed at the base's medical lab concerned themselves with a variety of physiological puzzles such as those related to submarine escape. In the early years of the U.S. submarine service, at the beginning of the twentieth century, there was no procedure, and virtually no hope, for submariners trapped in downed subs, even in relatively shallow water. Safety features in the early boats were primitive at best, and gradual asphyxiation or drowning were often the survivors' macabre options. But after three submarines sank in the 1920s, newer American subs were fitted with the equivalent of emergency exits, known as escape trunks. By the late 1930s a drum-shaped pod called the McCann Rescue Chamber was devised that could be lowered like an elevator from a surface ship and, with the aid of divers, locked on to the sub's escape hatch. But if the chamber couldn't reach a sunken submarine, the sailors inside needed something like an underwater version of a parachute to escape on their own.

Several methods evolved to enable a sailor to bail out of his doomed vessel and get himself to the surface. From the *Archerfish* Bond wanted to test one that had been borrowed a few years earlier from the British called "buoyant-assisted free ascent," or just "buoyant ascent." It was also known less formally as "blow and go." Holding your breath underwater might feel like holding on to life, but not if you're swimming to the surface after inhaling a lungful of highly compressed air. The final breath taken in the escape trunk just before leaving the *Archerfish* would expand from something like five quarts to nearly thirteen gallons en route to the surface, a phenomenon explained by Boyle's law, named for the seventeenth-century Irish physicist Robert Boyle. Assuming temperature remains the same, as pressure increases, volume decreases; as pressure decreases, volume increases. So it is that a lungful of compressed air inhaled at any depth will expand tremendously as a person approaches the surface and the surrounding water pressure gradually drops. A submarine escapee, therefore, had to learn to exhale forcefully and consistently, like an exuberant tuba player, all the way to the surface—a strange sensation that took some getting used to.

The gravest danger during an escape to the surface is from an arterial gas embolism, or "getting embolized." If not adequately exhaled, the expanding air will backfire into a pulmonary vein, sending emboli in the form of tiny bubbles to the brain. The consequences could include seizures, slurred speech, loss of vision, loss of muscle coordination, unconsciousness, even instant death. The human lungs aren't wired to the autonomic nervous system, so even if perilously overinflated they won't set off any synaptic sirens and zap the brain with pain signals the way other body parts do, like a finger on a hot stove. From three hundred feet down if you didn't blow and go just right, you could kill yourself as surely as if you stayed on a stranded submarine.

Prior to Bond's attempted escape from the *Archerfish*, about the deepest experience that he or anyone else had with the blow and go technique came from inside the escape training tank. At New London the tank towered above everything else on the wharf and was the main attraction on base tours. Esther Williams, "Hollywood's Mermaid," once came to have a look. For anyone going through submarine school, the tank held the promise and fear of a rite of passage. Eleven stories high, it resembled a silo that might have been transplanted from a farm in the surrounding Connecticut countryside. But it was taller than a silo and instead of a simple dome, it was topped with an octagonal cupola, which gave it the look of a misplaced airport control tower.

Built into the sides of the tank were several airlocks similar to actual submarine escape trunks. A sailor could pass from the airlock into the tank water as if making a real escape, and practice reaching the surface unharmed. The deepest entry point was near the bottom of the tank, at almost 120 feet. Most trainees made their required mock escapes from a lock at fifty feet, but the confluence of physics and physiology is such that a blow and go escape begun at that depth could be every bit as life-threatening as one from the bottom of the tank. During training sessions half a dozen instructors hovered in the tank water, trailing each trainee like pilot fish. A doctor was on duty, too, in case of an accident, and Dr. Bond became a proficient escape artist himself.

Sailors often asked whether it would be possible to blow and go from, say, 240 feet, the depth at which the USS *Squalus* sank during a test dive off the coast of New Hampshire in 1939. The best answer that Dr. Bond or anyone could give was that, yes, in theory, you should be able to safely blow and go in the open sea from a depth of at least three hundred feet or so—but no one had ever done it. Bond decided it was time to put theory to the test.

After receiving the approval of the rear admiral in command of the submarine force, U.S. Atlantic Fleet, Bond planned to stage a series of progressively deeper open-sea escapes, culminating with one from more than three hundred feet. Bond needed a buddy to make the escape with him and chose Cyril Tuckfield, a chief engineman recently reassigned to a submarine out of Key West after working for several years as a training tank instructor at New London. "Tuck," as the good-humored sailor was known, was thirty-eight and a World War II submarine veteran who was also trained as a diver. He had an infectious, face-crinkling grin and was unique among the enlisted men for his aversion to using foul language. Tuck's father once heard him cuss during a sandlot baseball game, gave his boy hell for it, and young Tuck never swore again.

As a former training tank instructor, Tuck understood the peculiar dangers of this deep escape as well as anyone but was delighted to be asked to accompany his friend Dr. Bond. The fact that a medical officer like Bond would put himself on the line was one of the things that endeared him to enlisted men like Tuckfield. Indeed any number of sailors would have gladly volunteered to join Bond for the escape from the *Archerfish*.

Once the two men locked themselves into the escape trunk, with the sub at a depth of just over three hundred feet, they would have at least as much to worry about as someone parachuting from an airplane. Their procedures and timing would be critical. Pressure inside the trunk would approach 150 pounds per square inch, about ten times as great as inside the submarine, where the air is kept close to 14.7 pounds per square inch, the same atmospheric pressure as at sea level. The elevated pressure would enable them to make the transition from the submarine's escape trunk into the sea. That pressure would also expose them to physiological dangers including one commonly known by deep sea divers as "the bends." Ugsome aches and pains, paralysis, even death await the diver who ascends too quickly to the surface—but only after having spent enough time at depth for gases to seep into the blood and tissues. To avoid getting "bent," as divers say, Bond and Tuck knew they would have strict time limits at each step of their escape. They would need about five minutes to prepare the trunk and flood it shoulder-high with seawater. Then they'd open an air line inside the trunk and within another minute, as air noisily whooshed in, the pressure inside the remaining air space would match the pressure outside. At that point, according to prior calculations, they would have a maximum of three minutes and fifteen seconds to reach the surface. Any longer than that and they would risk getting bent.

Throughout this process Bond and Tuck would also have to be alert to the symptoms of breathing nitrogen under pressure. Four-fifths of air is nitrogen, but when inhaled at high pressures the omnipresent gas could instantly put a diver into a giddy haze similar to drunkenness, and that could cause him to make fatal mistakes. Largely because their time at the full depth of 300 feet would be brief, Bond and Tuck were less likely to be dogged by the mental haze of nitrogen or the bends. Getting embolized posed the greatest danger in making this deep escape. Once they began their ascent, sufficient blowing, all the way to the surface, was the only thing that would keep them from getting embolized by tiny, backfiring bubbles from their lungs. In the training tank, to avoid potentially deadly cases of embolisms, instructors would routinely jab a trainee in the gut to make sure he was blowing forcefully enough, and if anyone's ascent ever looked problematic, they would pull the ascending man into one of several air pockets built into the wall of the tank. In the open sea, of course, safety pockets did not exist.

On board the *Archerfish* that morning, neither man ate breakfast. Bond thought it would be wise to keep their stomachs empty, in case of an accident. They drank only black coffee. Bond puffed his ubiquitous pipe and Tuck smoked a cigarette as they mulled over the plan for their impending escape. It would have been like Bond, a consummate raconteur, to recall his memorable introduction to the hazards of making an escape. It had come five years earlier, at the Pearl Harbor submarine base, where Dr. Bond was first assigned soon after completing his Navy schooling for submarine duty and deep-sea diving. At Pearl Harbor Bond spent much of his time around the escape training tank, a replica of the tank at New London that was built a couple of years later and survived the infamous Japanese attack.

There were no guarantees with escape training, as Bond learned one Friday in the fall of 1954 when he met a fellow Navy doctor named Charles Aquadro. Charlie Aquadro, whose very name suggested a love of the life aquatic, was the medical officer with an Underwater Demolition Team, the progenitor of the Navy SEALs. Aquadro was an experienced diver, fourteen years younger than Bond, lean and athletic, raised and educated in Kentucky and Tennessee. He was polite but playful, with a disarmingly gentle demeanor and a ready smile—a boyish Southern gentleman. The two doctors hit it off immediately, at least in part because of their common Southern roots.

Aquadro's UDT group was soon scheduled to make practice escapes from the bottom of the training tank at 120 feet. Bond watched the UDT drill as Charlie Aquadro took his turn, blowing a steady stream of bubbles, looking good all the way up. Aquadro reached the surface, climbed out of the water, stood up on the deck around the top of the tank, said a few words, seemed to be fine, and then collapsed, unconscious. Aquadro had somehow embolized himself. As Dr. Bond knew from his recent training, the surest remedy for quashing menacing bubbles, the cause of the embolism, is to put the ailing person back under pressure.

The group lugged Aquadro's limp body into the pressure chamber that had been installed for just such emergencies. Bond stayed inside with Aquadro. Those outside sealed the hatch and flooded the chamber with a whoosh of compressed air to raise the pressure to that at 165 feet, a standard treatment protocol. Aquadro regained consciousness, but couldn't see a thing. You could never be sure where a bubble blockage would occur or what effect it would have.

Aquadro's sight gradually returned. As the hours passed, the pressure inside the chamber was eased back to surface level. They were closing in on the surface when Aquadro went into severe convulsions, then stopped breathing. Those working outside the chamber cranked up the pressure again, hoping to quash the bubbles that were clearly short-circuiting Aquadro's brain. Temperature rises when a gas is compressed, so the air inside the little chamber was as hot as a desert summer. Bond sweated it out with Aquadro on the hyperbaric roller coaster for almost forty hours. By Sunday morning, the two men finally staggered out of the chamber, their camaraderie bolstered by the wild ride.

Their paths would cross many times. In October 1959 Lieutenant Commander Aquadro was the submarine squadron medical officer at Key West and among the safety divers assigned to watch as Bond and Tuck attempted their deep escape. Aquadro and the rest of the support crew sailed to the designated escape site, about fifteen miles southwest of Key West, on the USS *Penguin*, a submarine rescue ship equipped with a pressure chamber.

At nine o'clock in the morning, after a three-hour ride, the *Archerfish* settled near the bottom at 322 feet. That put the outboard door of the escape trunk at 302 feet, deep enough to prove the point—if it could be proved. A messenger buoy was released to let the *Penguin* crew know where Bond and Tuck would appear, barring any significant currents or catastrophes. The two men approached the forward torpedo room and the ladder leading up into the escape trunk. Tuck gestured for Bond to go first, part of a tongue-in-cheek ritual the two friends developed at the New London training tank.

"After you, Courageous Leader," Tuck said.

"Thank you, Brave Follower," Bond replied.

Each wore only shorts, a diving mask, and a Mae West—an inflatable life vest nicknamed by astute airmen who noticed that it gave them a silhouette similar to that of the buxom screen legend. Bond stashed his pipe, tobacco pouch, and a letter from his wife in a pocket of his Mae West. He and Tuck passed through the open hatch in the floor of the escape trunk, as if crawling up into an attic the size of a broom closet. They sealed the hatch behind them. Tuck then opened the valve that allowed seawater to pour in.

At sixty-eight degrees Fahrenheit, the water felt chilly as it rose around them, a palpable reminder that this escape would be for real. In the controlled environment of the training tank, the water was kept at a balmy ninety-two degrees. For several minutes the water noisily spilled in until it reached their chins. Then Bond shut the vent leading to the torpedo room. Tuck opened a valve and high-pressure air whooshed in, warming the remaining air pocket around their heads. The pressure rose fast, as planned. Bond was not surprised to feel pain in his eardrums and instinctively cleared his ears, as divers do, to spare his eardrums from sharp pain and possibly rupture as the pressure around his body rose. To clear your ears, you clamp your nose shut, close your mouth, keep the back of your throat open, and try to exhale with something like the force of a suppressed sneeze. The objective is to create internal pressure on the eardrum to match the mounting external pressure. Once the pressure is equalized, there's no net effect on the eardrum, no painful push from either side. After half a minute, Tuck turned off the air and the trunk fell silent. Now they were breathing under pressure equal to the water pressure outside, about 150 pounds per square inch. From that moment, the clock was ticking. To test for signs of a nitrogen-induced mental haze, Bond quizzed himself on his telephone number, his middle name, and the names of his kids—*Gail, George Junior, Judy, David.* Tuck seemed to be doing fine, too.

With the pressure equalized between the flooded compartment and the seawater outside, there was no longer any great resistance on the trunk's escape hatch. They inflated their Mae Wests and Bond pushed open the escape hatch, which was about waist high in the side of the trunk. The Courageous Leader took a last breath from the remaining air pocket. His Brave Follower did the same. Each man's lungful of air was now a time bomb, a massive gas embolism waiting to happen. As soon as they passed through the hatch into the darkened sea outside, they began to exhale vigorously.

Tuck grabbed hold of Bond's life vest belt so that they would stay together. In their inflated Mae Wests, they ascended briskly, like helium balloons set free on a breezy day. A valve on the Mae West allowed the expanding air to whistle out so the life jacket wouldn't explode on the way up. Like their lungs, their life jackets started out with ten times as much air as they would be able to hold on the surface. Bond raised his left arm, like Superman in flight, to help steer them in a straight line to the surface. They were rising at about six feet per second, in a great hail of bubbles.

Bond could feel his Mae West ballooning as the water pressure decreased. He worked his legs and arms slightly, checking for signs of the bends as he and Tuck made their swift ascent. They blew forcefully and steadily. The trick was to exhale enough to avoid embolizing themselves as Charlie Aquadro had done, but not so much that they ran out of breath before surfacing. They were used to the strange sensation of continuous blowing and from three hundred feet down, more than twice the depth of the training tank, that sensation was stranger still. Bond felt as though their ascent was taking an awfully long time, but he figured that was just because he had never risen through three hundred feet of water. Fortunately the current was light. At 180 feet the water turned noticeably warmer as they passed through a thermocline. Visibility improved and Bond could make out the underside of the waiting *Penguin*.

With about a hundred feet of water still between them and the surface, Tuck sensed that he had nothing left to blow. He had been exhaling like mad for more than half a minute. Experience had taught Tuck not to panic, and within another few feet the air in his lungs swelled anew and he kept on blowing. The final thirty feet or so, which should take less than ten seconds to cover, could be the most dangerous of the entire run. In those last thirty feet the volume of air remaining in the lungs would double.

Fifty-three seconds after Bond let go of the *Archerfish*, he and Tuck broke the surface. They gratefully gulped the fresh sea level atmosphere at 14.7 pounds per square inch. Of course as Charlie Aquadro's accident at Pearl Harbor demonstrated, Bond and Tuck were not necessarily all right. They might still be struck by an embolism or possibly a case of the bends, which could take hours to manifest itself. Doctors watched carefully as the two escape artists were plucked from still blue water and raised on a diving platform to the *Penguin* deck. Still dripping wet, Bond produced his pipe, smiled, and lit up as photographers snapped his picture.

George Bond and Cyril Tuckfield received widespread praise for their daring escape. They were awarded the Legion of Merit, and their feat was noted in *Time* magazine, *The New York Times*, and their hometown newspapers. Bond had a lengthy first-person account published in *True*, a reasonably respectable men's magazine. He was also a guest on NBC's *Today* show and on *Conquest*, a new Sunday evening science program on CBS.

This was not the first time that the press had taken notice of Bond. A year before the submarine escape, in November 1958, Bond had made himself the lead subject in a series of tests to analyze possible causes of Navy pilot deaths from carrier-based jets that crashed in the sea. Using the blow and go method to escape from a sinking aircraft gave these jet escape trials something in common with submarine escapes. Dressed in full pilot gear, Bond strapped himself into a jet cockpit that had been affixed to the afterdeck of a submarine. With the sub headed down at sink rates of up to ten feet per second, Bond extricated himself, took a last breath, and swam for the surface, blowing all the way. A lengthy story in the New London newspaper, the kind of publicity that irked some of Bond's Navy colleagues, was headlined: "Cmdr. Bond Escapes from 'Aircraft' Underwater to Come Up with Answers." Bond was already in his early forties and until now he had not looked like the obvious man to come up with answers like these.

Prior to joining the Navy, Bond had developed no special interest in either ships or the world underwater. He never acted like someone eager to sign up for any kind of military service, even when so many young men of his generation were marching off to fight in the Second World War. Dr. Bond was more poet than soldier, and for a time it appeared he might be more poet than doctor. The son of a well-to-do lumberman, George Foote Bond was born on November 14, 1915, in Willoughby, Ohio. He spent much of his privileged childhood in DeLand, Florida, in his family's magnificent home filled with Stickley furniture near the offices of the Bond Lumber Company. George was not quite ten years old when his father died late in the summer of 1925, but the family could still afford a nanny, summer camps, and private schools. As a teenager at Mercersburg Academy in Pennsylvania, Bond worked on the school's monthly literary magazine and became known as the class poet. He took up pipe smoking as a teen and developed what would become a lifelong fondness for bourbon. His nickname would not have shocked his Navy bosses. His schoolmates called him "Rebel."

After graduating from Mercersburg in 1933, a few years into the Great Depression, Bond enrolled at the University of Florida. His course list revealed his divided interest between letters and sciences, though he generally did better in classes like Old English and Imaginative Writing than in General Chemistry and Histology. His literary side won out, at least initially. Bond earned his bachelor of arts degree in early 1939 and went straight into a University of Florida graduate program for a master's degree. He studied English and for his thesis wrote about the Appalachian dialect spoken around Bat Cave, a hamlet in the Blue Ridge Mountains about ten miles from the family's summer home at Chestnut Gap, near Hendersonville in western North Carolina. Bond had gotten to know the speck of a town—named for a nearby cavern that was a seasonal home to migrating bats—when he went to summer camps in the area after his father died.

In those formative years George, his older sister, and brother had continued to live with their mother, Louise Foote Bond, a stern, college-educated woman, as the family moved between DeLand and the more modest home at Chestnut Gap. Bond loved the rocky and wooded backroads around Bat Cave and neighboring hamlets like Chimney Rock and Bear Wallow. He spent many lazy days hiking, horseback riding, and fishing with his two best friends, Lonnie Hill and Homer Lafayette "Fate" Heydock. Both were products of far less privilege and formal education, but that made no difference to Bond. Fate himself served as one of Bond's fifteen master's thesis subjects and helped Bond gather documentary material. Bond made notes on distinctive words, pronunciations, and such Appalachian pearls as "I'll take no back-sass offen you" and "My popskull mash went blinky on me."

Bond was inspired by the simple rustic ways of the mountain people and their independence. As a kid he once lived with Ben and Dooge Conner for the better part of a month at their ramshackle place on Bear Branch, a little tributary near the head of the Broad River. Ben Conner was quiet and removed, but Dooge, his wife, was wiry, snaggle-toothed, jubilant, and gregarious. It was Dooge—pronounced DOO-jee—who early on had said something like: "George, why don't you go to school, make a medical doctor, and come back here? We've never had nary one, and we need one bad."

After Bond finished the English thesis and got his master's he went to work as a laborer during the construction of the scenic Blue Ridge Parkway, no doubt pondering his future plans. By then he had married Marjorie Barrino, whom he had met in the summer of 1935 while she was on a church camping trip up around Bat Cave. She was a brunette with a cheerful face, two years older than Bond, and in his six-foot-plus company appeared more petite than she was. "Margit," as Bond took to calling her, was from nearby Marshville and a family of more ordinary means than the Bonds. In an act of social unorthodoxy, George and Marjorie eloped during the summer of 1938 while in New York City.

Bond decided he should go down the professional trail that Dooge had urged him to follow. He knew firsthand that there was a real need for a doctor in the mountains. As a boy he had seen two people die for lack of prompt medical attention. With his sights set on medical school, Bond enrolled at the University of North Carolina at Chapel Hill in early 1940 and spent the next fifteen months shoring up his science background— and got a good ribbing from fellow students for riding a horse to campus.

In the fall of 1942, nine months after the Japanese attack on Pearl Harbor, Bond started medical school at McGill University in Montreal. Marjorie and their firstborn child soon followed. Bond applied for a reserve commission in the Army's Medical Administration Corps and was assigned to Ready Reserve, subject to active-duty call on fifteen days' notice. That was as close to the wartime action as he got. At McGill he initially made average grades but by his fourth year had moved to the upper fifth of his class, ranked eighteenth out of ninety-eight. Bond found that he liked practicing medicine more than pondering it in class, and always seemed to have a penchant for grand plans. At one point during surgical training he complained about a lack of practical experience and brashly prepared an outline for overhauling the entire system of medical education "from grammar school up."

In 1945, Bond graduated from McGill and embarked on a year-long internship back in North Carolina, at Charlotte Memorial Hospital. Well-paid city jobs beckoned but money was never a great motivator for Bond. He cared little for fine furnishings or the kind of lavishly appointed houseboat his father once owned. He and Marjorie, along with their three young children, moved into a log cabin in Bat Cave and Dr. Bond, then thirty-one years old, went to work. Making house calls came with the territory, although Bond, ever the philologist, preferred to call his service "family practice in the home." He drove a surplus Willys jeep, bouncing and weaving over rocky roads that led to the homes of some five thousand souls squirreled away over five hundred square miles of Hickory Nut Gorge. During twenty-hour workdays Bond might easily cover two hundred miles.

As Bond would say, he often found himself trying to bring modern medicine into eighteenth-century settings. Not everyone welcomed his remedies. On one memorable occasion up at Reedy Patch, Granny Hill (no relation to Bond's pal Lonnie Hill) saw this young doctor fixing to stick her kin with a needle to inject penicillin and tetanus antitoxin and she aimed the barrel of a 12-gauge shotgun at him. Bond carried a .38 in his medical bag and brandished it while seeking to ease the old woman's fears. She put down the gun and Bond gave the injections.

Bond learned to perform delicate procedures in remote shanties, swatting houseflies by the light of kerosene lanterns. Hardly a baby was born in Hickory Nut Gorge that Bond did not deliver. Unorthodox methods were an essential part of his practice, in part because the nearest hospital was some twenty miles away. Sometimes, when fatigue and despair came knocking, Bond believed it didn't hurt to invoke the power of prayer. His patients invariably urged him to do so. Although religion had not been central to Bond's upbringing, he became an active Episcopalian and a great student of the Bible, much as he relished studying Chaucer and Shakespeare in school. In Bat Cave Bond served as a lay preacher at the Church of the Transfiguration, a little stone-walled sanctuary a stone's throw from the rocky Broad River. On many Sundays he would fill the little chapel with his sonorous baritone.

In the early months of 1947, an outbreak of influenza swept the gorge along with heavy winter snows. Those exhausting days and nights prompted Dr. Bond to expedite his plan to get a hospital built in Bat Cave. With a certain amount of cajoling and even tacit collusion, Bond raised money and rallied the community. Volunteers like his old friends Fate and Lonnie—who were much handier with hammers and nails than he— donated their time to transform the abandoned Bat Cave schoolhouse into the twelve-bed Valley Clinic and Hospital. They built the new hospital at a fraction of the normal cost and Bond believed the year-long effort could serve as a model for improving medical care throughout rural America. Young Dr. Bond and the Valley Clinic were even celebrated in an American Medical Association film and written up in *Reader's Digest*, among other publications. Bond's status as beloved country doctor was secured. Everyone knew him as "Doctor George."

When the Korean War started Bond got a telegram informing him that he was eligible for recall to the Army. But it wasn't until after the armistice in 1953 that Bond finally had to pull up his mountain roots to relieve a paratroop doctor. Soon after arriving at Fort Sam Houston in San Antonio, Bond found that he and several hundred others would be redirected either to the Air Force or the Navy, where more doctors were needed. Bond chose the submarine service for no particular reason other than his understanding that the hazard pay would be the same as it would have been with the paratroopers. Bond's next stop, as part of his undersea medical training, was the Navy Yard on the Anacostia River in Washington, D.C., the home of the Navy's school for deep-sea divers. This unplanned immersion into diving began his professional conversion.

After dive school Bond spent six months at the New London base studying submarines and related medical issues. By the summer of 1954 he had moved with his young family to Pearl Harbor to serve as a submarine squadron medical officer. The post included shifts at the escape training tank, and Bond's touch-and-go ride in the pressure chamber to revive Charlie Aquadro was not his only such act.

George Bond was by then a father of four and he relished having a much more manageable schedule than was possible during his eight years as a country doctor. He was about ready to leave his Navy post, as scheduled, in the spring of 1955 when Navy officials ordered him to Hollywood, supposedly to act as an adviser for a training film on submarine escape. Bond was sitting in a little waiting room at the NBC studios when suddenly, as cameras rolled, he became the latest unsuspecting guest of honor on *This Is Your Life*, the popular TV show. Dr. Bond was lauded for his exploits as a country doctor, and as the program ended host Ralph Edwards handed Bond the keys to a new Mercury station wagon that had been converted into an ambulance for the Valley Clinic. The television program served as a fitting epilogue to Bond's life as a rural physician.

While Bond was in the Navy, another doctor had taken his place at the Valley Clinic. The two planned to go into a partnership upon Bond's return, but Bond soon found that he disapproved of how his replacement was running the practice—sending out bills, for example. Such a formal demand surely came as a shock to the mountain people. Bond had always waited until his clients could rustle up the dollars—or he would gladly take payment in the form of a ham shank or a basket of vegetables. But there were more serious concerns weighing on Bond, too. His childhood friend Fate Heydock had recently died of leukemia. Lonnie Hill, who was among the surprise guests flown in to give testimonials about Dr. Bond on

This Is Your Life, had been depressed over Fate's passing, as Bond knew. Bond figured he could lift his old friend's spirits once back in Bat Cave, but by the time Bond got home Lonnie had hanged himself from a dog-wood tree near the house he had built with his own hands.

The passing of his childhood friends and the schism with the other doctor may have provided a catalyst, but Bond's professional passion had already shifted like the tide. Within six months Bond volunteered to re-turn to duty and soon got orders to report to the Medical Research Labo-ratory at the U.S. Naval Submarine Base in New London. In March 1957, he took over as the assistant officer-in-charge of the lab and moved into his office in a brick building on the upper base. More than working with submarines, it was diving—the "diving game," as Bond would sometimes call it, with folksy flair—that had captured his lively imagination. Soon after taking his new lab post he began plotting the sunken jet escapes and the unprecedented three-hundred-foot blow and go from the *Archerfish*. These acts were testaments to his love of the game, and mere prologue to Bond's grandest plan yet, a plan buoyed by his vision of a future in which man would live in the sea.

2

DIVING

Living in the sea had long been a staple of myth and science fiction. But what Bond learned during his early Navy years got him wondering how to bring Jules Verne–style fantasy into the realm of reality. There were practical reasons for doing so, and Bond could launch into quite a sermon on the subject. An undersea frontier equal in size to the continent of Africa could be opened for "exploration and exploitation," Bond liked to say, if only divers were able to live and work on the continental shelves of the world, the submerged fringes of the planet's landmasses. To reach this undersea frontier would mean diving to depths of at least six hundred feet, far beyond the standard diving limits of the day, and then figuring out how to stay there for a while.

Bond had in mind a whole range of undersea activities—scientific, humanitarian, military, industrial. The Navy, given its undersea expertise and wherewithal, not only seemed like the most logical leader of a quest to live in the sea, but stood to benefit militarily through a dramatic expansion of its aquatic acumen. Spin-offs into civilian life and science would undoubtedly follow as endeavors such as mineral mining, marine biology, and marine archaeology were brought within man's reach. Bond believed that undersea exploration would bring the next generation of antibiotics, and that massive supplies of fresh water that boiled up from the continental shelf could be tapped. He believed, too, that the very survival of the human species depended on our ability to take up residence on the seabed and learn to harvest the ocean's edible protein.

In a speech to an audience of honor students at Albany Medical

College, Bond said, in his lolling pulpit brogue: "Those of you who are close to the agricultural picture know that with the expanding populations which we have, should we avoid all-out war, and God willing we will, we are going to come up against the old enemy hunger, and it's gonna be serious. There's no reason to believe that we can take care of the combined hungers of the expanding populations a hundred years hence. The oceans around us can do it, however, the organisms in the oceans and the continental shelves. We had best be about the business of trying to find out how to get our food out of the seas, and I say this in all seriousness." In many of his speeches and lectures it was clear that Bond's plans had grown from saving the ailing souls of Bat Cave to saving the world, or at least a good part of it, by leading a pilgrimage into the sea.

The first nuclear-powered submarine had already been launched when Bond took his job at the New London base in early 1957. Yet to Bond underwater vessels weren't the complete answer. Man would never attain the ultimate goal of exploration and exploitation locked in a submarine or a submersible pod, like the bathysphere the naturalist William Beebe famously used to descend a half-mile down, tethered by a cable to the surface in the 1930s.

To fully explore and exploit the seas, man had to be able to swim freely, like a fish, or, for that matter, like his primeval self. With a specially designed sea dwelling as a base, divers could live and work on the ocean floor for days, weeks, even months at a time. From a pressurized sea floor base, with the proper dive gear, they could pop anytime into the surrounding water, day or night. They could then spend hours in the water, working or exploring, and return to their base as needed for food and shelter, like weary mountaineers retiring to a cozy, well-stocked tent, or astronauts to a moon base.

There were a couple of fuzzy spots in this grand vision. One was that no such sea dwelling had ever been built. The other made the whole concept of living and working in the sea seem even more farfetched than putting a man on the moon: Long-accepted deep-diving limits typically allowed for bottom times of scant minutes. The deeper the dive, the shorter a diver's stay. No one spent hours or days in the sea, nor did they spend long periods exposed to high pressure, either in the water or in a dry chamber. The depth limit for most Navy divers, even for brief sojourns, was far less than Bond's initial target of six hundred feet. But as a scientist Bond believed it could all be done. As a man of faith he believed it would, for it is written: "And God said, Let us make man in our image, after our likeness; and let them have dominion over the fish of the sea, and over the fowl of the air, and over the cattle, and over all the earth, and over every creeping thing that creepeth upon the earth." This divine directive, from the opening of the Book of Genesis, resonated deeply with Bond. He liked to say that mankind would one day "acquire dominion over the seas, and the creatures therein," a neat paraphrase that also deftly shifted the biblical emphasis from "the fish of the sea" to the sea itself. "Dominion over the seas" became for Bond the ecclesiastical version of his more pragmatic exploration-and-exploitation mantra. Either way, he intended to give the Lord a hand.

Not long after Bond settled in as assistant officer-in-charge at the New London research lab, he wrote up a characteristically grand plan that he called "A Proposal for Underwater Research" and submitted it to his immediate superiors around the time the Soviet Union shocked the world with its launch of the Sputnik satellite. The proposal was relatively brief, but bold in its argument that diving and deep-sea exploration needed a breakthrough like the one in space. Divers who could live and work on the ocean floor would gain access to a vast, invaluable submerged frontier. "In man's history efforts have been made, with varying degrees of success, to derive the benefit of this offshore treasure," Bond wrote. "To date, these efforts have resulted only in sporadic and drastically limited forays, which offer little promise of full utilization, of the continental shelves of the world."

The sound barrier had been broken a decade earlier, in 1947, ushering in the space age. But the depth barrier that dogged divers remained intact, although no one much thought in those terms. "Depth barrier" was not part of the lexicon, and there was nothing as specific as the speed of sound to be broken, nor anything as dramatic as a sonic boom to punctuate an equivalent underwater achievement. As Bond had learned during his Navy dive school training, no one had ready answers to what struck him as the essential questions: *How long can a man stay down? How deep can a man go?* In many ways, Bond would find, deep-sea diving methods and equipment had changed little more over the previous century than rural life in the backwoods around Bat Cave.

A case in point was the Mark V, the venerable "hardhat" diving gear—so called because of the bulbous copper helmet, which had become the icon of the deep-sea diver. The Mark V had been in use ever since the U.S. Navy became serious about diving during the Woodrow Wilson administration. The basic hardhat design hadn't changed that much since

Augustus Siebe, a German-born instrument maker and gunsmith who settled in England, first bolted a similar helmet to the collar of an airtight diving suit in the 1830s.

The helmet, the bulky rubber-lined canvas suit, the lead-weighted boots, and the hose connecting the helmet to an air compressor at the surface—hand-cranked in the early days—gave human beings a reliable way to spend more than a breath-hold's worth of time underwater. For centuries, breath-holding divers trained themselves to wring as much time out of a lungful of air as possible, going as deep as they could to grab sponges, oysters, or other quarry off the bottom. Evidence of rudimentary forms of diving gear intended to allow for longer underwater stays is scattered across the ages. Aristotle, writing in his *Problemata* in about 360 B.C., referred to divers supplied with containers of air, and in the first century the Roman naturalist Pliny made one of the first of a number of early historical mentions of divers using breathing tubes, often in warfare to elude enemies. Leonardo da Vinci sketched a few underwater breathing devices. A variety of apparatuses, some more preposterous than clever, were devised during the two centuries preceding Siebe's hardhat. But Siebe did for diving what the Wright brothers would do for flight in 1903. An early demonstration of Siebe's revolutionary design was made in the busy port of Portsmouth, England. For half a century the harbor had been obstructed by the famed wreck of the *Royal George*, which sank in 1782. The battleship was in no more than ninety feet of water, but nothing much could be done to salvage it until six decades later when divers put on Siebe's new gear.

As Bond would find out for himself, there was much more to hardhat diving than tramping along the ocean floor like an undersea gladiator, as might be inferred from the movies. The Mark V helmet alone weighed some fifty pounds, the lead-soled boots twenty pounds each. Together with the rubber-lined canvas suit and a belt of lead weights held up by leather suspenders, the hardhat diver wore nearly two hundred pounds of gear. On dry land he could hardly walk. Tenders had to help him dress and then monitored the compressed air supply delivered to the helmet through a hose—a lifeline fittingly called the umbilical. Improvements had been made since Siebe's day—notably the addition of voice communication—but the Mark V, and similar versions made in other countries, was like a Model T frozen in time.

Once a hardhat diver was lowered into the water, the gear's substantial weight diminished. Give or take a few leaks in the suit, the diver remained dry. He wore undergarments for warmth and often a wool cap.

Besides keeping the diver relatively warm and protected, the hardhat gear functioned as a pressure suit. The pressure of the air a diver breathes has to be about equal to the surrounding water pressure. If the pressure differential becomes too great—it doesn't take much—his respiratory mechanics quickly break down, as they had for Charlie Aquadro. If a hardhat diver's breathing gas is insufficiently pressurized, or if a nervous diver were to hold his breath while descending, the increasing water pressure could crush a rib cage and cause the under-pressurized lungs to collapse, a condition known as "lung squeeze."

Such diving hazards underscore the physical risks from the rapidly rising pressure at depth. In the everyday atmospheric pressure at sea level, the human body, and every object, is subjected to 14.7 pounds of air per square inch, the pressure that comes from the weight of the air at sea level. For every thirty-three feet of descent underwater, the pressure increases by another 14.7 pounds per square inch, the basis of a unit of measure called an atmosphere. The term atmosphere itself is a poignant reminder of just how very distant anything quickly becomes underwater. Thirty-three feet is about the height of a three-story building—just a few flights of stairs away. But once underwater, any distance is much farther from home. Pressure increases by nearly a half-pound per square inch with every foot of depth. At thirty-three feet a diver is at a depth of two atmospheres (sometimes counted as just one when excluding the pressure of the one atmosphere above sea level). The total water pressure on a body immersed in thirty-three feet of seawater is then about 29.4 pounds per square inch. At sixty-six feet, you're at three atmospheres, or 44.1 pounds per square inch, and so on down to the deepest canyons of the abyss, more than a thousand atmospheres away. In contrast, outer space is just one atmosphere away.

Even on the surface, the body is routinely under thousands of pounds of air pressure, but the pressure doesn't bear down in a single direction, like a boulder crushing a bug. Liquids and gases envelop an object and exert pressure in all directions simultaneously, and human bodies have the advantage of being mostly water. Liquids like water are essentially incompressible, as are solids like rock or bone, and the enveloping pressures cancel each other out, creating a state of equilibrium. But the air in the ears, sinuses, and mainly the lungs is compressible, so a diver has to have adequately pressurized air to maintain the equilibrium needed to breathe, and to prevent lethal embolisms or lung squeeze.

These were the kind of lessons in physics and physiology taught at the dive school at the Navy Yard in Washington, D.C. While much of the science might be familiar to medical officers like Bond, the experience of hardhat diving was usually not, until they got to practice in the turbid waters of the nearby Anacostia River. The brisk current and dismal visibility acquainted trainees with the kind of challenging conditions they could expect to find at sea. They couldn't see much beyond their faceplates. They had to learn to work by sense of touch and avoid tangling or kinking their umbilical—getting "fouled"—which could cut off their air supply, snag them on the bottom, or both.

A hardhat diver had to carefully manipulate the valves that regulated the flow of air in and out of the helmet. The main valves could be opened and closed like faucets—the one controlling air intake was positioned around the chest; an exhaust valve was on the right side of the helmet. If the diver's hands were occupied or he couldn't reach the exhaust valve, he could work a "chin button" from inside the helmet, pushing it open with his chin or yanking it shut with his mouth.

Besides regulating the flow of fresh air for breathing, the valves adjusted the buoyancy of the airtight suit. As per Boyle's law, the volume of air inside the suit increases and decreases with changes in depth and pressure. Even small changes in pressure affect the diver's buoyancy and the pressure equilibrium between the inside and outside of the suit. A skilled hardhat diver could reduce his buoyancy and plant his feet firmly on the bottom—or as firmly as a muddy sea floor would permit. If need be, he could also make himself neutrally buoyant and float like a weightless spaceman. He might shut off the air flow entirely for a few minutes to cut the whooshing cacophony so he could talk and listen over the telephone set in his helmet. All of this could make hardhat diving as artful and risky as flying a plane.

If the suit becomes overly buoyant, the diver starts to float upward. Boyle's law kicks in, the gas inside the suit expands still more, and the diver would find himself in a "blowup," rocketing toward the surface, limbs outstretched, looking like the Michelin man swept up in a tornado. Besides causing embarrassment, a blowup could give a diver a case of the bends from the fast ascent. A "squeeze," while less commonplace than a blowup, could be even worse. Pressure inside the suit suddenly drops below the pressure outside because of a ruptured umbilical or perhaps an unintentionally fast descent. The higher surrounding water pressure then closes in unrelentingly. Because the diving suit is flexible and the helmet solid, the body can get squeezed like a tube of toothpaste into the helmet with potentially torturous force and ghastly results. Even if the diver isn't killed, he could bleed from the lungs, mouth, nose, ears, a bloody mess.

In addition to experiencing hardhat diving for themselves, Bond and the other doctors were schooled in diving physiology and medicine, a specialized field filled with unsolved mysteries about breathing gases and the effects of high pressures on the human organism. One major topic was the bends, more formally known as decompression sickness. For years the condition was called "caisson disease" because it was best known for striking nineteenth-century workers who spent their days in pressurized workspaces, or caissons. Workers would climb down into a hollow shaft and pass through an airlock leading into the caisson, which had all the comforts of a muddy tomb. As long as the air pressure inside was at least equal to the pressure of the water or the oozing muck outside, the workspace would remain dry enough so that crews could do their jobs, jobs like excavating New York's East River to build the submerged foundations of the Brooklyn Bridge.

The puzzling thing about the caissons was that workers would generally be fine until they left their pressurized job site and returned to the surface. There were a lot of deaths, paralysis, and excruciating pain. Afflicted workers doubled over as if mimicking the "Grecian bend," a passing fancy among Victorian women who took their promenades bent forward from the hips, pigeonlike. Thus decompression sickness was nicknamed "the bends."

An explanation for the bends, and how best to prevent it, came in 1878 from the great French physiologist Paul Bert. Bert's experiments showed how, under pressure, the problem originated with the absorption of nitrogen gas into the blood and tissues. Nitrogen makes up nearly four-fifths of atmospheric air but is inert, meaning molecules play no part in metabolism. They merely tag along as oxygen molecules are carried through the bloodstream. Carbon dioxide, the body's exhaust, flows out—from tissues and organs into the blood, then into the network of tiny capillaries covering the translucent membranes of millions of air sacs in the lungs, the alveoli, and finally into the lungs to be exhaled through the nose and mouth. As oxygen is taken in, and carbon dioxide released, omnipresent nitrogen makes the trip, too, in and out, shadowing the life-giving gases.

Under increased pressure underwater or in a caisson, nitrogen begins to accumulate in the tissues, harmless and unnoticed. If the pressure rapidly drops, as when one returns to the surface, the nitrogen emerges as bubbles, a phenomenon often likened to opening a bottle of carbonated soda pop. Bert found that these nitrogen bubbles were the cause of the bends. They could fizz like soda pop almost anywhere, in varied size and number, and block the vital flow of oxygenated blood to tissues and organs, wreaking corporeal havoc—excruciating aches and pains, paralysis, unconsciousness, and death. To avoid the telltale signs of decompression sickness, the caisson worker or the diver had to be slowly returned to the lower pressure. His body could then gradually release nitrogen by the normal process of respiration. The time required to safely decompress became known as a decompression penalty—the price a diver had to pay for each dive.

If a diver did get bent, the best treatment was "recompression," which meant putting the victim into a sealed chamber where pressure could be raised so that errant gas bubbles would dissolve. Severe aches, pains, or other symptoms would then usually subside, and an orderly process of decompression could be resumed in the chamber's controlled atmosphere. The dive school at the Navy Yard had chambers for training and testing and as a student Bond likely got a first chance to be locked into the kind of chamber he would return to many times, often to treat bent or embolized divers like Aquadro.

Paul Bert's discoveries—enumerated in his thousand-page tour de force on barometric pressure—were a huge advance in unraveling the mystery of the bends. Still, his slow and steady approach to decompression did not always prevent decompression sickness. He might have refined his methods but died suddenly at age fifty-three. The unsolved mysteries of decompression sickness would linger for three decades, until 1905, when the British physiologist John Scott Haldane led a committee that codified Bert's discovery of the bends into the first diving tables, also known as decompression schedules.

The dive tables look a lot like bus schedules—a grid made up of rows of different diving depths and durations ("bottom times"), with corresponding columns showing the depths at which a diver has to make decompression stops on the way back to the surface. A certain amount of time is specified for each stop to allow the accumulated nitrogen to ease out of the system without bubbling. Haldane's tables took the guesswork out of decompression and gave divers a guide to making a safe return to the surface. The tables were tested up to 210 feet and became the foundation for decompression schedules worldwide, including those adopted by the U.S. Navy. But even by the 1920s, Navy hardhat divers were not typically qualified to dive much deeper than ninety feet.

The Haldane tables, while a major advance, were not foolproof and were continuously tested and refined over the years. There are many variables on a dive, including a diver's unique physiology and the degree of physical exertion a dive requires. There are also many different ways to adjust a decompression schedule, and significant depth barriers remained. Divers were strictly limited in both how deep they could go and how long they could stay down. One major problem was that at depths of more than about 130 feet, divers began to experience something as puzzling as the bends. They started acting as though they'd just stumbled out of an all-night martini party.

The phenomenon became known as nitrogen narcosis, or, as divers would say, getting "narked." The French called it *l'ivresse des grandes profondeurs*—rapture of the great depths. Under pressure, nitrogen short-circuits the human brain like a round of stiff drinks. The deeper a diver goes, the stronger and more menacing the effect. As with ordinary drunkenness, the effect of nitrogen narcosis varies from diver to diver, but universally judgment is seriously impaired.

In the 1930s, after identifying nitrogen as the culprit in narcosis, a team of U.S. Navy researchers, led by Charles Bowers "Swede" Momsen, a renowned submarine rescue advocate, went looking for a remedy. They found it in helium. The compressed air typically used as a breathing gas is just ordinary air, with its abundant nitrogen, that's been pressurized and thus compressed. By replacing the nitrogen with helium, divers could go deeper without experiencing the potentially deadly mental haze of nitrogen narcosis. Like nitrogen, helium is inert, and under increased pressures it accumulates in the blood and tissues, but at different rates than nitrogen, so researchers devised new decompression schedules for divers breathing a helium-rich mix. To handle the helium they also had to create a new version of the venerable Mark V dive helmet, known as the "helium hat," which required even heavier gear. The noticeably larger bulbous helmet weighed more than a hundred pounds, about twice as much as the standard air-fed Mark V. To compensate for the heavier hat the helium diver's lead-weighted boots were forty pounds each instead of twenty. A fully outfitted diver wore some three hundred pounds of gear.

The helium hat was just out of the experimental box when the USS *Squalus* sank off the East Coast in May 1939. The sub came to rest about 240 feet down in the freezing Atlantic, and as the historic rescue and salvage operation began, the first hardhat divers got to the bottom the old-fashioned way, breathing ordinary compressed air. One diver passed out. Another called out football signals. A third befuddled diver almost cut his air hose and lifeline. Such was the power of the potent narcotic haze. Swede Momsen, who ran the diving operations, decided they should try the recently developed helium gear, and its successful use became one of the most significant diving milestones in the century since Siebe had devised his hardhat system. The helium diving gear, along with the inaugural use of the McCann Rescue Chamber—an elevator-sized, drum-shaped pod run on a cable between sub and surface ship—made possible the rescue of thirty-three survivors from the *Squalus* crew of fifty-nine, a triumphant first in submarine rescue. After the celebrated rescue, helium divers went back to the bottom to lay the groundwork for the successful salvage of the sub, which was ultimately raised, repaired, and recommissioned as the *Sailfish*.

During the 1940s, between the *Squalus* recovery and Bond's arrival at the dive school, there was another major development in the long evolution of diving gear. It was self-contained underwater breathing apparatus, which would become known by its familiar acronym, SCUBA. Self-contained gear eliminated the need for an umbilical and was relatively lightweight. The diver could carry a portable supply of compressed air and swim freely, without any tethers to the surface for air. Rough-hewn scuba designs had been tried with varying degrees of success since the nineteenth century. Then, shortly after the *Squalus* rescue, during World War II, a French naval officer named Jacques-Yves Cousteau, working with the engineer Emile Gagnan, came up with the pièce de résistance for scuba: a regulator that automatically and reliably allowed a swimming diver to breathe the highly compressed air from his tanks at the same pressure as the water around him. Cousteau and Gagnan would patent their invention and call it the Aqualung.

The simplicity of the Cousteau-Gagnan brand of scuba made diving accessible to virtually anyone. If traditional hardhat diving—with its need for an umbilical, compressor, heavy gear, and surface tenders—was like a railroad system, the Aqualung was like a private automobile. The Aqualung's success, rather like the success of the car or any consumer durable, was at least partly due to good timing. Swim fins and masks had only recently been introduced. There was growing interest in diving as a sport and as a tool for oceanography. Before Cousteau, diving was mainly a means of doing an underwater job, often a difficult or unpleasant one. Salvage operations such as those carried out after the attack on Pearl Harbor,

inside the jagged remains of bombed ships, in oily blackness, could be terribly dangerous even if they weren't deep.

Jacques Cousteau played a starring role in popularizing and promoting
both recreational diving and the Aqualung. A few months before Bond
entered dive school in 1953, Cousteau published his first book, *The Silent
World*, which was filled with diving adventures and photographs from the
wet world that few readers had ever seen. The book sold briskly and Cousteau's name quickly became synonymous with the underwater world. The
National Geographic Society signed on as a major sponsor of Cousteau's
undersea ventures in the early 1950s and its widely read magazine gave
Cousteau a preeminent forum for his stories, pictures, and of course the
Aqualung. In early 1954, Cousteau made his debut on American network
television on *Omnibus*, the premier cultural program of the day. On the
Sunday night show, Cousteau embellished his account in that month's
National Geographic with ample film footage about his team's excavation
of an ancient Greek wine-transport ship off the coast of Marseille. With
foot fins and the Aqualung, Cousteau told his TV audience, "You get the
impression of being a spaceman, free to fly in any direction." No man had
yet flown in space, but this kind of enticing imagery added to Cousteau's
soaring popularity and that of recreational diving.

Scuba was proving to be a plus for more than just recreation—search
and rescue, for example—but the U.S. Navy, or at least many of its hardhat divers, did not immediately embrace it. Part of the reason, as Bond
may have detected at the dive school, was that working Navy divers were
accustomed to the traditional hardhat system and they liked it. They liked
staying dry underwater, sealed in their helmet and heavy suit. The suit
and helmet offered good protection on treacherous salvage jobs. Hardhat
divers liked to be able to communicate by the phone link inside their
helmets. Scuba did not present a welcome change. Having a mouthpiece
stuffed in their faces and being unable to speak annoyed hardhat stal-
warts. Scuba also meant swimming, often in cold water. Not all hardhat
divers were swimmers, and staying warm and protected without a bulky
dive suit introduced a new challenge. But despite the misgivings, scuba
was gradually more accepted in the U.S. Navy. In mid-1954 the Navy established a new, scuba-oriented diving school at Key West called the U.S.
Naval School, Underwater Swimmers.

George Bond liked the way scuba enabled a diver to swim and roam
more freely. But the self-contained gear did nothing to change the long-
standing depth barriers. Duration was also limited by the amount of air

that could be compressed into the tanks carried on a diver's back. And the deeper a dive, the faster the air in the tanks would be used up. The Navy favored scuba for relatively short, shallow dives but generally considered it unsuitable beyond 130 feet—about half the depth at which helium hat divers worked on the *Squalus*. Helium and the helium hat had made deeper dives possible, but 350 feet was pushing the standard Navy limit. Duration, too, remained limited to a matter of minutes. Rarely would anyone stay at a substantial depth for more than half an hour.

Against this backdrop, diving looked ripe for a major breakthrough, at least to Bond. In his "Proposal for Underwater Research," first submitted in 1957, Bond said that he, along with fellow scientists at the New London Medical Research Laboratory, should begin a series of experiments that could show whether longer, deeper dives were physically and physiologically possible. Bond believed they were, as some earlier researchers had suggested. Man could not live in the sea, even with a well-crafted sea dwelling, if it turned out that he could not survive prolonged exposure to pressure or safely breathe under pressure for periods of time longer than the conventional dive tables allowed. No one knew for sure, but Bond wanted to find out. *How long can a man stay down? How deep can a man go?*

Bond believed that even those who were skeptical about the grander elements of his proposal—building laboratories on the sea floor, turning divers into mineral miners, or finding an undersea solution to world hunger—would see the value in attempting to break depth barriers. Longer, deeper dives had great practical potential for the Navy's own salvage and submarine rescue operations. It had taken more than six hundred hardhat dives and almost four months to complete the rescue and salvage of the *Squalus*. A diver's typical bottom time had been twenty minutes. Nearly twenty years later, at the dawn of the space age, an underwater job similar to the *Squalus* recovery would have been handled much the same way.

The Bureau of Medicine and Surgery, under which the New London Medical Research Lab operated, was in no mood to approve Bond's proposed experiments. Bond later summed up the official response this way: "You are wasting time. In the first place, who wants to go to the ocean bottom, there's nothing down there worth fooling with. In the second place, it can't be done. And, thirdly, you were pretty good at delivering babies and looking at tonsils. Why don't you get back to doing that and knock off this wild dreaming?" Bond considered his BuMed boss to be a man of virtually no vision, and a personal enemy as well.

Having just reentered the Navy and started his new job at the Medical Research Lab, Bond might have chosen to back off, but that was not his way. He believed that it was time for the breakthrough that diving deserved and that providence prescribed. Undaunted by the rejection of his proposal, Bond was determined to conduct his undersea crusade clandestinely, but he was going to need some help.

3

GENESIS

Bundled inside the hefty Sunday papers of November 9, 1958, with their front-page news about atomic weapons testing and the botched flight of the latest Pioneer moon rocket, *The American Weekly* thudded onto doorsteps across the country. The widely circulated Sunday supplement had an eye for the sensational and on this fall day it featured a tantalizing glimpse of the plan that George Bond had outlined in his rejected "Proposal for Underwater Research." The renowned illustrator Alexander Leydenfrost had sketched a futuristic sea floor scene showing scuba divers working around an enclave of boxy two-story structures that matched the hypothetical description for an undersea laboratory in Bond's proposal. Two divers were depicted cruising along in a kind of undersea Corvette. The title splashed across the illustration and accompanying article read: "YOUR FUTURE HOME UNDER THE SEA," with the subheading: "Fantastic? Not at all, say Navy scientists—it's not only possible but we're a lot nearer to it than you think."

The lone Navy scientist cited in the "exclusive first look at this amazing project" was Commander George F. Bond of the U.S. Naval Medical Research Laboratory, New London, Connecticut, "one of the Navy's experts on underwater physiology." Bond had somehow managed to get his vision of life on the continental shelf featured in the *Weekly*.

"So far, no human being has explored our inclining coastal shelf for more than minutes at a time due to the need for slow, painstaking decompression on the return trip to the surface. Consequently, our knowledge of nearby underwater terrain is embarrassingly scanty," the *Weekly* said, adding an unequivocal statement from then Commander Bond: "But we possess all the necessary hardware to complete a rapid conquest of our oceans."

Such showy talk readily contributed to Bond's reputation, and his ability to promote this vision of living in the sea could sometimes be as significant as anything he did as a scientist. Bond understood the basics of the hardware that might be put to use for this undersea conquest, but he was no engineer and had nothing specific on the drawing board. Nonetheless, as the *Weekly* article showed, Bond eagerly promoted his exploration-and-exploitation ideas. He maintained a peripatetic schedule and could often be found on the lecture circuit, giving his lively brand of scientific sermon, and in Washington, looking for influential converts to his cause.

Despite the formal rejection of the "Proposal for Underwater Research" he was determined to proceed, sub rosa, and prove that the picture sketched for the *Weekly* was not something lifted from Buck Rogers—a favorite sci-fi icon of Bond's. The first step was to test animals and Bond enlisted the help of scientists on the Medical Research Lab's staff, much as Swede Momsen had an expert team around him in the development of the helium hat. One man in particular would become as dedicated as he was indispensable. He was Commander Walter Mazzone, a decorated veteran of some of World War II's most harrowing submarine patrols.

Bond first met Walt Mazzone late in the fall of 1958. Mazzone had recently arrived to run the School of Submarine Medicine at the New London base, where Navy doctors and hospital corpsmen received specialized training before being assigned to their boats. Mazzone had come to manage the curriculum, run academic review boards, oversee the students' required projects, and generally juggle the demands of a top administrator. But he could often be found serving as a valued mentor, mingling with the new medical personnel and offering practical pointers about Navy customs and the rigors of submarine life, which he knew well.

Mazzone was at the Medical Research Lab one day that fall when Bond strolled in. Mazzone had barely been introduced to Commander George Bond when Bond jauntily grumbled about his recent trip to Washington, home of the surgeon general and the headquarters of the Bureau of Medicine and Surgery. The bosses there had been none too pleased about the *American Weekly* article, but now it became the catalyst for a conversation during which Bond impressed Commander Mazzone as a charismatic visionary, and Mazzone was not easily impressed. Bond invited Mazzone to join his nascent animal research project, which he intended to pursue with or without official approval. As Mazzone would quickly learn, Bond believed God said it would be done and it was time to give the Lord a hand.

In Mazzone, Bond had found a true acolyte, but not an obvious one. Mazzone was not a diver. He was not devout. He was not on the research staff of the medical lab. His duties were largely administrative, not scientific. There were also the disparate routes by which these two middle-aged commanders—Bond had just turned forty-three and Mazzone would soon be forty-one—had attained their ranks. Unlike Bond, the son of a well-to-do lumberman, Mazzone came from a working-class family of Italian immigrants. Determined *not* to follow his father's footsteps into the Steinbeckian world of California's canneries, Mazzone went to San Jose State College, not far from his family home, and earned a degree in biological and physical sciences. His sights were already set on medical school when he graduated in 1941, but with the Japanese attack on Pearl Harbor he went to war instead.

Mazzone chose the Navy over the Army, mainly because a couple of relatives were Navy men. He also loved the ocean. As a kid growing up in San Jose, young Walter and his family, and often a few neighbors, would drive their Model A cars over the Santa Cruz Mountains to the beach. Mazzone would return many times during his youth, hitching rides when he had no car. During the Depression, the families could help feed themselves by fishing and, when the tide ebbed, digging up clams and plucking mussels and sea snails.

About the time that Bond began his studies at McGill, living Mazzone's dream of going to medical school, Mazzone was aboard the submarine *Puffer*, whose crew endured, among other things, a prolonged nightmare of a stalking around the eastern side of Borneo. For nearly forty hours Japanese destroyers threatened to send Lieutenant Mazzone and his shipmates to the bottom of the Makassar Strait. But the *Puffer* crew dodged the repeated depth charges and finally slipped away.

Mazzone was thickly built, like Bond, but shorter and more compact. He was a boxer in college and apt to express himself in staccato bursts—a manner well suited to a line officer who had to bark orders. Mazzone spit words like bullets, in sharp contrast to Bond's lolling pulpit baritone, and his vocabulary drew more heavily on the profane than the sacred.

After the war Mazzone left the Navy and enrolled at the University of Southern California for a degree in pharmaceutical chemistry. But pharmacy work made him restless. By 1950 he had rejoined the Navy, just in time for the Korean War and to become part of the new Medical Service Corps. This time around Mazzone was stationed in Occupied Japan at the newly commissioned U.S. Naval Hospital at Yokosuka. After the war Mazzone worked out of an office in Brooklyn and within five years he became a high-ranking manager in the Armed Services Medical Procurement Agency. It was a good job, with responsibility for big budgets and vast supplies of blood, blood derivatives, drugs, and chemicals, but submarines were in Mazzone's blood. When he got an order to report to the New London base, he welcomed it.

If he followed George Bond's lead, Mazzone would have an opportunity to do the kind of research that had made him want to study science and medicine in the first place. Besides piquing Mazzone's academic interest, Bond was offering an opportunity to learn to dive, which had an added appeal for an adventuresome soul who had lately been confined to an office. Mazzone's only experience with diving had been brief, cold, and courageous. Toward the end of the war Mazzone was on the USS *Crevalle* as part of a submarine group sent on a mission to the Sea of Japan. For a month the sub group dodged mines and attacked Japanese shipping. They eventually made their escape to the north, through La Perouse Strait and into the Sea of Okhotsk. On the way out, the *Crevalle* crew noticed its sub was making an irksome whining sound, enough of a noise to be picked up by Japanese sonar. A propeller may have been damaged, or perhaps the sub caught a mine cable and was dragging deadly cargo. Someone was going to have to dive in and have a look around. Lieutenant Mazzone volunteered.

The submarine surfaced and Mazzone crawled out the hatch of the aft torpedo room and put on a Jack Browne, a basic face mask fed with air through an umbilical for shallow-water dives. In the event that a fast getaway became necessary, a shipmate stood by with the grim task of cutting Mazzone's umbilical. If Mazzone felt his air supply cut off, he was to swim away from the sub and up to the surface. The sub would return for him, if it could. So, wearing only the Jack Browne and a pair of shorts, twenty-seven-year-old Mazzone jumped off the submarine's deck into the icy Sea of Okhotsk. He ran his hands along the idled propeller blades, noting some damage and looked for errant cables or other possible sources of the whining. At least the sub was dragging no mines. Once back on board, his reward was whiskey and, rarer still, a hot shower. Mazzone gave no further thought to diving until the day he met George Bond at the New London lab.

Bond arranged to have Mazzone sent to the dive school at the Washington Navy Yard. It was an unorthodox move, since there was no official need for a top administrator like Mazzone to take a month or two off to go through dive school, but Bond had a way of making things happen, something that Mazzone admired. Mazzone made an ideal student—sharp, willing, energetic. He read up on the history and science of diving, embracing a favorite maxim of Cicero along the way: *To be ignorant of what happened before you were born is to be ever a child. For what is man's lifetime unless the memory of past events is woven with those of earlier times?* Or, as Mazzone summed it up: *Don't reinvent the goddamned wheel!*

Not long after completing his dive school training, Mazzone was with Bond, Cyril Tuckfield, and the rest of the crew for the deep submarine escape trials from the *Archerfish*. During the first couple of days at sea, in preparation for Bond and Tuck's record buoyant ascent, Mazzone and a partner made successful blow and go escapes from 150 and also two hundred feet—impressive feats in their own right. Bond and Tuck made ascents from those depths, too, prior to the grand finale, their daring escape from three hundred feet. Mazzone would gladly have attempted the three-hundred-footer with Bond, but instead dutifully served as one of the safety divers, along with Charlie Aquadro. This was one of Mazzone's early lessons in stepping back to allow Bond to take center stage. Their working relationship might not have lasted if Mazzone shared Bond's appetite for publicity. Mazzone worked well without fanfare and preferred it that way. He was happy to let Bond be Bond, and considered it a small price to pay for the opportunity to work with a visionary.

Once the animal research experiments got under way in the late 1950s, Mazzone assumed a leading role, albeit on the moonlighting basis made necessary by his official duties. The main physiological issues were, first, to find the right gaseous recipe for an artificial atmosphere that would sustain life under pressure for indefinite lengths of time. Second, there was the matter of devising a decompression schedule. Nitrogen would have to be either reduced or eliminated—its narcotic haze would become deadly long before a diver set foot on the continental shelf, six hundred feet down and nearly twenty atmospheres away. Oxygen, too, would have to be carefully regulated. Oxygen toxicity, or oxygen poisoning, could quickly be brought on by the same laws that govern all gases. When pressure on a gas is increased, the concentration of the gas—a measure known as its "partial pressure"—undergoes a proportionate increase. Under the pressure of two atmospheres, a depth of just thirty-three feet, the concentration of pure oxygen doubles, and that's enough to cause the twitching, dizziness, and convulsions associated with oxygen poisoning, which in turn could be fatal to a diver. Jacques Cousteau found this out the hard way when he tried to dive with an oxygen-only breathing apparatus but went into convulsions forty-five feet down, lost consciousness, and nearly drowned.

To prevent the concentration of oxygen, or any gas, from rising to dangerous levels under pressure, you had to put less of it into the mix the divers breathed. Two hundred feet down, for example, where the pressure is seven atmospheres, you need one-seventh the amount of oxygen found at sea level. So to equal the healthy surface equivalent of 21 percent oxygen in an artificial gas mixture for diving you would need an oxygen concentration of about 3 percent. While depth and pressure are the major variables in determining the composition of breathing gases, others that come into play include water temperature, the diver's individual physical and mental characteristics, and the degree of exertion an underwater job requires.

Much of this was familiar science. Ever since the *Squalus* rescue, it was known that helium was a good substitute for nitrogen because it eliminated the haze of narcosis. Some researchers experimented with other gases, mainly hydrogen and neon. Whatever the gas recipe, most experimentation followed the paradigm that deeper dives would be of very limited duration. The standard decompression schedules did not go beyond dives of about four hours, but even that boundary was untested in practice. The unmistakable message in the standard dive tables was that the deeper you go and the longer you stay, the longer your decompression penalty. In other words, beware the depth barriers. It was not uncommon to spend more time decompressing than working on the bottom. At 240 feet, the depth of the *Squalus* operation, a half-hour of bottom time would require about two hours of decompression. Even if a rescue diver lasted a full hour on the bottom, the decompression penalty would approach three hours. This double whammy of limited bottom times and long decompression penalties made underwater work inefficient and added to its expense.

But what if diving were approached differently? What would happen to a man breathing an artificial atmosphere, under pressure, for much longer than anything outlined in the standard dive tables? Some researchers had suggested that longer-duration deep dives might be feasible. Among them was Albert Behnke Jr., the renowned Navy scientist who discovered that nitrogen was responsible for narcosis. Working with Swede Momsen in the 1930s, it was Dr. Behnke who pioneered the use of helium in time for the *Squalus* rescue. Although Behnke retired from the Navy just as Bond entered the scene, he remained active in the field of hyperbaric medicine and became a supportive elder and friend to Bond.

Behnke and a few others had taken some preliminary and relatively shallow steps over the years into longer-duration dives, mainly by testing their subjects in pressure chambers. But those experiments didn't break depth barriers the way Bond believed they could and should be broken. One leading researcher, Dr. Edgar End at Marquette University medical school, concluded from a one-hundred-foot test he ran in a chamber that twenty-seven hours was about as long as a human being could stay under high pressure without risking his life. Until Bond came to New London in the 1950s, no one was actively exploring the concept of long-duration deep dives, a concept known as "saturation diving." Saturation diving, a term bandied about in the mid-1940s to describe some of the rare longer-duration dives, was never part of everyday diving lexicon until it became the holy grail for George Bond. The basic concept of saturation itself, however, had been understood for years.

Much as a sponge absorbs liquid, bodies absorb gases. The human body becomes "saturated" as the gases a person breathes disperse into the blood and tissues. At the ordinary pressure of one atmosphere, everyone on earth is saturated, mostly with nitrogen, the major component of air. But saturation levels can rise or fall, depending on the depth or altitude pressure at which a person is breathing. At greater depths, for example, a diver has to breathe higher-pressure gases, so the degree of potential saturation goes up—sort of like a dry sponge absorbs more liquid the longer it's held underwater. As the diver breathes, he gradually becomes more saturated, up to the point at which the concentration of gases absorbed in his body reaches the same level as in his pressurized breathing gas. At that point of equilibrium—whether at one atmosphere, two, three or more—a body would be considered saturated. Like a soaked sponge, it could absorb no more unless depth and pressure were increased.

But it wasn't completely clear how long it took for a body to become thoroughly saturated. Furthermore, neither Bond nor anyone could be sure that allowing a diver to become saturated, the key to saturation diving, would open the door to breaking the old depth barriers. Bond's team would have to experiment with artificial atmospheres, learn to maintain a healthy breathing mix for extended periods, and also devise a decompression schedule to bring a saturated diver safely back to the surface.

Decompression and its underlying theories were not simple matters.

Saturation decompression theories, being new and untested, created yet another layer of complexity. Human bodies are, after all, far more complex than a sponge as is the process of gas absorption in the body. Inert gases like nitrogen or helium can be absorbed more quickly into tissues with a rich blood supply, like the brain and spinal cord—two areas where bubbles from the bends are especially dangerous. These more readily absorbent, "fast" tissues, as they became known, get saturated much more quickly than tissues less well supplied with blood, like the joints, which were categorized as "slow" tissues, and where the bends were notorious for causing severe aches and pains, but not necessarily death.

Depth and duration were of course the primary variables of decompression theory, but there were others, like gas solubility. Nitrogen is about five times as soluble in fat as in water, for example. So fatty tissue can absorb significantly more nitrogen and thus take longer to saturate than tissues with a higher water content. That difference also applies in reverse, during decompression. Fast tissues "desaturate" more quickly than slow tissues.

Mathematical models representing the full range of tissue types were the key to predicting saturation, desaturation, and decompression rates. But just as musical notes on a page cannot be heard and fully judged until played by musicians, these models had to be tested on real bodies to find out whether they produced a harmonious result. The optimal schedule was the one that could be completed in the shortest time possible while doing the diver no harm. If someone got bent, paralyzed, or killed in open-sea dives or in controlled chamber tests, a decompression schedule would have to be altered. This trial-and-error science of decompression, along with the technological and mathematical tools to pursue it, had been evolving since the first diving tables at the turn of the twentieth century. George Bond was a beneficiary of that evolution, and to help puzzle out lingering mysteries in the physiological interplay between math and medicine he called on his friend and fellow medical officer Dr. Robert Workman.

Soon after George Bond took over as officer-in-charge of the Medical Research Lab in mid-1959, Dr. Workman succeeded Bond as assistant officer-in-charge. Bob Workman was studious and genial, but a bit reserved, especially compared to his gregarious boss, who was by then planning his daring three-hundred-foot submarine escape. Workman had neatly combed dark hair, earnest eyes, a broad face, and a dimpled chin like the actor Robert Mitchum. At six feet he was a couple of inches shorter than

Bond and six years younger. Like Bond, he was a pipe smoker. He grew up in a small town in central Iowa and had always wanted to be a doctor. After college and medical school at the State University of Iowa he put in a couple of required years with the Army, and then started a rural practice in northwestern Iowa, much as Bond had in Bat Cave.

In the early 1950s, for some reason, Workman went scuba diving. Perhaps he got the idea from *The Frogmen*, a 1951 movie about underwater commandos in which scuba plays a central role. Or he may have been among the myriad readers captivated by Jacques Cousteau's first magazine articles and *The Silent World*. Whatever the case, it was Cousteau's new Aqualung that had made it relatively simple for almost anyone to dive. Workman and his buddy donned the gear and explored the silent world beneath the surface of Iowa's Lake Okoboji. Dr. Workman underwent some kind of conversion. It helped that he, like Bond, had come to know the unrelenting rigors of a rural medical practice. Not long after his first diving experience Workman changed career course and joined the Navy. A couple of years after Dr. Bond, he, too, went through dive school en route to becoming a submarine medical officer.

Workman soon distinguished himself as an outstanding and well-liked Navy scientist. He was a virtuoso with a slide rule and had a keen feel for the mathematical models underlying the physiology of diving and decompression theory. In 1958, the year before Workman joined Bond's staff in New London, he won a year-long postgraduate fellowship at the University of Pennsylvania and studied with Dr. Christian Lambertsen, a leader in the field of diving physiology and equipment development. Lambertsen's many contributions included his amphibious respiratory unit, a breathing device used by combat swimmers during World War II. Lambertsen even shared credit for coining the SCUBA acronym.

Like Bond, Workman was naturally curious about the concept of saturation diving. Bond needed a scientist of Workman's caliber to help do the physical calculations necessary to determine whether saturation diving was a viable way to break depth barriers. At Bond's behest, Workman wielded his slide rule and confirmed a critical point: If a diver stayed under pressure long enough to become fully saturated, his decompression requirement would become fixed. The decompression curve flattened out. On paper, it appeared that a saturated diver's decompression penalty should be the same whether he stayed down a few additional minutes, a day, a month, or possibly even longer. This would mean that a diver could stay underwater indefinitely with only one decompression—a key to living in the sea and a stunning advance over the standard practice of making multiple short dives, each one requiring a time-consuming decompression. The animal experiments would be the first step in finding out whether the hypothesis held up. Bond called these experiments Genesis. They marked a beginning, as the name implied, but mostly the name was Bond's way of paying tribute to the first book of the Bible, the spiritual origin of Dominion over the Seas.

Down by the Thames River, in a dreary building adjacent to the bottom of the escape training tank, Bond had access to a pressure chamber, much like the one at the top of the tank, where trainees suffering from an embolism could be locked in for emergency recompression. Like other such steel chambers, including the one in which Bond treated Charlie Aquadro, this one was cylindrical and looked like a midget railway tank car. It was roughly twenty feet long, with a circular hatch on one end. That hatch led into a compartment that served as an airlock for entering and exiting an adjacent second compartment separated from the first by another hatch. Both chamber compartments were just large enough around for a man to stand up. The chamber wasn't designed for experimental use and didn't have much in the way of instrumentation—Bond's lab still awaited the installation of a new, state-of-the-art test chamber—but this old chamber's dual compartments at least allowed the researchers to maintain a desired test pressure in the second compartment while tending the animals inside by using the first compartment as an airlock.

Bond, Mazzone, and Workman, the Genesis trinity, would take turns peering through little portholes to observe the animals they led into this pressurized Ark, which was about a ten-minute walk from the medical lab on the upper base. Two dozen rats would be the first of many critters to get locked in. The initial goal was to keep them for fourteen days under a pressure of seven atmospheres—more than one hundred pounds per square inch—the same as would be experienced at a depth of two hundred feet. Two hundred feet was considered a substantial depth, and fourteen days was far beyond the durations of a few hours outlined in the standard dive tables. This struck Bond as a good starting point for collecting some benchmark data. A good long saturation dive like this would give them ample opportunity to identify any signs of physiological trouble. It also gave them a chance to practice the art of prolonged atmospheric control. They would begin by filling the chamber with ordinary compressed air. They knew that if nitrogen narcosis didn't kill the animals, oxygen poisoning probably would. Sure enough, by the thirty-fifth hour, all two dozen male rats were dead. Some amount of helium was going to be needed to create a healthy artificial atmosphere.

Over the course of about two years, Bond, Mazzone, and Workman took turns watching over dozens of animals—some two hundred, by Bond's unofficial count. The need for constant monitoring during a Genesis experiment could add long hours to the researchers' days. Rats, then guinea pigs, and eventually little squirrel monkeys were locked in the chamber and put under pressure in a helium-infused artificial atmosphere for up to two weeks at a time. As the hours passed, the researchers would watch for any signs of animal sickness or discomfort, such as changes in appetite, lethargy, dragging a limb. In some guinea pigs they observed a heightened friskiness. Perhaps artificial atmospheres would join the long line of purported aphrodisiacs. No one could be sure. Later, in checkups and dissections at the medical lab, they examined the animals' organs for any signs of trouble attributable to living under pressure.

At times the lab evoked a high school science experiment. Inside the chamber the experimenters laid out trays of lithium hydroxide, a carbon-dioxide-absorbing chemical that looks like coarse sand. They positioned fans to help circulate the chamber atmosphere and added some boric acid powder to the trays to eliminate the growing stench of excreta. Because they were working with an old pressure chamber, sub rosa, in their spare time, without any formal budget or backing, this kind of jury-rigging was a must. To try to improve their setup at one point, Mazzone and Workman—who were both much handier than Bond—retrofitted the chamber with a homemade version of the kind of gas-filtering device used on submarines, known as "scrubbers," to better "scrub" carbon dioxide out of the enclosed atmosphere. Bond and Mazzone were spending money out of their own pockets to buy animals, animal feed, or whatever else couldn't be "borrowed" from the Medical Research Lab or some other department around the submarine base, such as costly helium.

The Genesis troika eventually settled on a breathing mix that was about 97 percent helium and 3 percent oxygen at the trial pressure of seven atmospheres. Nitrogen was virtually absent, and Bond knew that some scientists had expressed doubt that the human body could function indefinitely without any nitrogen to breathe. Yet that did not seem to be an issue, the troika was finding. The tests on smaller animals were a good start, but if they were going to make the case that human beings should be allowed to try prolonged saturation dives, they needed to test some larger animals. Goats would be among the last animals into the pressurized Ark. Goats had been a top choice in experimental diving work for years because their anatomy more closely resembled that of humans. If the goats could survive the saturation plunge and subsequent decompression, chances were good that men would, too.

Bond got some of his bleating test subjects from neighbors around the old farmhouse where he lived in nearby Ledyard with Margit and their four children. There were also goats enlisted from the submarine base, where they were left to graze to keep the grass trimmed. The scientists saturated a number of goat pairs, paying special attention to the decompression schedules. At certain predetermined depths, the decompression stops ranged from thirty-six hours to a few minutes. With each stop the gas recipe also had to be altered to keep the concentration of oxygen from becoming poisonous. This was the same sort of trial-and-error decompression testing that had been done with human subjects to refine the schedules for conventional dives. When those divers felt the telltale aches of the bends coming on, they would say so and adjustments could be made to the gases and pressure. But with the goats, and all the other animals, the three researchers had to watch for nonverbal warning signs. In the case of bends, the goats might gnaw at themselves as if they had a bad itch, or take the weight off a leg and let it hang limp.

The Genesis experiments were already under way in early 1960 when Bob Barth began a stint as a training tank instructor. Barth was a few months shy of his thirtieth birthday and had a dry, wisecracking personality. He had a proclivity toward pranks and often spoke with an inscrutable smirk. He was of medium height and build, with thinning dark hair and a full, rounded face prone to ambiguous expressions. He had enlisted in the Navy at seventeen and had worked his way up to quartermaster first class. For the past few years Barth had ridden submarines and trained as a diver. When it came time for shore duty, the prospect of spending hours a day in a fat column of balmy water as a training tank instructor looked paradisiacal. Bob Barth could not have guessed that his days at the New London submarine base would include goat herding and the beginning of a long association with a medical officer named George Bond.

Bond would come down to the tank to put in shifts as the doctor on duty and didn't just watch or pace around in a lab coat. He liked to get in the water. Bond would join Barth and other instructors in making breath-holding dives from the top of the tank, or "bottom drops." On a single breath, they would drop all the way to the bottom, nearly 120 feet, and then return to the surface, much like free diving sponge and pearl divers. When visitors came to the tank, Barth and Bond took part in demonstrations that featured bottom drops and a carnival atmosphere. One favorite stunt was to toss a pair of tennis shoes into the tank. Then a breath-holding diver would follow, "flying" to the bottom, as Barth would say. Once on the bottom he would put on the white sneakers, with studied nonchalance, taking time to lace them up, letting the clock tick, before finally resurfacing.

Between the showy demonstrations and the sober job of watching over potentially deadly practice escapes, a good deal of camaraderie developed around the training tank. Cyril Tuckfield returned to submarine duty when Bob Barth arrived, and soon Dr. Charlie Aquadro, who had lately been chief medical officer of a submarine squadron in Key West, joined Bond's staff at the Medical Research Lab. The boyish Southern gentleman gave the training tank's carnival side a boost when he started bringing his recently acquired pet Kampuchean monkey to work. Aquadro named the monkey Scuba and doted on him. Scuba liked to watch from the deck around the top of the tank as instructors and trainees came rocketing from the depths of the water column in a hail of bubbles. Before long, Aquadro had a machinist create a Mark V–style dive helmet for the monkey. Scuba would don the gear and Aquadro taught the little primate with the prehensile tail to dive. Scuba eventually made it all the way to the bottom of the tank in his little hardhat. Bob Barth and the other instructors grew fond of their unofficial mascot, and Barth even took Scuba home one weekend to monkey-sit while Aquadro was out of town.

Once the goats were locked in the test chamber they required more work than the smaller rodents and primates. Bond also wanted the goats taken for a brisk trot as part of their post-decompression checkup, to see that they were running on all fours and not limping or displaying any other signs of ill health, like a slow-acting case of the bends. Cleaning up after goats and other animals in the chamber was one thing. It might be unpleasant but it could at least be done in relative seclusion. Running goats around the base required a certain nonchalance. Had anyone other than Dr. Bond asked Barth to volunteer as goat herder, Barth might have responded with a prickly answer and a dismissive smirk. But for Bond, and for the cause of Genesis and saturation diving, Barth did it willingly.

Barth put the goats on leashes, two or sometimes four at a time, and herded them from out of the bowels of the escape training tank. The goats did not always want to go in the same direction, but Barth would do his best as he ran them along the nearby wharf, in full view of docked submarines, ships, and their crews. One can only imagine the merciless epithets hurled at the hapless sailor trying to keep his bleating ungulates in line. If anyone could weather the verbal assaults, it was Barth. But even he could get tongue-tied when trying to explain why he was running goats along the wharf, as happened one day when he was stopped by a base patrolman. Barth might have just said that his goats were like Able and Baker. Everyone knew Able and Baker, the monkeys famously launched into space on an American rocket. Unlike their simian counterparts, however, the Genesis goats would not get banner headlines or make the cover of *Life* magazine. Yet those goats were doing for diving what Able and Baker did for space travel.

During the final phase of Genesis, Bond, Mazzone, and Workman focused on refining decompression schedules. Several pairs of anonymous goats were locked into the chamber and allowed to become saturated at the equivalent of two hundred feet. At this point, having gotten the hang of controlling the artificial atmosphere, the team of researchers began with a fairly conservative approach that gradually brought the goats back to the surface over seventy-two hours, an ascent rate of about three feet per hour. For half that time the goats were held at a pressure equivalent to a depth of eighty-four feet, for the other half at twenty-six feet, with the concentration of oxygen being increased along the way. A decompression lasting several days, as this one did, was long by traditional standards, but if paying this kind of decompression penalty meant a diver could live and work in the sea indefinitely, it would be time—and money—well spent. So far the signs were looking good.

On another try, a larger goat weighing 120 pounds, closer in size to an adult human, exhibited symptoms of the bends. They raised the chamber pressure back to seven atmospheres, following the traditional method of using recompression to stop the fizzing soda pop effect of the bends. It worked. After three hours the goat's symptoms subsided. This was good news for humans who might someday get bent while surfacing after a long saturation dive. Recompression still seemed to be effective medicine. They then successfully brought the goat back to the surface with a more conservative decompression schedule, pausing for full-day stops at one hundred feet, forty-three feet, and ten feet before returning the goat to the surface. The moonlighting hours could really add up during Genesis. Following the goats' ritual run along the wharf with Bob Barth and a day of observation, none exhibited any symptoms of the bends.

The goat tests marked the culmination of the groundbreaking animal experimentation and in early 1962, Bond, Mazzone, and Workman published an official report for the Bureau of Medicine and Surgery outlining the methods and results of Genesis: "Prolonged Exposure of Animals to Pressurized Normal and Synthetic Atmospheres." The forty-page report, made publicly available, reflected the sober influence of Workman and Mazzone in that it was straight scientific fare. Gone were the grand flights of fancy found in Bond's "Proposal for Underwater Research." But the report did point out the potential for "operational exercises" that might include underwater construction, deep salvage, and marine research.

To anyone who wanted to hear it, the report's message was clear: What worked for the animals could work for human beings. A formidable barrier had been cracked, and an undersea future that had existed only in science fiction was seeping into reality. Just how long a man could stay down and how deep he could go still remained a mystery. Also off in the Buck Rogers realm was the engineering of a sea dwelling—an undersea laboratory or shelter something like the boxy structure depicted in *The American Weekly*. But it was beginning to look like man could actually live and work in the sea, if only someone dared to try.

4

FRIENDLY RIVALS

In the years leading up to Bond's Genesis experiments, there were sporadic and often sensational efforts to dive deeper, if not to stay down for very long. Several record-breaking dives were attempted in the 1940s, including one by a Swedish engineer, twenty-eight-year-old Arne Zetterström. Breathing a special gas mixture through traditional hardhat gear, the lanky Swede was lowered into the Baltic Sea on August 7, 1945, the day after the atomic bomb was dropped on Hiroshima. Zetterström successfully reached a depth of 528 feet, but his diving platform was raised too quickly. By the time he reached the surface, he was dead. In September 1947, a month before the first supersonic jet flight, one of Jacques Cousteau's most experienced divers, Maurice Fargues, made a dive to 385 feet in the Mediterranean. Fargues was breathing ordinary compressed air, testing the depth limits of the Aqualung. Upon reaching his target depth—a record for an air-breathing diver—he apparently passed out. His safety line went slack, alerting his buddies at the surface that something was wrong. They hastily hauled him up and tried to revive him, but Fargues was dead.

A year after Fargues's accident the Royal Navy staged a series of deep dives in Loch Fyne in western Scotland and passed the official U.S. Navy helium diving record of 450 feet. On the final dive Wilfred Bollard, a petty officer diver first class, spent five minutes at 540 feet. He suffered a minor case of bends while undergoing nearly nine hours of decompression—but lived to celebrate his record. That record stood until the mid-1950s when another Royal Navy diver, Lieutenant George Wookey, was lowered in hardhat gear to six hundred feet in a Norwegian fjord, in October 1956.

With the water temperature near freezing, he did a brief task with his numbed hands, simulating a procedure for submarine rescue, but then ate up precious minutes struggling to free his badly fouled umbilical. He was lifted back to the surface, dizzy and shaken, but alive.

Deep-water marks like these were fine achievements when they didn't turn tragic, but Bond wasn't interested in depth alone, as he would make clear to anyone who would listen. He wanted divers to remain at depth long enough to perform useful work. While some in the Navy either doubted his abilities or disliked his unorthodox ideas, Bond found receptive audiences elsewhere. In the spring of 1959 he presented the keynote speech to the Boston Sea Rovers, a dive club formed in the early 1950s when the Aqualung was gaining popularity. Bond found himself in good company. The Rovers' featured speaker the year before was Cousteau, who in 1957 had won an Academy Award for *The Silent World*, his film based on his popular book of the same name. That was also the year he took over the prestigious directorship of the Oceanographic Museum in Monaco. By then Cousteau had retired from the French navy and transformed an American-made, 140-foot minesweeper into the *Calypso*, the research ship that would become synonymous with Cousteau, diving, and undersea exploration.

The Rovers, unlike some Navy traditionalists, were captivated by Bond's ideas, as were members of similar groups that formed around the growing interest in scuba. A man who heard Bond speak before the Illinois Council of Skin Divers around this time wrote to him to say that his presentation was the best he had ever heard by a doctor or by a naval officer. "Your message was important, and your sense of humor delightful. Thank you for coming. I know 600 others feel the way I do about your wonderful talk." By 1960, the year after the daring three-hundred-foot submarine escape, the Boston Sea Rovers named Commander George Bond their Diver of the Year.

The Rovers took understandable pride in having introduced Bond and Cousteau to each other during this period. Bond would fondly recall that his first significant meeting with Cousteau took place in the fall of 1959, following a seminar on undersea activities at Town Hall in New York City. He joined Cousteau after the event at the home of Cousteau's good friends the writer James Dugan and his wife, Ruth. Jimmy Dugan had been a correspondent for *Yank*, the U.S. Army magazine during World War II, when he first met Cousteau, and his 1948 article in *Science Illustrated* introduced Cousteau, his underwater films, and the Aqualung to

America. Dugan's role in Cousteau's rise to prominence deepened almost immediately. He helped transform the Frenchman's logs and notes into *The Silent World* and from then on would have a hand in almost everything Cousteau published, including the lyrical scripts for his films.

The conversation at the Dugans' went deep into the night. Bond spoke a little French, picked up over the years in school and during some adolescent adventures abroad. Cousteau, the son of a businessman, had lived on 95th Street in Manhattan for a couple of years as a boy, attended an American school and a Vermont summer camp, where he hated riding horses but loved frolicking in lakes. The possibility of undersea living would likely have sparked Cousteau's vibrant imagination. But Cousteau had no formal scientific training, so talking to Bond, a fellow diver and U.S. Navy doctor, must have been inspiring. Bond explained the concept of saturation diving, told Cousteau about Genesis, and no doubt assured him that Dominion over the Seas was just around the corner.

Bond was pleased to be able to share ideas, and perhaps a few watts of Cousteau's limelight. Walt Mazzone couldn't help noticing that Bond got into the habit of saying "Jacques Cousteau and me." Other divers, notably the accomplished Austrian Hans Hass, preceded Cousteau in writing books about undersea adventures and in developing the new art and technology of underwater filmmaking, but Jacques-Yves Cousteau would become something more: the first celebrity diver. In the spring of 1960 he was on the cover of *Time* magazine. A year later, Cousteau joined President John Kennedy for a ceremony in the White House Rose Garden. The president proclaimed him "one of the great explorers of an entirely new dimension" and presented him with the National Geographic Society's Gold Medal, a tradition since the presidency of Teddy Roosevelt. Cousteau's ready smile and Roman nose, his lean physique and distinctive French accent, were becoming familiar across the United States and to millions around the world.

"JYC"—pronounced *zheek*—as Cousteau was known to friends, was a private operator, unfettered by military bureaucracy and backed by sponsors like the French government and the National Geographic Society. He could also tap the proceeds from his roles in profit-making corporations, notably U.S. Divers, founded in the 1950s and based in Los Angeles, which was growing into a major supplier of diving gear, including, of course, the Aqualung. An inveterate promoter like Bond had to have recognized that Cousteau was uniquely equipped to advance—and to advertise—the cause of saturation diving. If the celebrity diver could somehow provide a good showcase for the quest to live in the sea, then Bond might have an easier time appeasing the saturation skeptics and getting the U.S. Navy on board. Being first to put a man on the ocean floor could have wide-ranging political, social, or economic ramifications, but the quest to live in the sea was still nothing like the space race, which had turned the heavens into the prime arena for a great showdown between Us and Them, West and East, Americans and Russians, Good and Evil.

The Russians also had been engaged in diving activities, inspired by American experimentation with helium and by the advent of the Aqualung. Yet their advances, often shrouded in Soviet-style secrecy, appeared to lag behind those in the United States and elsewhere. The British, like the French, were longtime leaders in diving but the Royal Navy was focused on improving conventional methods and striving to reach deeper depths, as Lieutenant Wookey's experimental dive to six hundred feet had shown.

Perhaps more lively competition, even from a military ally like Great Britain, might have been a boon to Bond, but in the days of Genesis he had few known rivals. Those he did have tended to be friendly, like JYC, and another emerging and enterprising competitor in the quest to live in the sea, the American industrialist Edwin Albert Link Jr.

Ed Link had the kind of inventive spirit that had characterized Thomas Edison a couple of generations earlier. Like Edison, Link didn't have much use for traditional schooling. His parents sent him to several private high schools but in 1922, at the age of seventeen, Link dropped out of school and started working full-time at his father's business, the Link Piano and Organ Company, in his hometown of Binghamton, New York. It was a good fit. As a boy Link had been fascinated with gadgets. He was especially fond of his electric trains and happily tinkered away for solitary hours, which he often had to do because his brother was seven years older. Working on player pianos, nickelodeons, and theater organs, Link learned the pneumatics of organ mechanisms and compressed air. He also loved to hang around dusty airfields, lending barnstormers a hand in order to glean lessons about planes and flying. His father strongly disapproved— some pilots had shady reputations—and he threatened to disown his son. But Edwin Junior made his first solo flight in 1926, the year before Charles Lindbergh's solo transatlantic flight in the *Spirit of St. Louis*.

As Link would learn, aspiring aviators had little choice but to fly by the seat of their pants. Many were killed in training. This morbid fact gave Link the idea of developing a machine that would allow people to begin learning to fly without leaving the ground. A few such devices had been built but none was able to simulate the sensations of real flight. By fusing his knowledge of stops and props, Link created a prototype made from organ parts and pumped with compressed air to mimic the movements of a flying plane. His Kitty Hawk of a flight simulator would be called the Link Trainer. In 1929, twenty-five-year-old Link formed his own company to sell his invention. At about the same time he met Marion Clayton, who was twenty-two, fresh out of journalism school at Syracuse University, and working as a newspaper reporter in Binghamton. The two were set up on a blind date and Marion then lined up an interview so she could write an article about this intriguing young local pilot. A couple of years later, in 1931, she married her best story, as she liked to say, and soon took an active role in her husband's activities and enterprises.

Link's new flight trainer business stalled in the wake of the stock market crash and because the U.S. Army Air Corps, a presumptive major customer, wasn't interested. With their nine-foot wingspans, the early trainers looked something like a carnival ride, and during the Depression Link found a sales niche in amusement parks. In the meantime he persevered and by the mid-1930s attitudes and economics began to shift in his favor, especially once the military finally bought in. Over the next two decades an estimated two million airmen, including half a million World War II pilots, would learn to fly using Link's flight simulators.

Link began to ease out of the trainer business in the 1950s after his company, Link Aviation, merged with General Precision Equipment Corporation, a large holding company in New York City. He established the Link Foundation to support research and education in the fields of aeronautics and oceanology, and channeled more of his seemingly bottomless reserves of energy into the sea. Already a self-made millionaire and a holder of twenty patents, Link had grown restless and taken up sailing. In 1951, Link tried the Aqualung and joined the growing legion of scuba enthusiasts. He was in the Florida Keys with his wife and two young sons that summer on what would be the first of many expeditions in search of shipwrecks and sunken treasure. Link often coordinated his forays into marine archaeology with the Smithsonian Institution and other reputable organizations, and he became enthralled with the history underlying the artifacts he found. High school had never been like this.

Link's discovery of a four-hundred-year-old cannon a couple of years later got him hooked on researching and retracing the Caribbean routes of

Christopher Columbus. He even scoured the north coast of Haiti for the lost *Santa María*. This blossoming field of undersea archaeology, fostered by scuba, also gave Link fresh opportunities to exercise his inventive spirit by designing needed tools—like a jet air hammer and an underwater metal detector. When Ed Link caught wind of Bond's Genesis experiments he was quick to see the potential of saturation diving. The greater efficiency of staying down longer had a natural appeal to a businessman with a sharp eye for the ocean bottom and for the bottom line.

Link had made numerous influential friends and industry contacts over the years, including some in the U.S. Navy, and he did what he could to convince the Navy brass to embrace and develop saturation diving. He argued that it held great promise and ought to be pursued more actively than just through Dr. Bond's unofficial experimentation. The Navy didn't seem to agree. If Link wanted to see real progress anytime soon he would have to take matters into his own hands. Unfettered by official red tape, he had his own financial and inventive resources.

So in February 1962, shortly after Bond's trinity had wrapped up the animal phases of Genesis and published its favorable results, Link presented a project he called Man-in-Sea to the National Geographic Society's Committee for Research and Exploration. He soon had the society's backing for his proposed experiment in undersea living, and in addition to the money, he could count on a story in *National Geographic* magazine. Link still had to invest substantial sums of his own, as he had for other expeditions. He didn't mind. His earthy pragmatism and simple tastes lent credence to his frequent refrain about having little interest in money, except as a means of survival. In Europe he and his wife got around in their battered Volkswagen Bug. Yet Link had the wherewithal to put up $500,000 in the late 1950s to replace a couple of his first boats with a new, specially outfitted expeditionary yacht that he designed with marine archaeology in mind. *Sea Diver*, which might be considered Link's version of Cousteau's *Calypso*, was ninety-one feet long, steel-hulled, and equipped to accommodate a dozen crew members. It had navigational systems to rival those on a large ocean liner. It had a built-in crane and winches for raising heavy relics from the sea floor, and a diving compartment crowded with gear. *Reef Diver*, an eighteen-foot auxiliary cruiser, was slung in davits on the stern. That still left ample space on the open deck to carry Link's twin-engine plane, *Widgeon*.

Link met with Jacques Cousteau a number of times in Monaco, where Cousteau had an office at the Oceanographic Museum, housed in

a luminous limestone edifice built a half-century earlier by Prince Albert
I. The building blended majestically into the Rock of Monaco and overlooked the Mediterranean's azure expanse. From the renowned museum,
it was a short walk past the Royal Palace—home to JYC's friends Prince
Rainier III and Princess Grace, the former American actress—and then
down the steep embankment to Monaco's exclusive harbor, where Link
kept *Sea Diver* berthed in winter. Cousteau's team had a craft moored
nearby.

While it's not clear how and when the paths of the inventor and the
celebrity diver first crossed, it couldn't have hurt that they were practically neighbors for at least part of the year. Link's reputation may have
preceded him, too, because Cousteau had always dreamed of flying. In
the mid-1930s Cousteau was in training to become a navy pilot until he
almost lost his right arm in a late-night car crash. The twenty-six-year-old
French lieutenant was then forced to work at sea instead. He was assigned
to gunnery duties on a battleship and took up swimming to strengthen his
injured arm. His fascination with the silent world soon followed. George
Bond was apparently the only one of the three men who didn't yearn to
fly before learning to dive. Any fancy for flight on Bond's part may have
died with his older brother, Bobby, a pilot, who was killed in a plane crash.

Late in the fall of 1961, Link had lunch with Cousteau and his wife,
Simone, at their Monaco home and the two men discussed their mutual
interest in staging a prolonged deep dive. They also talked about the possibility of teaming up for an experiment the following spring. Upon Link's
return to Monaco a few months later, and with his Man-in-Sea project
taking shape, he met again several times with Cousteau. It was clear that
the U.S. Navy was in no hurry to pick up where Bond's animal experiments left off, so Cousteau and Link decided to put saturation diving and
undersea living to the test. Cousteau's team would build a "house" big
enough and well-equipped enough to shelter at least a couple of divers for
the duration of the Man-in-Sea experiment. It would have the cylindrical,
railway tank car appearance of a pressure chamber. The air pressure inside
this sea floor habitat would be kept equal to the water pressure outside.
That meant the sea would not enter the pressure chamber. There would
be a pool of water in a circular opening about the size of a manhole in
the floor. A diver in the habitat could pass through this liquid looking
glass freely, from the comforts of his dry shelter into the uncertainties of
the silent, wet world outside, and then he could come back again. Ideally,
to best show off the potential advantages of living in the sea, the divers

should be able to spend ample time in the water, not just lounge inside the habitat.

Link's contribution to the project would be an aluminum cylinder he had designed and built with backing from the Smithsonian Institution. It looked like a section of shiny sewer pipe, capped on top and with several portholes around the midsection. In many respects it was similar to a diving bell, a piece of underwater equipment that had been around for centuries, beginning with simple forms that weren't much more than upside-down barrels with an open end at the bottom. When lowered into the water, the air pocket trapped inside the bell provided nominal shelter and the means to stay down longer than would be possible on a single breath. Alexander the Great is said to have used a diving bell during the siege of Tyre in 332 B.C.

Much more recently, in the 1930s, the British Navy improved upon the basic concept with what was called a Submersible Decompression Chamber, which gave Link some ideas for his cylinder. Prior to the SDC, a diver had little choice but to take his decompression stops hanging in mid-water on a "shot rope," a line between the bottom and the surface. In cold water and strong currents, a long decompression could be incredibly tedious and exhausting. The SDC was like a pressurized elevator, with a diver's tender on duty inside. At a depth of about sixty feet, after having made some deeper stops along the shot rope, a hardhat diver could climb a ladder into the open bottom of the SDC, passing through a liquid looking glass from the water into the bell's dry, pressurized interior. With the help of the tender, the diver could take off his helmet and continue decompressing in relative comfort and safety. Link's cylinder, which was eleven feet tall and three feet in diameter, would serve a similar purpose, as Link and Cousteau discussed. They could use the cylinder as a pressurized elevator to shuttle divers or visitors between the surface and Cousteau's undersea dwelling. If habitats were ever going to be placed deep on the continental shelf, this type of diver delivery and retrieval system could be as essential as an elevator in a skyscraper, even more so, considering the water and the pressure. Of course the success of all this newfangled Man-in-Sea hardware assumed that human saturation divers would fare as well physiologically as the goats had in Genesis.

The plans for a joint undersea venture with Cousteau were left to simmer. Link sailed on *Sea Diver* from the Côte d'Azur for southeastern Sicily. With the new cylinder on board for preliminary testing, Link and his crew went to work with an archaeological team to help dredge up tons of Byzantine artifacts lost twenty-five centuries before—columns of white marble, composite cornices, foundations, mosaic fragments, a ritual saucer engraved with religious motifs, even a pulpit carved in green marble. All this human history had been hidden in less than forty feet of seawater. Other treasures yet to be discovered lay much deeper. Later that summer, Link went to examine an ancient Greek shipwreck north of Sicily near the Lipari Islands. The wreck was at 150 feet, deep enough to make excavation a time-consuming and expensive prospect by "bounce" dives, as conventional dives were sometimes called because of their brief bottom times. With saturation diving, if it proved feasible, divers stationed on the sea floor could carry on with their underwater work uninterrupted for days, almost as if they were on dry land.

When Link returned to the south of France, during the summer of 1962, it became clear that Cousteau was bailing out of the Man-in-Sea project. Link was told that the undersea house wasn't ready, but he believed that the celebrity diver was only using that explanation as a convenient excuse. In February Link had arranged for him and Cousteau to meet with one of Link's many high-ranking friends, Vice Admiral John T. Hayward, a deputy chief of naval operations, in the hope of getting some Navy participation in Man-in-Sea. Link believed that the real reason Cousteau had backed out was that he had discovered that there was considerable money to be made in selling the story and the television rights about an underwater house—so best not to have to split the proceeds.

"Frankly I am a little sore about this double-crossing by Cousteau since it was we who interested him and started him off and he agreed to work with me," Link wrote to his friend Mendel Peterson, a Smithsonian curator. Link told Peterson he wasn't interested in a public squabble with Cousteau. "When queried, I am simply announcing that Cousteau did not get his part of the program ready to go along with us at this time, and we are proceeding as scheduled." It was by then the end of August, and Link planned to start his experiment in less than a week. He wouldn't have the benefit of Cousteau's underwater house, but he could use his cylinder instead, even if it was more pup tent than bona fide dwelling.

That same week, George Bond sent a letter he hoped would persuade Link to hold off on his attempted saturation dive. Bond worried that Link was putting too much stock in the animal experiments and that a rush to put humans under prolonged pressure could be dangerous, even deadly. Experimental human exposures, under controlled and carefully monitored laboratory conditions, were needed before anyone took the concept of saturation diving deep into the open sea. Furthermore, Bond said he was on the verge of receiving Navy approval to proceed with a new phase of Genesis in which he would be allowed to test human volunteers. Sub-rosa animal experiments were one thing, but even George Bond couldn't lock people into a chamber for days at a time without proper approval.

Link sensed a hint of selfish interest on Bond's part, and he may have been right. If Bond savored being the first to demonstrate a three-hundred-foot submarine escape, he might also want to be the first to show that the concept of saturation diving—*his* concept—could enable man to live in the sea. Bond even raised red flags about national security—not a characteristic concern of his, and one that neither Link, who was hardly cavalier about such things, nor even the cautious U.S. Navy brass, shared. In any event, Link said he wasn't trying to steal Bond's thunder. Were it not for Genesis and Bond's animal experiments, Link knew he would not attempt such a long, deep dive. Bond deserved all due credit, Link would say. In fact, he would have preferred to have the U.S. Navy and Dr. Bond take charge of the experiment he was about to do. When he lobbied Vice Admiral Hayward in the hope of convincing the Navy to actively pursue saturation diving, he encountered many of the same fusty attitudes Bond had. So one of Link's motivations for going ahead on his own was to prod the Navy into action, show them it could be done, should be done, and the sooner the better. And if the Navy took up the cause of saturation diving, Link would gladly drop his own experiments. But until then, the intrepid inventor was not going to sit on his capable hands and wait.

Bond's expressed desire to have Link drop his experiment, or at least postpone it, likely stemmed in part from a concern that any serious mishap or fatality could make it that much harder to sell the Navy on saturation diving. One tiny bubble on the brain of one diver could ruin everything. If Navy skeptics had their worst suspicions confirmed at this early stage, the quest to live in the sea could end almost before it began. Rockets could blow up and test pilots could die in spectacular fireballs, but the space race and the cosmic tug-of-war between Us and Them would go on, regardless. Not so with saturation diving, at least not yet, despite George Bond's inspired sermonizing, and Ed Link's determined prodding.

As it turned out, Link didn't receive Bond's entreaty until after his experiment had ended. Link would later say that if he had gotten the letter sooner he probably would have heeded Bond's warning. But that summer found Link decided that his split with Cousteau might have been for the best. Cousteau and his team were planning what seemed to Link an overly prudent venture. They were building an undersea habitat, a worthwhile effort, but they were going to place it at two atmospheres, a depth of just thirty-three feet. The week-long duration envisioned for the Cousteau team's dive certainly had great publicity value, Link thought, but at that minimal depth the experiment would represent mere baby steps onto the continental shelf. Cousteau would prove only that man could live in a swimming pool. Link set his sights at two hundred feet, the same trial depth Bond had simulated with his Genesis menagerie in New London.

From the port of Monaco, Link sailed ten miles or so west into a sheltered bay at the medieval citadel of Villefranche-sur-Mer, which offered a variety of possible depths. Villefranche was also the Mediterranean home port of the U.S. Navy's Sixth Fleet. Link's prodding of top military brass had not been entirely in vain: The Navy took charge of delivering Link's cylinder to France and planned to provide him with the costly helium he would need. The Navy also granted Link's request to have a submarine rescue ship, equipped with expert diving teams and decompression chambers, available in the bay during the planned two-hundred-foot saturation dive.

Link had no medical specialist on *Sea Diver* but through his various Navy contacts specifically requested that Charlie Aquadro be assigned to his project. Aquadro had befriended Link while stationed at Key West and had been among the Navy divers assigned to two of Link's recent high-profile underwater archaeological expeditions, in Jamaica and Israel. Aquadro would have been happy to join Link again for Man-in-Sea, but he was about to serve as medical officer aboard a Polaris submarine, so Link's request for medical assistance went to Dr. Robert Bornmann, an able lieutenant commander with the kind of know-how Link needed.

Bornmann had come to the Navy by way of Harvard and medical school at the University of Pennsylvania, and had been through the Navy's deep-sea diving school in Washington, the submarine school, and the training tank in New London. When the assignment with Link came along, Bornmann had been the medical officer and an instructor at the Navy diving school in Key West for the better part of a year. Prior to that he had spent a couple of months on Bond's staff at the Medical Research Lab. Although not directly involved, Bornmann was certainly aware of Genesis, and he struck up a friendship with Dr. Bond. A cordial letter from Bornmann to Bond that included an update on Link's progress is what had prompted Bond to write back, to Bornmann, with the warning that Link should hold off until Bond could run his tests on human subjects in chambers. Bond and Link surely followed each other's underwater activities, but the two Americans rarely had direct exchanges and in that sense they were more rivals than friends.

Bornmann had no specific experience with the peculiarities of saturation dives—outside of Bond's immediate circle no one really did. Nonetheless, calculating decompression schedules and treating cases of bends were part of his training. Among the items Bornmann packed for the trip was the published report on Genesis. In a sense the report was as much a talisman as anything because its gas mixtures and decompression schedules, while perhaps safe for a goat, probably wouldn't be exactly right for a man, as Bond had stressed in his letter.

Not everyone found Ed Link easy to work with, but Bornmann, a bright and affable medical officer, got along well with the inventor, who had a stubborn streak and a bit of a temper, as Bornmann discovered. Link was an intense worker and devoted little time to diversions. Apart from daily newspapers, and perhaps *Reader's Digest*, he typically read only what pertained to his projects. The only television he watched was the news. At fifty-eight, Link was tanned and as active as someone years younger, despite a graying wreath of hair around his head. His confident grin would deepen his facial crags.

Toward the end of August, Link moored *Sea Diver* in a shallow section of Villefranche Bay and began a series of test dives, starting with one in which Link would lock himself into the cylinder. Link, like Bond, would never ask anyone to do anything he wouldn't do himself. Link's aim was to test his prototype in an airtight mode, as if it were a mini-submarine or a bathysphere like William Beebe's. With a hatch in its lower end sealed shut, the cylinder maintained a pressure of one atmosphere, the same as at the surface. Link sat in his cramped contraption like a man in a phone booth, tinkering and running through some checks. He was about sixty feet down. Suddenly he told his topside crew to bring him up. The hatch beneath his feet had apparently given way. The sea surged in, equalizing the pressure. Link was soon up to his neck in chilly saltwater. The boom and lifting line on *Sea Diver* were designed to handle the cylinder's weight of more than two tons, but when the cylinder was filled with water, it became too heavy to handle. If the boom or the cable snapped, Link's cylinder would sink to the bottom of the Mediterranean and become his casket.

The crew cranked up the supply of compressed air, delivered by umbilical to the top of the cylinder, and forced the seawater out. But as the cylinder began to rise, Boyle's law kicked in, the air inside expanded, and the over-buoyant cylinder shot out of the water like a great silver torpedo, clearing the surface by a good few feet before dropping back into the sea. Fortunately, Bornmann and the quick-thinking crew cranked up the air again before the cylinder could take on more water, and got it buoyant enough to reel in.

Once safely back on deck, Link said they had done enough for the day. The supreme fix-it man then disappeared below deck, making his way through the wood-paneled lounge, past the fireplace of coral encrusted bricks and the mermaid figurine near the galley door, and downstairs into his private cabin. His biggest test was soon to follow.

On August 28, Link attempted his longest, deepest dive: sixty feet for eight hours. Similar depths and durations had been achieved, but mostly by caisson workers or in dry test chambers. The standard U.S. Navy diving tables maxed out at a bottom time of between three and four hours, depending on whether air or a more complex helium-oxygen mixture was used, so there was no information about whether Link might suffer the bends. Even at this modest depth of less than three atmospheres, Ed Link was taking his chances, in a prototype that had just threatened to drown him. The pressure inside the cylinder was raised to equal the surrounding water pressure so Link could open the hatches in the bottom-facing end of the cylinder and swim freely in and out, as if using a diving bell. He donned scuba gear and spent much of the afternoon working in the water.

A lunch of macaroni and cheese was delivered in a sealed container by the younger of Link's two sons, twenty-one-year-old Edwin Clayton Link, known as Clayton, an experienced diver who was becoming a regular participant in his father's expeditions. Clayton also brought his father a few letters. The prompt and pragmatic businessman in the cylinder dictated responses by telephone from his seat inside the silver cocoon. Link was relaxed enough to find time for a nap, which he took while hunched over a small tray table, like a dozing schoolboy. He later received a hot roast chicken dinner as the cylinder was gradually raised through the water and the pressure inside decreased. With two of the planned six hours of decompression still to go, Link's crew hauled the sealed-up cylinder onto the deck of *Sea Diver* and laid it on its side, on a base that Link built to hold it in place. Once horizontal, Link was able to lie down for the first time in about twelve hours. He fell asleep almost immediately. Dr. Bornmann crunched some decompression figures and extrapolated from the Navy decompression schedules for "exceptional exposures." These schedules were considered somewhat risky, but they were better than nothing if a diver ever got stuck on the bottom for much longer than the standard tables allowed.

About eleven-thirty that night, Dr. Bornmann opened the hatches at the end of the cylinder and peered inside. The inventor was asleep. Wasn't he? Other than hearing his voice over the intercom, there was no way to be sure he was all right, nothing monitoring his heartbeat or other vital signs. For a moment, like a nervous parent who tiptoes into a room to check on a sleeping child, Bornmann experienced a fillip of adrenaline as Link lay motionless in the aluminum cradle. Then Link suddenly stirred, crawled out, and kissed his wife and expeditionary partner, Marion Clayton Link. He had survived in the sea for eight hours at sixty feet!

The cylinder had proven to be a viable shelter, at least down to three atmospheres. It enabled Link to do a few tasks in the water and had maintained the desired gas mixture and pressure for a total of nearly fifteen hours, including decompression. Link seemed uninterested in his having broken a diving record that was more than two centuries old. In the 1690s, Edmund Halley, the English astronomer known primarily for tracking the comet that would be named for him, built a barrel-shaped diving bell out of wood and lined it with lead. It had greater girth than Link's cylinder but about the same interior volume. Halley devised a system of barrels and leather tubes that channeled fresh air from the surface to give the bell's occupants more breathing time.

During one trial, Halley wrote that he and four others stayed in the bell for an hour and a half at a depth of about sixty feet. The men couldn't do much while submerged, although Halley had ideas for supplying air so that divers could work outside the bell. No one suffered any ill effects, but Halley and his crew had no way of knowing that with the depth and duration of their dive they narrowly averted getting bent, or worse. Nonetheless, his innovative bell and record stay in the sea made diving history, even if Halley's Comet would become better known than Halley's bell.

So it was that Link, whom many considered a Renaissance man, broke a Renaissance record. But his day at ten fathoms was only meant as the final test before dropping the cylinder down to two hundred feet for a few days. Because Link felt he should oversee the experiment—and because the earlier trials were a reminder that at fifty-eight he was getting a little old for this sort of thing—he decided not to make the next saturation dive himself. Instead he gave the starring role to a volunteer half his age, an experienced Belgian diver named Robert Sténuit. Link had met Sténuit at Vigo Bay, on the northwest coast of Spain, where Sténuit was with an expedition searching for a fleet of galleons and cargoes of gold sunk during Halley's day. The Belgian treasure hunter impressed Link as a skillful, even-keeled diver, and Sténuit further proved himself on Link's expedition to Sicily that spring. After returning home to Brussels, Sténuit received Link's invitation to make the two-hundred-foot dive in the cylinder. He readily accepted. It was as if he'd been invited to be the first man on the moon. Sténuit, twenty-nine, dark-haired, svelte, and mustachioed, rejoined *Sea Diver* in Villefranche for a few days ahead of the experiment to get acquainted with Link's underwater shelter. This would be his first experience breathing a mix of helium and oxygen. If all went well, he might also be the first man to live in the sea.

5

DEPTH AND DURATION

On September 5, 1962, *Sea Diver* anchored within view of the shoreline near Cap Ferrat, which juts like a thumb into the Mediterranean to form the eastern rim of Villefranche Bay. Like other parts of the Côte d'Azur, this scenic little bay was ringed by rocky hillsides spread with a mosaic of pastel residences with tile roofs. The sea depth just beyond the bay dropped to more than two hundred feet, perfect for Robert Sténuit's trial dive in the cylinder.

As Link had just learned from his experience at sixty feet, the cold was going to bother Sténuit as much as anything, That was one drawback to substituting helium for nitrogen in order to avoid narcosis. Helium is a fantastic conductor of heat, whisking it away about six times faster than air. And not only would Sténuit be bathing in an atmosphere that was almost pure helium, he would be breathing the gas. Helium would sap his body heat. While Sténuit got a good night's rest before the dive, Link was busy installing an additional heater inside the cylinder.

Winds whipped up big swells and threatened to delay the dive, but calm returned the next day. The USS *Sunbird*, the submarine rescue ship Link had requested, arrived at dawn as promised. Sténuit was locked into the cylinder for the ride down to two hundred feet, which would put him about forty feet above the muddy bottom. He was to stay down for two, maybe even three days. Cousteau's undersea house might have made the Belgian's accommodations more comfortable, but Link's cylinder would have to suffice.

Keeping warm inside the cylinder was one challenge, but Sténuit also had to stay warm in the cool water outside. Even with a protective wet suit, itself a recent innovation, Sténuit quickly learned that the numbing cold of the deep Mediterranean made it difficult to spend more than a few minutes in the water. He was able to dive long enough to pose for photographers from *Life* and *National Geographic* who made brief bounce dives from the surface. Sténuit simulated some simple tasks and fetched a dinner that was sealed in a container and lowered like fish bait to the cylinder. The well-intentioned heaters built into the top of the cylinder fairly broiled Sténuit's head, but the rest of him nearly froze. No prototype is perfect, but these added risks due to excessive cold, or equipment failures, were part of the reason Bond had tried to discourage Link from running his test at sea. In addition, Link didn't have the equipment to monitor and collect detailed physiological data. Without it, little could be learned about the actual effects on a diver of a prolonged exposure to pressure, cold, helium, and the rest.

Sténuit went to sleep by nine that night, seated and hunched over the little tray table with his head on his arms. Largely because of the cold, he spent most of his time inside the undersea turret, rather than out, reading books he brought along like Céline's *D'un château l'autre—Castle to Castle*. The goal had been to demonstrate that a saturation diver could do a day's work in the water, but in the end Sténuit made just two ten-minute dives, including the one to pose for the photographers. Still, Sténuit had put in twenty-four hours under pressure and at sea, which was believed to be longer than any diver before him. A barrier had been broken, even if it didn't seem to be as striking a human achievement as the miracle of flight, or even the first ascent to the summit of Mount Everest a decade earlier.

Sténuit was supposed to stay down for another day or two to extend this new duration mark, and to make a more convincing case for living in the sea. But as the first day came to an end, Link called off the experiment. Sténuit protested over the intercom, but a leak in the homemade ventilation system was sapping their limited helium supply and they couldn't risk running out before Sténuit had completed his decompression. There was also a predicted return of rough weather that could make the raising of the two-ton cylinder hazardous. Despite the curtailment of the experiment, Link considered it a real breakthrough. He believed they had proved that man can safely live at significant depths for long periods of time. Back in New London, however, Bond and Walt Mazzone were much less impressed. Perhaps Link had just been lucky, like Edmund

Halley a few centuries before. And the experiment was not over yet: They still had to bring Sténuit back to the surface, safe and bubble-free.

The *Sunbird* came to the rescue with enough additional helium to last through a long, steady decompression. Dr. Bornmann factored in both the standard Navy diving tables for helium and the Genesis results. After twelve hours at one hundred feet, the pressure inside the cylinder was dropped to the equivalent of fifty-five feet. It was about then that Sténuit felt a groaning ache around his right wrist that soon crept up his forearm— a telltale sign he was bent. As the pressure was reduced by another few pounds per square inch, the pain got worse, leaving little doubt that bubbles were affecting his wrist and arm, and there was always a fear the bubbles could spread. Bornmann promptly cranked up the pressure to take his Belgian subject back to a depth equivalent of seventy feet. As Bond's group had found with the saturated goats, recompression still seemed to be the best remedy for the bends, even when coming out of a full saturation. The added pressure relieved Sténuit's pain. Bornmann cautiously held the diver for another twelve hours at seventy feet before resuming the schedule that would have him back on the surface in a couple of days. Sténuit had no choice but to remain sealed inside the oversized aluminum can, passing time and peering out the portholes, assuming all the while that he would be able to rejoin those on the outside.

The crew waited for another patch of rough weather to pass before hauling the vertical cylinder out of the water and laying it on the deck of *Sea Diver*, as they had when Link was inside. After the twelve-hour pause Sténuit was eased closer to sea level pressure by about ten feet every few hours. There also had to be a methodical increase in the amount of oxygen in the breathing mix to compensate for the decreasing gas concentration as the pressure gradually dropped. The cylinder's oxygen level had been kept at just 3 percent while Sténuit was at seven atmospheres, but the laws of nature gave that 3 percent a seven-fold boost, thereby creating a healthy sea level equivalent of about 21 percent oxygen. There was some wiggle room, but too little or too much oxygen could spell trouble, and ultimately death.

At twenty minutes past six o'clock on a Monday morning, September 10, Sténuit finally ended his pressurized confinement after a grand total of three full days, twenty hours, and thirty minutes. For almost twenty-six of those hours he was at a depth of two hundred feet, followed by the sixty-six hours it took to reel in the cylinder and ease him back to the benign pressure at sea level. A decompression lasting more than twice as long as a diver's time on the bottom was not the way saturation diving was supposed to work, but this was an experiment. Even if Sténuit stayed only a day, it was fair to say, as Link and his team did, that he had become the first man to live in the sea. A year earlier, of course, in May 1961, Alan Shepard had made the first space flight, a suborbital trip that lasted just fifteen minutes—and he was hailed as the first American astronaut. The Russians had put the first man in space a month earlier than that when Yuri Gagarin orbited the earth on April 12—a considerably greater achievement than the Shepard flight. Both trips were indelibly etched in history.

Link's friend and admirer John Godley, the third Baron Kilbracken, had been on board *Sea Diver* to witness the cylinder dives and wrote about them for *National Geographic*. The issue in which the article appeared— spread across fourteen pages with color pictures—wasn't published until eight months later. But it said that Sténuit's dive "would be an advance as complete—and as important in its own way—as the orbits of the space- men." Lord Kilbracken declared: "There can be no doubt whatever that Robert Sténuit's dive was an outstanding achievement" and "an immense step forward."

This belated assessment was perhaps compensation for the meager story that appeared in the back pages of *The New York Times* the day after the cylinder was raised—two noncommittal paragraphs under a small headline: "Diver Up After 34 Hours." No mention of immense steps forward, nor of an inner space advance to rival those in outer space. This would not be the last time that barrier-breaking dives received barely a drop of ink in the newspapers. The quest to live in the sea may not have been the space race, but it did have Jacques-Yves Cousteau, and the benefit of being the world's first celebrity diver could be seen in the finer treatment of a story about Cousteau's impending experiment on the very same page of the *Times*. Accompanied by a picture of the undersea explorer's familiar face and a catchy headline—"Two Frenchmen to Spend Week Living in House on Floor of Sea"—the article dwarfed the Link story. It also made Cousteau's venture sound much more impressive.

Cousteau's prototype was not exactly a house, and one of his divers lightheartedly dubbed it Diogenes after the truth-seeking Greek philoso- pher who was said to live in a tub. The dwelling was set up in a protected cove near the island of Pomègues, a strand of jagged, glittering limestone a short boat ride from the sprawling harbor at Marseille, where Cousteau's research group had its headquarters. It was a steel chamber shaped like a stout railway tank car, as envisioned when Cousteau had considered collaborating with Link. Seventeen feet long and eight feet wide, it was certainly roomier than Link's cylinder. And instead of hanging vertically from the surface like an elevator it would be positioned horizontally, more like a house, and held in place by thirty-four tons of pig iron that hung in ungainly clumps from four heavy chains affixed along each side of the tank. With the habitat adequately pressurized, the sea would stop at the circular opening in the floor, just as it did at the bottom of Link's cylinder. But this "house on floor of sea," while it might appear to be a big advance, was really more about spectacle than science, as Link saw it.

Thirty-three feet, or two atmospheres, the planned depth for Cousteau's house, had long been considered the safe depth for a dive that would require no decompression stops, regardless of bottom time. Down to thirty-three feet, a diver could stay as long as he liked and swim straight back to the surface without getting bent. Such "no-decompression" dives, often called "no-D" dives for short, could also be made to greater depths, but you had to be careful not to exceed strict time limits. You could dive to sixty feet for sixty minutes, for example, or to ninety feet for a half-hour, and then come straight back to the surface without paying any decompression penalty. But from the surface down to about thirty-three feet, there was theoretically no time limit.

Cousteau and his team were being intentionally prudent, reducing the risk of a widely publicized disaster that would besmirch the nascent reputation of saturation diving. Cousteau's divers would also breathe ordinary compressed air, a much simpler approach than mixing helium and oxygen. At thirty-three feet they would be far too shallow to have to worry about *l'ivresse des grandes profondeurs*—nitrogen narcosis. While the project might not yield major scientific breakthroughs, Cousteau could nonetheless put on a good show by spotlighting duration, if not depth.

Sixty people, including fifteen support divers, would set up and maintain Diogenes, and Cousteau picked two favorites from his team to be his leading men. They were Albert Falco, thirty-five and swarthy, and Claude Wesly, who was stocky like Falco, but a few years younger, with sandy hair and a more effusive personality. They were good friends and Cousteau sensed that they would make compatible undersea roommates.

Cousteau's project was known in English as *Conshelf One*, short for Continental Shelf Station Number One, or *Opération Pré-Continent No. 1*. The operation began shortly after noon on September 14, just a few days after Link had ended his experiment at Villefranche—timing that seemed to lend credence to Link's suspicions about why Cousteau had dropped out of Man-in-Sea. A hundred miles to the west of where Link had placed his cylinder, Falco and Wesly climbed up a ladder through the liquid looking glass in the belly of Diogenes, which hovered about seven feet over the Mediterranean seabed. Compressed air, water, and power were piped in from the *Calypso* and *Espadon*, Cousteau's smaller boat. A command center with backup systems was set up on Pomègues where Cousteau and other observers, including members of the press, could watch the unfolding drama live on closed-circuit TV.

Once settled inside Diogenes, Falco and Wesly were free to don their Aqualungs and jump through the open hatch into the sea. They didn't quite reach the goal of spending a total of five hours a day in the water, but they did put in a lot more time than Robert Sténuit. The two divers performed some perfunctory tasks, such as assembling patterned cubes as part of a psycho-technical test. They pieced together cement block fish houses and set up what Cousteau called a "fish corral"—basically a tent made of netting. The two divers also poked around a shipwreck discovered nearby. On a few occasions they swam down to eighty feet, a depth that, according to standard dive principles, should be safely within reach of their base at thirty-three feet. Back in the habitat, Falco and Wesly could take hot freshwater showers. To while away the hours inside their habitat, the two divers had a radio and television on the cluttered shelves over their sleeping cots. Wesly, a violinist, enjoyed listening to Bach and Vivaldi. Falco passed time by building a model ship. Two coffin-sized pressure chambers were also built in, opposite the cots, although at this depth there would presumably be no emergency need for them.

The *Espadon* chef sent down the main meals, some with regal sauces and pastries. The two divers were ravenous but by the third day suffered nasty bouts of gastritis and upset stomachs. They eased off the grains, animal fat, and milk in favor of more readily digestible dishes. Cousteau brought caviar and a bottle of Bordeaux on one of his visits. Doctors came down for several hours each day to observe the divers and do checkups. Cousteau even had a dentist and a barber drop in. At times it got so busy that Falco and Wesly just wanted to be left alone. Omnipresent photographers and the filmmaking crew added to the commotion. It was like a movie set, thirty-three feet down.

On the seventh day, September 21, Falco and Wesly were set to return to the surface. Although no decompression stops should be needed,

Cousteau's doctors agreed on a precautionary schedule intended to hasten the release of whatever minimal amount of nitrogen had seeped into their tissues. So the divers each put on a mask and lounged on their cots for a couple of hours while breathing an oxygen-rich gas mix. By one-thirty that afternoon, on a calm and clear day, Falco and Wesly took a final plunge through the looking glass and then swam to the surface. Cousteau's chief doctor kept an eye on them for two days afterward. As hoped, neither diver showed any signs of bends or other illness.

Link privately remained adamant that Conshelf One proved little more than that man could live in a swimming pool, but it did make a good week-long show. Marion, Link's journalist wife, lamented that Cousteau's "exhibition," as she called it, had thoroughly eclipsed their worthy Man-in-Sea project in the newspapers. Cousteau would acknowledge that Conshelf One was an experiment more logistical than physiological, and a timid one at that. But the following year in *The Living Sea*, his first book since *The Silent World*, he proclaimed that his divers were "the first men to occupy the continental shelf without surfacing for a significant period of time." Falco and Wesly, too, took pride in being the world's first human sea dwellers, even if theirs was a shallow venture. As Wesly saw it, Robert Sténuit's dive, although much deeper, was simply too short to be considered a true example of living in the sea.

One team had depth, one had duration. Whatever each may have lacked, George Bond could hardly have picked better preachers for the gospel of undersea living. A month later, Cousteau and Link were both in London spreading the word at the Second World Congress of Underwater Activities, a landmark gathering of diving specialists and dignitaries. Ed Link, one of two dozen featured speakers, paid brief tribute to Bond, Dr. Workman, and their animal experiments as he told of his eight hours at sixty feet in the cylinder. Mostly Link heralded Sténuit's record dive— he didn't hesitate to use the word "record"—of twenty-six hours at two hundred feet. Sténuit was in the audience and Link had him stand up, just to let everyone see that the dark-haired, lanky Belgian was indeed "a healthy specimen" and none the worse for his record number of hours under pressure and under the sea. Within a matter of months, Link said, he would push further into the depths with another long-duration dive, this time to an astonishing four hundred feet, a depth rarely attained for a few minutes, never mind days.

While Ed Link made quite a splash, Cousteau had opened the prestigious conference with a flamboyant keynote speech in which he waxed poetic more than scientific about Falco and Wesly's recent undersea experience and about the prospects for a manned undersea future. The celebrity diver announced that his team was already planning to build an undersea "village" that should be ready by the following year. He never mentioned saturation diving, or the target depth of his planned village, but fired the imagination of his appreciative audience by predicting the coming of a new kind of man, a surgically altered, deep-swimming under-water breed he called "Homo aquaticus." *Homo aquaticus!* American news services earnestly reported Cousteau's divination. JYC's friend and scribe Jimmy Dugan breathed further life into this hypothetical species with an article in *The New York Times Magazine.*

At the dawn of the 1960s, the mushrooming nexus of high technol-ogy and hubris held boundless potential. Man had lately split the atom, broken the sound barrier and the four-minute mile; he had produced the revolutionary transistor, cracked the DNA code, and was on his way to the moon. Living in the deep sea, on the continental shelves of inner space and deeper still, should be well within reach.

6

EHPERIMENTAL DIVERS

In the fall of 1962, as Jacques Cousteau and Ed Link made their bold pronouncements in London, George Bond was getting ready for a new phase of Genesis. The combined effect of Link's prodding of Navy brass, the favorable outcomes of Man-in-Sea, and Cousteau's showy Conshelf One no doubt helped set the stage as Bond sought the necessary seals of approval to experiment further with saturation diving. Experimenting on animals had been one thing, but now he wanted to run tests with human subjects. For that he would ultimately need approval from the secretary of the navy. The fact that he got it was a tribute to his own persistent sermonizing, because along the way Bond befriended some influential supporters, perhaps none more important to his cause than Dr. James Wakelin, the Navy's first assistant secretary for research and development.

The Navy created the administrative post in 1959 and it gave Wakelin wide authority over R&D programs. A naval officer during World War II, Wakelin earned a doctorate in physics from Yale and then spent a decade in business before President Dwight Eisenhower appointed him to the new Navy post. Wakelin soon developed a reputation as an advocate for expanding the Navy's oceanographic research, and he was troubled by the technical limitations that prevented man from delving very deeply into the sea. Wakelin was among Ed Link's friends in high places and very likely had a hand in getting Link the Navy assistance he wanted for Man-in-Sea. How Bond managed to meet the secretary is unclear—he once recalled getting a visit from Wakelin at New London—but the two men developed enough of a rapport that Bond became a welcome visitor at the secretary's Pentagon office. In addition to their mutual marine interests, they were personally compatible. Wakelin, like Bond, was an affable sort—one aide fondly called him "Gentleman Jim." Because of Wakelin's top role in setting R&D priorities, his support for Bond's proposed human experimentation was critical, probably even decisive.

Bond arranged to run a first human test called Genesis C—the major animal phases had been called Genesis A and B. The test would have been run at the New London lab but Bond was awaiting the installation of a suitable new chamber, so the experiment moved south to the Naval Medical Research Institute in Bethesda. NMRI was part of the sprawling Naval Medical Center, its landmark headquarters a lone skyscraper in the Maryland countryside. President Franklin Roosevelt had chosen the home of the new medical center when the land was no more than a cabbage patch on a run-down farm, but the president was moved by the presence of a lovely spring-fed pond. It reminded him of the pool of Bethesda, a place of healing in the Gospel of John. Now Bond would come to Bethesda with Genesis and his diving gospel. Upon seeing the chamber, however, Bond thought it looked neglected and unworthy of the elegant breakthrough in biological science he was hopeful would take place.

For this first human test, Bond and Mazzone intended to make themselves two of the guinea pigs, but the Navy didn't want the principal supervisors doubling as test subjects. Dr. Workman was away in California as an observer at another experimental dive so to free up Bond and Mazzone, Bob Barth and others familiar with Genesis eagerly volunteered. For a test as risky as this one, participation had to be voluntary. Barth's stint as goat herder had stoked his interest in Bond's concept of saturation diving. Furthermore, he had come to trust and admire Commander Bond in a way that made him willing to be locked into a test chamber for six days over the upcoming Thanksgiving holiday. Barth's embrace of Genesis was also shaped by the diving games he played growing up in Manila, during the peaceful twilight before the Second World War.

Barth was an only child, the son of an Army officer. His parents divorced when he was young and Barth spent much of his itinerant youth with his mother, who managed an American shoe store in Manila, and his stepfather, a businessman. Through the age of eleven, when Barth wasn't poolside at the Army and Navy Club, swiping peanuts off the grown-ups' tables, his playground was the Philippine shoreline. The boys who lived around the bay were strong swimmers, a natural by-product of their island upbringing. One day Barth and another Army brat got hold of a gas mask and tried to fashion it into a makeshift diving rig. They climbed down the ladder at the deep end of the club pool, then continued their underwater antics at the officers' boat landing, within sight of Pan Am Clipper flying boats arriving from the Orient, skimming onto Manila Bay.

The mask and breathing tubes Barth and his pals used looked something like the gear that Leonardo da Vinci once sketched. As the boys dived, breathing surface air through their hose, they wondered why the simple act of inhaling and exhaling would become so laborious, even just a couple of feet below the surface. Years later, in dive school, Barth would remember his youthful hose-diving experience and immediately understood what was meant by lung squeeze. The year Barth turned seventeen he was living with his mother and stepfather in Durban, South Africa, and he wanted to get back to the States. He found himself a berth on a ship called the *Westward Ho*. He was taken on as an ordinary seaman, the lowest rank for the unskilled sailor, and worked his way back to the United States to join the Navy. More recently, besides getting his diver's training, he had served on submarines before taking his shore duty at the escape training tank.

On November 27, 1962, Barth and two young doctors from Bond's staff at the Medical Research Lab, John Bull and Albert Fisher, stepped into the Bethesda chamber. Like a lot of pressure chambers, it resembled a stout railway tank car, about seven feet in diameter and fifteen feet long, with a circular entry hatch at one end that led into the outer lock, a kind of foyer a few feet long that could be sealed and used as an airlock. Through a second hatch at the opposite end of this foyer was the inner lock, a considerably longer space of ten feet. This would be their main living quarters. The chamber was an older model, housed in a basement amidst a clutter of industrial fixtures.

While living in the chamber, the doctors would run myriad tests, poking and prodding Barth and each other. There would be regular pulse taking, temperature checking, and bloodletting. There would be daily electroencephalograms. There would be a lot of huffing and puffing into a "donkey dick," as they called the spirometer. The chamber was stocked with food and supplies that made it all look like a campout of crazed medics—syringes, vials, bandages, spirometers, rectal thermometers. As part of the experiment, Dr. Bull planned to examine the heat-sapping effect of helium and how it might play havoc with the human thermostat under pressure and over time. For Barth it all just felt like being trapped in an endless physical exam, and in a very cramped doctor's office.

With this first human trial Bond focused only on the effect of breathing an artificial atmosphere over a prolonged period. The chamber pressure was kept a little higher than the benign 14.7 pounds per square inch at sea level. In case of leaks, this slight pressure differential would ensure that no outside air could seep in, altering the artificial atmosphere. But they didn't need to worry about decompression schedules or potential cases of the bends. If anything went wrong—if Barth or the others got sick—the medical watchdogs could open the hatch and yank the men out, no decompression required.

Once Barth, Bull, and Fisher were locked inside, a gaseous cacophony signaled the filling of the chamber with 80 percent helium and 20 percent oxygen. They were ridding the chamber of virtually all nitrogen, although some scientists at the time worried that nitrogen, although inert, played a hidden role in human function. Going a week without breathing the element, so plentiful in ordinary air, might be as harmful as going a week without food or water or essential vitamins. An absence of nitrogen hadn't seemed to hurt the animals. Now they would find out whether that was also true for people.

The experiment hadn't been under way for very long when they noticed that the old chamber was leaking. Within a few hours it became clear that helium was slipping out so fast that they were having trouble maintaining the chamber's slightly elevated pressure. Had they been operating at much higher pressure, the leak could have given the subjects a bad case of the bends—or worse. As it was, the leak threatened the integrity of the experiment.

At times like this Bond was fortunate to have a right-hand man like Mazzone to handle the details. The erstwhile wartime submariner had the mechanical acumen that Bond lacked, and that was just one way in which the two men's different personalities could be complementary assets at work. Since Mazzone did not have Bond's appetite for publicity, he was happy with the lower profile that came from working night shifts. That left daylight hours, and any limelight, to Captain Bond—Commander Bond had recently made captain. During Genesis the two began to establish a routine in which Mazzone stood watch and manned the controls from dusk until dawn—or until Bond moseyed in to relieve him. The boss did not necessarily show up in the morning at five o'clock sharp as scheduled. Mazzone learned that his night shifts could stretch into the day by an hour or two, perhaps longer. It might depend on how heavily Bond had indulged in a favorite bourbon the night before, or how late he had stayed up writing, or both.

Mazzone sometimes bristled at Bond's self-assured obstinacy, probably honed when Bond was a lone country doctor accustomed to calling
all the shots. There were moments when Mazzone got so frustrated with
Bond he could have thrown a typewriter at him. But the same qualities in
Bond that could try Mazzone's patience were also those that Mazzone had
admired from the start, and had made possible the opportunity to carry
out pioneering scientific research. So let Bond be Bond was Mazzone's
basic attitude. And when a problem like a leaking chamber arose, or just
about anything else, Bond could always count on his capable friend and
colleague.

Mazzone busily dabbed suspect points around the chamber with a
sponge and soapy water, watching for bubbles as if he were trying to find
the leak in a bicycle tire. Wherever pipes or valves or electrical wiring
pierced the chamber's steel skin were the most likely areas for gas to escape. Barth could see Mazzone through the little portholes, dancing around
the chamber, dabbing and fretting. Any spot that bubbled, Mazzone would
slap it with monkey shit—sailor slang for caulking paste. The outside of the
chamber began to look like a fecal foray in abstract expressionism.

Helium leaks had prompted Link to cut Sténuit's cylinder dive short,
and now Bond and his group were getting a reminder that helium could
be much more difficult to contain than air. Helium is the second lightest
gas after hydrogen. The lighter the gas, the less dense it is and the smaller
its atomic structure; the smaller the gas's atomic structure, the more easily
it can permeate a seal, slipping like sand through a sieve. Helium might
solve the problem of narcosis, but its containment posed another challenge. To keep one particularly leaky hatch sealed they ended up using
dental cement, obtained from a nearby lab and cured with a blow-dryer
from a nurse's station. It worked. This would not be the last time they had
to improvise.

Bob Barth was not the kind of guy who liked to sit still, which made
him an interesting choice for a test like this. Apart from receiving his daily
pokes and prods, Barth didn't have a whole lot to do other than read the
sports pages or another chapter of pulp fiction. He never played cards, but
his inscrutable smirk could have been the centerpiece of a killer poker
face. He and his fellow Genesis subjects sat around stewing in 90 percent
humidity and similarly high Fahrenheit temperatures. Water droplets
rolled down the inside of the chamber walls. One constant source of
entertainment came from breathing the helium-rich atmosphere. As anyone who has ever taken a breath from a helium-filled balloon knows, the

ultralight gas makes the human voice sound like a falsetto cross between Donald Duck and the Chipmunks. Divers call it "helium speech."

A little bored one night, Barth found himself staring at the manifold inside the chamber. The oxygen masks hung there that he and the doctors put on when the chamber atmosphere was being emptied and refilled with the desired helium-oxygen mixture. The gauge indicated that oxygen could still be breathed through the masks. Commander Mazzone, whom Barth knew was anxious about helium leaks, was on the topside watch, so with a practical joke in mind, Barth put on a mask and took a few breaths of oxygen. He spoke a few words to himself. Sure enough, his helium speech all but vanished for a few phrases. Perfect! Barth then inhaled deeply from the mask and shouted at Mazzone over the intercom: "Hey topside! I think maybe we have an atmosphere problem in here." The sound of Barth's voice in its natural register prompted an immediate spike in Mazzone's blood pressure. Barth watched with glee through a porthole as Mazzone leapt from his chair, primed to wage holy war with monkey shit and dental cement. Just before the battle began Mazzone heard telling guffaws from inside the chamber. Mazzone could be tough on those he worked with but he also had a wry sense of humor. It was, as much as being a consummate doer, a requirement of his job, working as he was between the iconoclastic Captain Bond and whimsical white hats like Barth. Once Mazzone realized this supposed leak was only a practical joke, his pulse returned to normal and he gave Barth a mischievous order: *Try that one during Dr. Bond's shift!*

Dr. Charlie Aquadro, now back in Washington, working at the Bureau of Medicine and Surgery, would drop by the Bethesda chamber to see how his friends were doing with their latest chapter of Genesis. He was there as Barth and the others ended their week-long exposure to the artificial atmosphere. Barth emerged from the chamber breathing from a scuba tank filled with the same nitrogen-free mixture as in the chamber, as Dr. Bond had told him to do. Bond wanted Barth to be the one to continue breathing the artificial atmosphere while undergoing some follow-up blood tests, pulmonary exams, and electroencephalograms at the nearby naval hospital. Aquadro volunteered to push Barth in a wheelchair over to the hospital, with the scuba tank cradled in his lap like a beloved family pet. Unshowered, unshaven, having just emerged from a week-long helium steam bath, Barth felt like one of the goats who preceded him, reeking and hairy. As Aquadro wheeled him through the endless corridors, Barth could only gurgle pathetically into his mouthpiece, a severe handicap for someone fond of sardonic rejoinders.

As Barth was getting his final checkup, a well-publicized deep-diving exercise was under way off the coast of Southern California. It would become a chilling reminder of how risk factors multiplied when experimenting at sea. Near Catalina Island, about forty miles south of Los Angeles, a dashing and ebullient Swiss mathematician in his late twenties named Hannes Keller staged what was believed to be the first attempt at a thousand-foot dive—a big leap down to a whopping pressure of thirty atmospheres. Keller had gained widespread attention in recent years by doing deep dives and promoting the secret breathing gas recipes he devised with the aid of an IBM computer and the advice of Dr. Albert Bühlmann, a specialist in physiology and respiration at the University Hospital in Zurich.

Keller, like Bond, had come to diving as an outsider of sorts. He was teaching math to engineering students in his hometown of Winterthur, not far from Zurich, and wanted to learn to fly but found flying too expensive to pursue on a teacher's salary. In the late 1950s a friend introduced him to the growing sport of skin diving. Keller did his math and science homework and came to believe, much as Bond had, that deep-sea diving was entrenched in old methods and in need of a historic breakthrough. Within a few short years Keller had attracted enough attention with his secret breathing gas recipes to be invited to join Jacques Cousteau, Ed Link, and other noted experts on the prestigious program of the Second World Congress of Underwater Activities. At the London meeting Keller enthusiastically described his impending plan to "make history" by touching the continental shelf a thousand feet below the surface, and then swimming around at that depth for five minutes.

Keller's was a different approach from saturation diving and living in the sea, more along the traditional lines of a brief bounce dive from the surface. But he claimed that his secret gas mixtures would drastically cut the long decompression times believed to be necessary after a very deep dive, even a relatively brief one. The U.S. Navy was among those to take notice. If Keller's methods proved successful, they would revolutionize the dive tables and make deeper dives safer and more practical. The results of Keller's preliminary chamber tests and several shallower open-sea demonstrations looked promising. The Navy allowed Keller to make one of his test dives at its main experimental facility, next to the Washington dive school, and signed on as an underwriter of Keller's climactic thousand-foot dive.

To observe the dive, the Navy sent Dr. Workman, one of its leading decompression experts, along with another officer during the same week Bond and Mazzone were running Genesis at Bethesda. A few days after arriving at Catalina in late November, Workman joined Dr. Bühlmann, the rest of Keller's crew, and a number of observers on board *Eureka*, the experimental offshore drilling ship made available by the Shell Oil Co. Like other oil and gas companies, Shell was interested in any methods that might improve the efficiency of commercial divers, as offshore drilling operations were moving into deeper water off the California coast and in the Gulf of Mexico. Keller's high-profile plunge would begin inside a diving bell called Atlantis, a prototype similar to Link's cylinder, or a space age version of Halley's seventeenth-century bell. Atlantis looked like a big steel soda can, seven feet tall and just over four feet around, with a pad eye on top, a hatch in the bottom, and a dozen or more protrusions of pipes and valves.

Keller's diving partner, Peter Small, was a trim thirty-five-year-old British journalist, a well-known diver and a principal organizer of the Second World Congress of Underwater Activities. Small planned to write a first-person account of the record-breaking dive. On December 1, as a final test dive, he and Keller were lowered in Atlantis to three hundred feet and spent an hour diving outside the bell. While decompressing in the bell, both divers got relatively mild cases of bends—Keller in the belly and Small in the right arm. You could never be sure where bubbles might fizz.

Keller recovered on his own that night but Small's pain persisted until he was treated with recompression. But Keller insisted on proceeding as scheduled, rather than running further tests at progressively deeper depths before going straight down to a thousand feet. His decision was due at least in part to the presence of a boatload of journalists and other observers assembled to watch the historic dive. Time, money, and equipment were on the line, adding pressure to go ahead with the experiment. So around noon on December 3, a Monday, Atlantis dipped below the Pacific surface, its two divers sealed inside. The descent took less than a half-hour. Then, during a few dark and troubled minutes at the target depth of a thousand feet, Keller left the bell to plant a Swiss flag and an American flag on the ocean floor. His breathing hoses got tangled up with the flags and after he climbed back inside the bell, he passed out. The gas mixture had gone bad. Peter Small blacked out without ever having left Atlantis. Several support divers swam down to meet the Atlantis bell as it was hurriedly raised to within two hundred feet of the surface. One of the support divers sent down to meet the bell, Christopher Whittaker, who was just nineteen years old, vanished. He would not be the experiment's sole casualty.

Keller regained consciousness about a half-hour later and Small came around, but not for almost two hours. He asked Keller some lucid questions about what had happened. Small felt cold, and although he could speak, see, and hear, he couldn't feel his legs. He told Keller he was in no pain. Too weak to stand, he leaned against his Swiss partner and fell asleep as their decompression inside the bell continued. Several hours later, as Atlantis was shipped back to shore from the dive site near Catalina, Keller noticed that Small had stopped breathing and had no pulse. Keller tried mouth-to-mouth resuscitation and cardiac massage. Nothing. Small was cold and pale. The remaining pressure inside the bell, about two atmospheres, was hurriedly released to get Small to a hospital after eight hours locked in Atlantis. He was soon pronounced dead.

The Los Angeles County coroner said the cause of death was decompression sickness. Small contracted a fatal case of the bends; his tissues and organs were riddled with gas bubbles. Keller insisted that other factors were to blame, including a possible heart attack and the panic Small exhibited upon reaching their thousand-foot destination. Death due to any cause during decompression would prevent the proper elimination of absorbed gases, Keller would argue. Either way, the disastrous dive to thirty atmospheres and its two fatalities caused a sharp decline in interest in Keller's sensational methods. The prospect of experiencing this magnitude of failure had worried Bond when Ed Link took his cylinder to sea. Saturation diving had lukewarm Navy support as it was. A Keller-style tragedy, especially at this delicate stage, could put an end to Navy saturation experiments and the quest to live in the sea. Bond could not afford any such tragedy, and certainly not on his watch.

Within six months of the test at Bethesda, Bond arranged to lock Barth and two other volunteers into a chamber for another week, this time at a simulated depth of one hundred feet, or four atmospheres. Genesis D, as they called this next step, was set for the end of April 1963 at the Experimental Diving Unit. "The Unit" was a mecca for Navy diving, and the top testing zone for diving techniques and technology. It was where Swede Momsen had led the experimentation with helium that made clearheaded deeper diving and the *Squalus* rescue possible. The Unit shared Building 214, a brick complex at the Navy Yard in Washington on the turbid Anacostia River, with the Navy's deep-sea diving school. It had a pair of specialized chambers for running tests, and its own staff of divers who were accustomed to getting locked into those pressurized railway tank cars, often to find out whether the latest refinements and recalculations made on paper translated into improved, more streamlined decompression schedules, or whether they caused a bad case of the bends.

Much as being a test pilot was not for every pilot, so the job of experimental diver was not for every diver. There was always a chance of getting bent, and you could never know where those soda pop bubbles would lodge—joints, spine, brain. Peter Small had suffered all sorts of physiological short circuits before dying, but harmful bubbles could usually be squashed and proper blood flow restored by cranking up the pressure, as Dr. Bornmann did when Robert Sténuit's wrist began to ache. Those extra hours spent in decompression could make for a long, stressful, and possibly painful experience. EDU divers prided themselves on enduring this sort of thing, but life as a professional guinea pig was not for everyone.

Although chambers like those at the Unit were designed to simulate dives, there was a stark difference between chambers and flight simulators. In a flight simulator, the kind that Ed Link invented, a pilot never really left the ground. So if he got a bad case of motion sickness or for any other reason wanted to bail out, the simulator could be shut off and a training session aborted. Not so for the diver in a pressure chamber. Even if he wasn't in the ocean, a test diver was really under pressure. Once the hatches were sealed shut and the pressure was on, there was no bailing out—regardless of how sick or troubled a test subject might become. The experimental diver might as well be in orbit. For this reason Bond had intentionally kept the pressure off at Bethesda, but now they would be cranking it up to four atmospheres, and holding it there for an unprecedented six days.

The Unit chambers at the Navy Yard were similar to the one at Bethesda—divided by a bulkhead into two sections, with a smaller outer lock that led through a hatch into an inner lock, the main living area. There was one difference. The inner lock had a second hatch, a back door of sorts, that opened into a short tunnel just wide enough to crawl through. The tunnel led into the "igloo," a circular, domed adjacent section, about ten feet in diameter. A hole in the igloo's floor allowed a diver to drop down into the "wet pot," a vertical tank filled with about ten feet of water. There researchers could simulate a dive that was not just under a certain pressure but also underwater.

Over the course of the next week Barth and his two fellow volunteers would spend time in and out of the wet pot, replicating the experience of living in a pressurized, sea-based habitat and working in the water outside. Bob Barth would be teamed with two other instructors from the escape training tank, Sanders "Tiger" Manning and Ray Lavoie. Both Manning and Lavoie were hospital corpsmen—Navy-trained medics—so they could do the required poking and prodding of Barth and themselves. Dr. Bond also sent in a canary. It seemed like a wise precaution. Canaries had been put into service with caisson workers, just as they had in coal mines. If there was something foul in the breathing gas and the bird keeled over, that would serve as a backup warning system. A friend of Barth's likely spoke for others when he said: "They're gonna hurt you with experiments like that."

There was a palpable air of skepticism about this week-long test dive. *A hundred feet? For a week?* Some Unit personnel couldn't help but wonder why if this test was so important, a group from the New London submarine base had to run it. Others scoffed at the very notion of saturation diving. Whatever skepticism hung in the air may have been thicker, too, because the Unit had only recently played host to a Swiss wunderkind with similarly grand ideas about how to revolutionize deep diving—and everyone knew how his thousand-footer turned out. Now here was this erstwhile hillbilly doctor and his gospel of undersea living. Yet at least a few people around the Unit seemed receptive to the concept of saturation diving, and even the Jules Verne–sounding notion of living in the sea. The New London researchers also had the reassuring presence and slide-rule savvy of Dr. Workman, who had recently been assigned to the Unit after three years as Bond's assistant officer-in-charge at the Medical Research Lab. If all went well this week, they could begin to bridge the depth and duration gap of the Cousteau and Link trials—not just by racking up a record number of days under pressure, although Genesis D would in fact be a historic first for saturation diving, Bond reckoned. With careful physiological monitoring they hoped to better understand both the limits and possibilities of saturation diving.

Once locked into the EDU chamber, Barth found that daily life was much as it had been at Bethesda—a campout with crazed medics. He and the others eased into their routine of pokes and prods. The atmospheric control system had its limitations, so conditions in the chamber fluctuated between wintry and downright steamy, with stretches of relative comfort. Amenities for long-duration tests like this were lacking, as they were at Bethesda. There was a bunk for just one person inside. The other two had to squeeze their bedding in elsewhere. There were no showers, and toilets consisted of a portable, flimsy folding seat over a plastic bag. Now that they were living under pressure, they had to be careful not to tie up the potty bags too tightly when passing them out through the airlock. Otherwise, as Boyle's law worked its magic, the gases trapped inside the bag would expand and, very likely, explode.

The wet pot may not have been as thrilling or scenic as Villefranche Bay, and there would be no story in *National Geographic*, but they were accumulating valuable data on a diver's ability to work underwater and under pressure, in this case prolonged pressure. So Barth or one of the others would crawl through the short connecting tunnel into the igloo, put on breathing gear, and drop through the open hatch in the floor into the wet pot. A number of physical tests were performed on something like a swimmer's treadmill, a kind of counterweighted trapeze the diver could hang on to while swimming in place fully submerged, so that the doctors could track the effects of breathing harder and burning more oxygen.

As the days and hours crept by, Bond and Mazzone took turns keeping watch outside the chamber. Locked on the inside, visible only through a couple of small portholes, Barth and the others continued with the poking, prodding, diving, waiting. They seemed to be holding up well, although Ray Lavoie's blood pressure shot up about midway through the experiment. That sort of thing could happen under pressure, even on conventional dives. Different bodies respond differently to pressure, breathing gases, and the rest. The thousand-foot Atlantis dive provided a chilling case in point: Under similar circumstances, in a shared diving bell, Hannes Keller lived and Peter Small died. Dr. Bond was fairly convinced that Lavoie's high blood pressure was nothing to worry about. He recommended that Lavoie take tranquilizers three times a day.

As the hyperbaric campout approached its end, Bond gave his subjects a rundown of their forty-eight-hour decompression schedule. In the future they might be able to shorten the schedule, but for now they were still erring on the side of caution, taking it slow, as Dr. Bornmann had with Robert Sténuit. Considering the stakes, and the common knowledge that even a standard decompression is risky, everyone sounded remarkably relaxed. Someone joked that Bond ought to consult a Ouija board.

At one hundred feet, a depth of about four atmospheres, the breathing gas contained only about 5 percent oxygen, but the pressure had the effect of creating a four-fold increase. That made the oxygen concentration roughly equivalent to the 21 percent level of the gas found in ordinary sea level air. One practical advantage of this phenomenon was that the presence of relatively little oxygen in the chamber's atmosphere eliminated the threat of fire. A match would not have lit inside the chamber. Smoking, even if allowed, would have been impossible. But that would change as the helium was flushed out, the oxygen concentration raised, and the pressure reduced.

The potential for fire was on Bond's mind toward the end of the decompression as the chamber was to be filled with ordinary compressed air. Any appliances that might cause a fatal spark were turned off—the coffeemaker, electric fans that helped circulate the atmosphere, and a portable carbon dioxide scrubber, the cauldron-sized, electric-powered kind used for emergencies on submarines. It had hummed away during Genesis D to bolster the atmospheric control system and help prevent a poisonous CO_2 buildup during the unusually long experiment. The refrigerator could be left on, they decided. Bond joined the night watch during the final hours of the six-day dive and seemed almost tipsy with delight as he indulged in some characteristic repartee with his desaturating divers.

"Well, I hope you all feel good and brave. Pretty soon I'm going to start you up. I got your oxygen up to about 11.7 now," Bond announced over the intercom. "You see this piece of paper? This tells the whole story."

"Oh, yeah. Uh huh," a Chipmunk voice said with exaggerated uninterest. Bond went on rhapsodizing about the sheet of paper, clearly savoring his gag.

"I found it over here in the wastebasket. I thought it was a shopping list at first but it turned out to be your decompression schedule! It's all right, though. I can read it. Let me show you the curve." With that he put the paper to the porthole for all inside to see. It was a graphic rendering of their route back, a plotting of times, depths, and gas mixtures—a landing plan, in effect, for a safe return from four atmospheres to one. Then Bond joked again: "Well, I don't know which way that curve goes, but we'll get on it and ride it."

A Chipmunk muttered: "I thought you were at thirty-five feet now, right?"

"Ah, yeah. We're somewhere around there," Bond said.

"Well, that's nice," said the Chipmunk.

Early on the last morning of the experiment Bond wanted to hear what his subjects sounded like as helium was removed from their artificial atmosphere at around ten feet.

"Notice how your voices are changing. Talk a little bit now each of you," Bond said. "Say a few words, each of you." One after the other, each wise-ass Chipmunk responded: "A few words, each of you."

As the ultralight helium gas was replaced with air, Bond facetiously asked whether the canary felt any heavier.

"Well, we're going to have roast Tom Turkey tomorrow but other than that, no evidence," said a Chipmunk.

With less than ten feet and just a couple of hours to go, there was a sudden surge of anxiety when Barth and Manning detected smoke in the roof of the igloo. A motorized winch appeared to emit a gray puff. Electricity was shut down and Bond ordered the men to get out of there and secure the hatches to the igloo, sealing it off from the rest of the chamber as a precaution. The topside crew could then let off the pressure inside the igloo only, open an exterior hatch, and crawl inside to investigate the mysterious puff. A faintly acrid smell lingered in the domed section, they noticed. All the electrical systems were checked. None was found to be excessively hot, including the suspect winch. The smoke apparently came from the exposure of grease in the gearbox to the artificial atmosphere. In addition to physiology, there were lessons still to learn about optimal equipment design. The igloo remained sealed off as the three men rode out the rest of their decompression in the main chamber. The compressed air came whooshing in.

Bond drawled into a tape recorder as the experiment concluded: "The subjects are being held now at six feet on compressed air. I expect to have them out in about twenty minutes." It was three-thirty in the afternoon on April 30, 1963. "Their general condition appears to be quite satisfactory." Bond confirmed this soon after the three scruffy aquanauts emerged from their chamber.

To celebrate, Bond and Mazzone took the trio out for a seafood dinner at a nearby restaurant. They then climbed into their Navy truck to make the long drive back to New London. Mazzone dropped Bond off at his house in the wee hours of the morning, and Bond sat up until daybreak, sipping bourbon in the kitchen while telling his wife about the successful experiment. He would soon have another chapter of the Genesis story to tell.

7

DEEP LOSS, DEEPER THINKING

The changing atmosphere in the Unit chamber made a timely metaphor for a changing atmosphere throughout the Navy Department. As Genesis D successfully ran its course, the United States was reeling from the devastating tragedy of USS *Thresher*. The nuclear-powered sub was a stealthy, state-of-the-art Cold War weapon capable of diving to thirteen hundred feet, deeper than any Navy sub before it. The submarine had been in service less than two years when, on April 10, 1963, it was making a test run about two hundred miles off the New England coast. Its communications inexplicably fizzled and the worst was soon confirmed: The *Thresher* had gone down at more than eight thousand feet, taking all 129 men on board with it, including seventeen civilian observers. The disaster sent shock waves through the Navy, jolting some crusty attitudes in the process.

When the *Thresher* was first reported missing, a couple of weeks prior to Genesis D, Captain Bond received a call asking that he prepare his emergency divers for a possible rescue, but long before the sub reached the bottom, more than a mile down, the tremendous ocean pressure cracked its hull like an egg shell, causing an implosion that scattered the sub and everything on board like shrapnel across the ocean floor. The New London base was *Thresher*'s home port and the loss was a personal one for many sailors, including Bob Barth. He had just seen the *Thresher* crew members, alive and well, when they came through the escape training tank for a required review course. A handful of them had been close friends and shipmates—and Barth had almost gone down with them. He had requested and received orders to join the *Thresher* crew, but Bond had asked him to stick around for the upcoming hundred-foot Genesis trial at the Unit. Barth agreed and Bond arranged to have Barth's training tank duty extended. If not for Genesis, Barth would have been on the sub's final, fatal dive.

The best hope of locating the remains of the *Thresher* lay in a clever if somewhat quaint deep-diving vessel, the bathyscaphe *Trieste*. Designed by the Swiss physicist and high-altitude balloonist Auguste Piccard, the *Trieste* worked a lot like a hot air balloon in reverse, drifting down toward the ocean floor and then rising back up to the surface. Its steel "balloon," filled with jet fuel for buoyancy control, looked like a runt of a submarine. On the underbelly was a steel personnel sphere, attached in the center like a pea outside the pod. There was room enough for just two pilots, as the operators called themselves. The pod was built to withstand enormous pressure outside while maintaining a benign surface pressure inside, like the bathysphere the naturalist William Beebe had used in the 1930s to go a half-mile down. But the bathysphere was lowered by cable from a mother ship; *Trieste* could make its dives without tethers to the surface, and it could go much deeper.

The Navy bought the *Trieste* for research purposes and the vessel made its mark on January 23, 1960, with a harrowing record drop to the deepest known point on the planet in the Mariana Trench, southwest of Guam. The Challenger Deep is almost seven miles down, deeper than Mount Everest is high and more than a thousand atmospheres away. The bathyscaphe pilots achieved the inner-space equivalent of a moon landing, except that they couldn't leave their sealed sphere, plant a flag, or be seen on national television. They could only peer into the gloom through two tiny portholes and take pictures from mounted cameras. The *Trieste* wasn't designed to hunt for downed submarines—its battery-powered propellers gave it a limited roving ability. But with North Atlantic depths of more than a mile, and with the *Thresher* remains scattered across at least two square miles of ocean floor, the bathyscaphe was the best tool the Navy had.

Being ill-equipped for the *Thresher* search was one problem. Another was the impossibility of a rescue had the sub crash-landed on the seabed and remained intact but beyond the limited reach of divers and the standard rescue equipment. The Navy would have been helpless to save any trapped, suffocating survivors. To address these and other undersea concerns the *Thresher* tragedy had underscored, the Navy established the Deep Submergence Systems Review Group—just as Genesis D got under way at the Experimental Diving Unit. As the all-encompassing moniker implied, the group would reexamine many facets of the Navy's deep-sea operations, not only submarine rescue—while crucial, it was rarely needed—but salvage capabilities and diving. About sixty specialists in oceanography, engineering, and submarines were called on to develop a five-year blueprint to nudge the nation's undersea know-how into the space age.

Ed Link was among the first asked to join the group. Link had just returned to the picturesque Monaco harbor from a two-week cruise on *Sea Diver* when he was summoned by Rear Admiral E. C. Stephan, head of the Navy's Oceanographic Office, who was put in charge of the Deep Submergence Systems Review Group. Link flew to Washington the next day to meet with the admiral, and the two stayed up late at Stephan's home discussing the matter. Link was initially reluctant to get involved because it would mean putting off the extraordinary four-hundred-foot attempt at undersea living that he had announced the previous fall at the Second World Congress of Underwater Activities. Stephan did his best to twist Link's arm. Link's far-flung connections were such that the admiral believed that if he could get Ed Link on board, he could get almost anyone else. The hope was to assemble a group not only of military experts, but leaders from the realms of science, academia, and industry. After talking it over that night with Stephan, Link agreed to participate. He believed that the review group had important work to do, and could see that by spending weeks meeting with top underwater experts he would undoubtedly learn a great deal, probably more than he could contribute.

The review group was organized into a number of sections—operations, oceanography, research, engineering, and a variety of knowledgeable naval officials and scientists were ultimately appointed, but Captain George F. Bond, officer-in-charge, Medical Research Laboratory at the New London submarine base, crusader for saturation diving and undersea living, was not among them. Link headed the industry division. Bond might have been a logical choice to head the medical division, but the Bureau of Medicine and Surgery, where Bond still had a prime detractor in the boss who rejected his "Proposal for Underwater Research," tapped Charlie Aquadro for the job. Nonetheless, Admiral Stephan sought out Bond to do some work with the group, even if he wasn't a designated section leader.

By June, Link sailed *Sea Diver* to Washington and was given a berth at the Navy Yard. *Sea Diver* would remain docked there for the next six months. Link was provided with an office on the grounds and could frequently be seen around the Unit, a good place for the inquisitive inventor to soak up valuable information. On board *Sea Diver* that summer, with the assistance of a medical student, Link conducted some animal saturation experiments by locking mice into a shoebox-sized pressure chamber he built himself. More important, ultimately, was what Link and his team, including Robert Sténuit, who rejoined *Sea Diver* in Washington, could learn from the Unit and its medical chief, Dr. Robert Workman. In another show of Navy support for Link's plans, Workman was running some day-long deep dives in one of the Unit test chambers, with two Unit divers as subjects. They hoped to verify a decompression schedule that might be the key to Link's next crack at undersea living.

Link and the other members of the deep submergence group were just getting started with their review that summer as Jacques Cousteau's team geared up for a big second act: Continental Shelf Station Number Two, "the first human colony on the sea floor," as Cousteau showily described it. He again had the support of the National Geographic Society, a major Cousteau sponsor for a decade. This time the French national petroleum office also put up major funding, a further sign of industry interest in methods that might help divers work on deeper offshore drilling operations.

In June 1963, two months after the *Thresher* loss, Cousteau led his team of some forty-five men to the Red Sea. About twenty-five miles northeast of Port Sudan, on a spectacular coral reef known as Sha'ab Rūmi, or Roman Reef, Cousteau's team set up a pair of eye-catching undersea habitats. Starfish House, known in French as *L'Etoile de Mer*, was Conshelf Two's main sanctuary. Ten yards in diameter and painted a golden yellow, Starfish House had four compartments about the size of the EDU chamber that extended, like the legs of a starfish, from a pentagonal central living area and control room. Two compartments were sleeping quarters, one housed the kitchen, laboratory, darkroom, and toilet, the fourth compartment had a shower and dive station with a hatch in the floor open to the sea. An ultraviolet lamp in the sleeping quarters allowed each Starfish House resident to lie down and tan for ten minutes a day—not a real necessity, as far as anyone knew, but the sort of thing that would look neat in a *National Geographic* pictorial. A pair of windows the size of car windshields gave the central living room a magnificent view of the crystalline world outside. A gray steel control console looked like a set piece from *Voyage to the Bottom of the Sea*. The chef living in the underwater abode—yes, a chef, Pierre Guibert, who manned the galley's electric stove and appliances—prepared fine meals like *Bifteck sauté marchand de vins*. Bach, Mozart, and Vivaldi played on a reel-to-reel tape deck. A dozen support divers worked around the house and shuttled supplies and equipment to and from *Calypso* and the main utility ship, *Rosaldo*.

For all its stylistic improvements over the austere Diogenes tank of Conshelf One, Starfish House represented a similarly cautious step, with greater emphasis on duration than depth. No one wanted a repeat of the Keller disaster. Starfish House, like its predecessor, was placed at a no-decompression depth of about two atmospheres and pressurized with ordinary compressed air fed by umbilical from the surface. While accidents can happen at any depth, Cousteau's published accounts dramatically asserted that if any of these saturated divers were to swim directly to the surface, they would do so "only at peril of their lives." If they so much as drifted above Starfish House's roofline, he wrote, they "would die there in the shallows of massive decompression effects." Death by direct or partial surfacing was an unlikely fate from a living depth of thirty-three feet, but if any diver got bent en route to the surface, the recompression chamber on board *Rosaldo* would most likely prevent injury and death.

Starfish House served as center stage for a film Cousteau's team was shooting to be called *Le monde sans soleil*, or *World Without Sun*, as it would be known in English, an oddly ominous title considering that their world that summer was fairly drenched in sunshine, even underwater. Claude Wesly, one half of the Diogenes duo, said the setting would make for "the most beautiful" of Cousteau's adventures. Wesly's counterpart in the first Conshelf experiment, Albert Falco, led the selection of the Red Sea site for Cousteau and called it "a diver's paradise." Location was important for the film, of course. Another big pictorial for *National Geographic* and a book were also in the works.

The Red Sea water around the reef was clear and in the mid-eighties Fahrenheit, warmer than many swimming pools. Five "oceanauts," as Cousteau took to calling his sea dwellers, including Wesly, planned to spend a full month diving and working out of their house—a veritable five-star hotel compared to the Diogenes dorm. The oceanauts could freely smoke their Gitanes, and when Cousteau swam down for a visit and a meal, he could light up a Tuscan cigar. Instead of a canary as a backup warning device, they had with them a handsome parrot named Claude, after Wesly.

Outside Starfish House the scene included a shed for storing fish and other sea creatures. An onion-shaped dome, completely open at the bottom but kept dry inside by compressed air, served as a hangar for Cousteau's hydro-jet mini-sub, a two-man craft shaped like a mythical flying saucer—a vessel previously featured in *National Geographic*. During Conshelf Two it would star as the first submarine boat to operate from an undersea base. With filmmaking and photography in mind, the oceanauts wore unusual silver-hued wet suits, instead of standard black, with matching air hoses and gloves. The silver suits made them look like a cross between oversized sardines and spacemen from a Buck Rogers–era movie.

The most significant attempt at cracking depth and duration barriers came from Starfish House's neighboring undersea dwelling, called Deep Cabin, a vertical structure resembling a silo. It was twenty feet tall, seven feet in diameter, and nicknamed "the rocket." The dwelling was lowered down the side of Roman Reef, onto a ledge about eighty-five feet deep, and pressurized with an atmosphere that was a mix of helium and oxygen. Deep Cabin was three-tiered, with living quarters at the top, and a shower and dive station just below, in the midsection. Seawater filled the bottom section up to the floor of the dive station, at a depth of about three and a half atmospheres, forming a liquid looking glass in a circular opening. Two oceanauts spent an uncomfortable seven days in the stiflingly hot and humid structure. They perspired "like fountains," as one of them said.

Deep Cabin was in many ways a testament to the challenges of making an undersea habitat habitable. The cabin dwellers lost their appetite but forced themselves to eat and drink. One had a nagging earache. Among the more troublesome technical difficulties was a persistent gas leak—helium was the likely culprit. The Red Sea would rise into the dive station by more than a foot overnight, and the next day they would force the water out by adding more pressurized gas—the undersea equivalent of bailing water from a ship. But what if the leak got worse and flooded the entire rocket? Such unnerving thoughts made it difficult for the two oceanauts to sleep soundly, as did the oppressive cabin climate.

Relief came from swimming outside their sweltering cabin, and most days they spent several hours in the water. Cousteau had them make some dives to 165 feet, and on the sixth of their seven days, the two oceanauts swam past a thicket of black coral along the reef and down to a depth of 330 feet. They used their Aqualungs with compressed air, which would have been considered too dangerous if they had started from the surface. But they were starting out from eighty-five feet—after several days of breathing a helium-rich artificial atmosphere. They had a hunch that this would make it possible to dive deeper with an air-filled Aqualung without getting knocked silly by the narcotic haze of nitrogen—*l'ivresse des grandes profondeurs.* The divers later informed Cousteau that they accidentally went all the way to 363 feet, an additional atmosphere of depth, and just shy of the 385-foot mark, from which Maurice Fargues was hauled up dead sixteen years before. They reported that they felt fine but they did not stay down for long. Squeezed by the pressure at eleven atmospheres, their lungs sucked up about eleven times as much air with each breath. Their Aqualungs were quickly depleted, even though each diver had four tanks.

When the Deep Cabin dwellers completed their week-long stay at eighty-five feet, they swam up to Starfish House and spent the night there, using the time at thirty-three feet as a decompression stop. They surfaced with the other oceanauts the next day. This approach to making the transition from a deep habitat to the surface resembled an idea that Bond described in his "Proposal for Underwater Research." Bond suggested that sea floor habitats be placed at depths corresponding to appropriate decompression stops, like stepping-stones or way stations, so that divers could continue to live and work as they eased their way back to the surface—a better use of time than idly decompressing in a chamber.

This second Conshelf venture, like the first, was more about logistics than producing physiological data. Nonetheless it did its part for the cause of undersea living. The documentary *World Without Sun* came out the following year and garnered Cousteau his second Oscar. Cousteau wrote another major story for *National Geographic*—forty-two prestigious pages of pictures and text, including a friendly, if fleeting acknowledgment of Captain Bond. That article became the basis of his fourth book, a photograph-filled work also called *World Without Sun,* written with Jimmy Dugan.

There would be no film or book deals, but on August 26, 1963, about a month after Cousteau's team had packed up Conshelf Two and left the Red Sea, Bond and Mazzone began a third chamber test with human volunteers, Genesis E. It followed the previous phase, Genesis D, and Bond liked the coincidence that E stood for Exodus, the book of the Old Testament that comes after Genesis. With the parameters of this test the New London researchers planned to double the depth and duration of their previous test at the Unit.

For Genesis E, three people would be locked into an artificial atmosphere, at a pressure equivalent of two hundred feet, and *live* there for nearly two weeks. As in the wet pot at the Unit, they wanted to demonstrate that at this doubled depth a saturated diver would be physically capable of doing more vigorous deeds than planting flags or posing for pictures. "Useful work" was the catchphrase in Navy diving: A diver had to be able to do *useful work*. In this way they might build on the more cautious success at the Unit to produce a kind of supersonic moment—two weeks at two hundred feet—to convince the U.S. Navy, once and for all, that it ought to make the exodus from experimental chambers into the sea itself. Of course there was always a chance that this ambitious experiment might fail miserably, even fatally, and give skeptics a reason to end the entire effort.

The researchers conducted this third human experiment at New London's own medical laboratory, on the upper submarine base where the long-awaited pressure chamber had finally been installed. Although a modern machine, with advanced instrumentation for maintaining the desired internal atmosphere, the chamber had no wet pot. But they would make the best of it by using a variety of physical tests—such as push-ups and step-ups—to study a saturated diver's cardiovascular function and ability to handle the kind of useful work the Navy might want divers to do. The new chamber, like the older ones, was still not designed and equipped to accommodate such long stays, but by now Bond's subjects were accustomed to coping with an absence of plumbing and other comforts of home. Bob Barth, the original Genesis subject, again volunteered. Tiger Manning, the hospital corpsman, Barth's chamber mate at the Unit, and Dr. John Bull, who was with Barth for the first test at Bethesda, would also reprise their roles.

Bond and Mazzone would spend their daytime and nighttime shifts staring up at a variety of gauges and meters, including a couple of strip charts that produced data on rolls of paper, like a seismograph. These were part of the control console along one side of the new chamber's steel exterior, with a pair of built-in desks, one at each end of the console. Between the desktops was an assortment of shiny knobs and colored lights the size of half-dollars, part of the new and improved automatic control system. A porthole over each desk gave the investigators a view inside the chamber. Someone stuck a foot-long sign over the middle of the console with the handwritten words: GENESIS SUITE.

Once the men were locked inside the chamber, the ritual poking and prodding was more gut-wrenching than during the two previous Genesis tests. At one point, Barth and the others sported a rectal probe, an ear probe, wires in their arms, and a rubber hose that was threaded through their noses down into their stomachs. The best way to suppress the nagging nausea and generally minimize discomfort was to sit still, like monks in meditation. They were a long way from tanning beds and idyllic dives in the Red Sea, but the indignities of Genesis had to be tolerated so they could generate the data Bond needed to make his case for saturation diving. In the future that data could be the difference between success and catastrophe.

Outside the tank, Bond and Mazzone had to work out the bugs in their state-of-the-art machine. About a week into the experiment, Bond arrived for the daytime watch and was met with a litany of special instructions from Mazzone: "Before crossing emergency jumper 210 to post 18, check humidity readout. If this reads 92 on the dial (subtract 8 for true value), then this jumper cross is unsafe, and the problem must be reevaluated . . ." As more gauges and electronic controls malfunctioned, the chamber console became cluttered with red "out of order" tags that reminded Bond of Christmas tree decorations. But the experiment went on, with Barth, Bull, and Manning living and breathing and bantering like Chipmunks in their pressurized atmosphere made up mostly of helium and just 4 percent oxygen. The seven-fold increase in the concentration of oxygen at seven atmospheres made it close to a surface equivalent of 28 percent instead of the usual 21 percent. The researchers had observed during the course of Genesis that a slightly elevated oxygen content seemed to improve the human test subjects' overall performance under pressure. This was precisely the kind of insight they hoped to gain in the laboratory before facing the many complications that would come with attempting a similarly prolonged dive at sea.

A couple of days into the experiment, in honor of Bob Barth's thirty-third birthday, Bond and Mazzone passed a cake through a small supply airlock. Despite the chamber's technical difficulties, the experiment was going relatively smoothly. The test subjects showed some minor changes in their physiologic functions, but nothing alarming. Through it all, George Bond recorded the events of the day in a journal, as was his habit, filling pages with his distinctive loopy script. When he set pen to paper, after hours, his vivid imagination sometimes took over, buoyed by bourbon, like so many writers before him.

A few journal entries written during Genesis E read a lot like the secret life heroics of Walter Mitty, the daydreaming James Thurber character. In one, as Bond and Mazzone were running a preliminary pressure test, Bond described a sudden blast as "shattering glass in the room windows and rocking the chamber itself." The new chamber "had failed, blown the cork!"

Test subjects Barth, Bull, and Manning were not yet inside, fortunately. If they had been, Bond somberly wrote, they all "would have been dead." In a similarly dramatic episode, Bond described what sounded like a "sonic boom." The noise startled him out of a near slumber in his seat beside the chamber and then "all hell broke loose!" The pressure gauges were falling precipitously! The men inside were in danger! He rushed to the rescue, desperate to fix the cause of a potentially deadly leak. He and Mazzone assessed the situation and, soon enough, with the help of the subjects inside, things were brought under control.

Captain Bond might never get a forty-two-page spread in *National Geographic*, but he liked to tell a good story and often had his narratives typed up and distributed to friends and colleagues. Everyone was accustomed to creative flourishes like the Mitty heroics and understood that Bond didn't necessarily mean to be writing impeccable history, he was just . . . writing. Hyperbole was a frequent and entertaining ingredient in Bond's otherwise true stories, but the chamber blasts were apparently outright fiction. Witnesses to those experimental days recall no sonic booms or glass-shattering explosions. If there had been, Bond's saturation experiments might have been shut down for good. But as Bond also wrote in a more sober and factual manner, Genesis E came to a fairly uneventful end in early September 1963. After nearly twelve days living in the helium-rich atmosphere, including a decompression schedule lasting twenty-six hours, subjects Barth, Bull, and Manning showed no signs of having been done any permanent physical harm. With the Navy's blessing, the grand exodus into the sea Bond had dreamed of leading could begin at last, five years after the first animal tests.

8

TRIANGLE TRIALS

Before getting formal Navy approval to build a prototype sea dwelling, but hopeful that they might soon be taking their experiments from the lab into the sea, Bond and Mazzone, in typical moonlighting fashion, got hold of a surplus "survival pod." A sturdy steel structure, circular and domed, like a slightly larger version of the igloo at the Unit, it had originally served as an escape chamber for storm-battered crews of a Texas Tower. These towers had been built at sea, along the East Coast, to house Air Force radar stations. The old pod was as good a framework as any for a first home in the sea—and it was cheap, a good thing given their nonexistent budget.

The pod was delivered to the New London submarine base and sat in a vacant lot near the escape training tank, down by the Thames River. Bob Barth and a few other industrious volunteers were among those who spent a few months doing improvised hacking and welding in their spare time. It did not go as well as they had hoped. Looking over the botched structure one day, Mazzone grabbed a black marker and scrawled on its side: "Bond's Folly—Sealab I." The pod didn't survive their remodel attempt, but the name "Sealab" did.

In early December 1963, three months after the successful two-hundred-foot Genesis E experiment, and just a couple of weeks after the assassination of President Kennedy, Bob Barth got a call from Captain Bond. Barth was visiting his parents in the Florida Panhandle and Bond asked him to drive over to meet him at the Navy Mine Defense Laboratory, a base at Panama City situated around an inlet of St. Andrew Bay called Alligator Bayou. The reason, Barth learned, was to inspect a pair of old minesweeping floats, two hollow steel cylinders shaped like torpedoes that were big enough to stand in and nearly sixty feet long. The surplus hulks were in the base's salvage yard and easily the largest objects in the scattered purgatory of spare parts. As they strolled around, Bond explained to Barth that workers at the Mine Defense Lab were somehow going to transform the floats into America's first undersea habitat. Formal Navy approval was coming soon. Barth also learned that he would be part of the first crew to live in it.

After all the hours spent in experimental chambers, Barth was more than happy to try living on the ocean floor. There would still be plenty of poking and prodding as part of the continuing physiological studies, but he would be free to come and go from his home in the sea, living out a genuine Jules Verne fantasy. Barth's experience as a diver and as the original saturation diving guinea pig made him a logical choice as a first habitat resident. Captain Bond would find four others to join him. Of the four, one man seemed to come from out of the blue, although Barth certainly recognized his name. The whole world knew his name. Less than two years had passed since he had become a national hero as the second American to orbit the earth. He was Malcolm Scott Carpenter, a Project Mercury astronaut who wanted to become an "aquanaut," as Captain Bond liked to call a saturation diver equipped to live in the sea. Bond would now have his own celebrity diver, and he'd been sent by the original celebrity diver himself, Jacques Cousteau.

Like so many others, Carpenter had followed Cousteau's writings and films. The American space hero considered the French explorer a hero, and not long after Carpenter's historic orbit he met Cousteau at the Massachusetts Institute of Technology, where they were both attending an event. Carpenter took the opportunity to express his interest in the Conshelf experiments and offer his services. He had some experience with scuba and told Cousteau that he might be able to channel some valuable NASA know-how into Conshelf if he could join the French team. Carpenter also knew he had a fear of the deep ocean, and he wanted to overcome it. Cousteau said that he couldn't pay much, expressed concern that Carpenter didn't speak French, and suggested that Carpenter look into working with his own Navy. Until then, Carpenter had never heard of Captain George Bond or the new project the Navy was calling Sealab. The name Mazzone scrawled on the survival pod had attained official status, although it's not quite clear how. "Sealab" certainly made as much sense as any name, and had a purposeful, pioneering ring to it.

Carpenter was lent back to the Navy in the spring of 1964, just in time to train in Panama City with the Sealab team. The publicity value in adding an astronaut to the Sealab roster was obvious, although Barth wondered what someone with scant diving experience could bring to the program. But Bond wanted Carpenter, and they would train him. If Carpenter could pull off a feat like orbiting the globe, he could probably get along all right on the sea floor.

Carpenter was self-confident, competent, unflappable—making the cut to become one of the seven Project Mercury astronauts was ample proof of his personal qualifications. But Carpenter knew he would have to prove himself to Barth and the rest of the divers at the Panama City base. Even though Carpenter had a long list of lifetime hobbies and achievements— horseback riding, skiing, wrestling, gymnastics, studying engineering, flying planes, getting launched into space—Barth and some of the other divers couldn't resist doing a little hazing of their new recruit. They had an arsenal of pranks to choose from, like sneaking up on the spaceman and shutting off his air valve, or tugging on his face mask so it filled with water. Carpenter took it all in stride.

What Carpenter ultimately found most startling was how underfunded the Sealab program was, especially compared to the space program. The new undersea project was being run by the Office of Naval Research, an agency formed after World War II to coordinate the funding and developing of projects for possible Navy use. The ONR budget for Sealab was a relatively measly $200,000, and in early 1964 engineers at the Mine Defense Lab got word from ONR that they would have three months and about $35,000 to create a Sealab habitat out of those old floats in the salvage yard. There were few specific guidelines—the Navy had never built anything quite like this. The engineers were not divers themselves, so they solicited suggestions from Bond, Mazzone, and others to come up with a viable prototype for the inaugural U.S. quest to live in the sea. Mazzone, in his usual energetic fashion, had been shuttling between the New London submarine base and the Harvard School of Public Health, where he was earning a master's degree in environmental physiology to better serve the cause.

From his office window at the base, Bond could see Sealab being pieced together on a sun-splashed tarmac at the water's edge. By the time Carpenter arrived in the spring of 1964, the two old floats had been cut up and welded together to form a forty-foot-long, cigar-shaped capsule, nine feet in diameter. The capsule stood about five feet off the ground on four pairs of angled struts. The struts connected to two parallel troughs as a support base, like the pontoons on a seaplane, except that these pontoons were open at the top. When the time came, they could be filled with the ballast needed to sink the hollow Sealab and hold it securely on the bottom. The lab looked like a giant insect and stood out even more after it was painted a radiant orange. On either side were two small portholes and the name SEALAB I painted boldly in black, the roman numeral optimistically suggesting that more labs would follow.

To cut costs, Mine Defense Lab engineers bought quite a few fixtures from Sears instead of from Navy stockpiles—dehumidifiers, electric heaters, a hot plate, an intercom system, and a thermoelectric refrigerator. The engineers wanted to avoid an ordinary fridge cooled by Freon because any leak of the deadly gas could poison the Sealab atmosphere. To select the best conduit for fresh water, several varieties of hose were filled with drinking water, then laid out on the tarmac to bake in the Gulf sun. A Sears garden hose was the cheapest and provided the best-tasting water. The hose would become one strand in a thick umbilical bundle, a lifeline of communication cables, power, and gas lines linking the habitat with its support ship.

No one in Bond's camp was complaining about the need for penny-pinching. They had made a quantum fiscal leap from the days of out-of-pocket expenses for sub-rosa animal tests. An improved national attitude toward ocean exploration didn't hurt their cause, either. Shortly before President Kennedy's historic May 1961 speech in which he set the nation's sights on the moon, he asked Congress to nearly double the budget for ocean research and to spend more than $2 billion over the next decade on "the Earth's 'inner space.'" As Kennedy said: "Knowledge of the oceans is more than a matter of curiosity. Our very survival may hinge upon it." Two years later, after the *Thresher* sinking jump-started a review of the nation's deep-sea capabilities, Kennedy addressed the National Academy of Sciences at Constitution Hall in Washington: "We know less of the oceans at our feet, where we came from, than we do of the sky above our heads. It is time to change this, to use to the full our powerful new instruments of oceanic exploration, to drive back the frontiers of the unknown in the waters which encircle our globe."

Despite the inspiring words, Sealab was not showered with money, and the whole project might have become little more than Bond's folly if not for an unsolicited push from an obscure figure in the Office of Naval Research, Captain Lewis Melson. Melson came up through the

ranks of naval engineers and was more than a little miffed when he got
orders to head ONR's Naval Applications Group. His assignment to the
Office of Naval Research meant that he would be further removed from
the kind of practical shipyard work he preferred. But Melson followed
orders and went to Washington in late August 1963, when Bond was
getting started on Genesis E.

In his new job heading the Naval Applications Group, Melson looked
for whatever fruits of research seemed ripe enough to be transformed into
practical tools for useful work. He decided to act on an idea that occurred
to him a year earlier, after he had witnessed Typhoon Karen's trashing of
Guam's Apra Harbor. Winds of nearly two hundred miles an hour had
left a multimillion-dollar mess that required raising a dozen sunken ships
from depths of about a hundred feet. As head of ship repair, Melson saw
firsthand how time-consuming and cumbersome conventional diving
could be—so many dives, such brief bottom times. The Navy ought to
have better ways of getting work done underwater, he thought. Now, at
ONR, he had an opportunity to do something about it. Melson put his
top engineer, Henry A. "Al" O'Neal, in charge of creating a project that
would make divers into more efficient, effective, and therefore useful underwater workers.

Melson tried to gain the support of the Navy "diving establishment,"
as he scornfully called it—essentially the same skeptics Bond encountered—but found little interest. Neither Melson nor O'Neal was a diver,
which may have caused some Navy brass to doubt them, but soon O'Neal
found out about Captain Bond and Genesis and invited Bond to Washington. Melson still had doubts. On one hand, Bond was a good doctor
and an advocate for taking diving to new depths and durations. But when
Melson heard of Bond's unorthodox ways, he tried to recruit Dr. Workman instead. Workman's boss at the Experimental Diving Unit wouldn't
let him go and Melson decided to take his chances with Captain Bond
as the medical officer and "principal investigator" for the Office of Naval
Research project to be called Sealab.

Melson got the blessing of his ONR superiors to proceed with Sealab,
and Jim Wakelin, the influential Navy secretary for research and development, lent his support and asked for regular progress reports. The
bureaucratic blessings did not add up to much of a budget, however, so
Melson and O'Neal pooled funds from other projects, estimating that
about $200,000 would be needed to complete the job.

May 16 was Armed Forces Day and although the lab wasn't quite ready for its inaugural dip, they put Sealab I on view during the Mine Defense Lab's open house. Captain Bond stood on the tarmac in the hot sun fielding questions as visitors trotted up the stairs set up so they could take a peek into the portholes. A few thousand people got a look at America's prototype sea dwelling that day, although this first public display was more like a county fair sideshow than the unveiling of a pioneering vessel worthy of write-ups in newspapers across the land.

Within a week, and only a few weeks behind schedule, Sealab I was ready to undergo its first tests at sea. After adding enough ballast to sink the buoyant thirty-ton lab so that only its top half broke the surface, they towed it to a designated site a couple of miles from the glittering white sands of the Panama City shoreline. Bobbing along the surface the lab looked like a bright orange whale chasing after the tugboat that pulled it into the Gulf of Mexico. The support ship that would serve as Sealab's mission control was there, moored just as it would be during the actual experiment. It was a big ship, almost the length of a football field, fifty feet wide and boxy like an ocean freighter. Its deck, largely flat and open, was about two stories above the waterline. The ship was officially called the YFNB-12, the Navy designation for a steel-covered lighter, but before long the Sealab crew was calling it "Your Friendly Navy Barge." Built at the end of World War II as a freight barge, it had been modified several times and with some further modifications it would do the job—and it fit the budget.

Support divers and riggers attached Sealab to the barge by a line that ran from a large windlass down the middle of the ship, over the fantail and through a pulley at the stern. It was as if the barge were acting as a giant, floating fishing pole, reeling out the thick line to the lab in the water. The entry hatch in the belly of the lab—about the size of a manhole, through which the divers could freely pass into and out of the sea—was open, allowing the sea to form a small pool in the floor as it had in the Conshelf habitats. As the lab went deeper, the crew raised the internal pressure to keep the sea from rushing in, but lowering a large, buoyant habitat wasn't easy.

As Sealab was being lowered and the barge rose and fell on the passing swells, the thick line to the lab stretched. The combined effect of the stretching line and the bobbing barge caused a persistent yo-yoing and they lost control of the descent. The lab flooded and crash-landed about sixty feet down. Bond feared that they had just sunk the whole program. Lieutenant Commander Roy Lanphear, whom Melson had put in charge of Sealab operations, was summoned to Washington and the project was effectively threatened with termination. The fate of Sealab I was hanging by a hair, Lanphear reported upon his return, and skeptical authorities wondered whether this experiment in undersea living was safe. The Navy Department allowed the project to continue, but any further mishap could be their last.

The team raised Sealab to the surface and towed it back to the Mine Defense Lab. Within a few days the soggy lab was cleaned, dried, and readied for further tests. The next descent was more dignified. The crew raised and lowered the lab a couple of times for good measure. It was then left on the bottom overnight, about sixty feet down, to make sure everything was working properly and that there were no gas leaks of the sort that had forced Ed Link to cut his cylinder experiment short—or that had kept Mazzone hopping with monkey shit.

Following the preliminary tests in the Gulf, Sealab I was lifted onto the YFNB-12 and shipped more than a thousand miles to the U.S. Naval Base in Bermuda, the chosen base of operations for the actual experiment. From there, after further preparations, the ultimate destination would be about twenty-seven miles southwest of the British colony and nearly two hundred feet below the surface. The site was selected in part for its relatively warm and clear water. Most of the ocean depths are much colder, and shrouded in greater darkness, but that added challenge could wait. For this maiden trial, keeping five men on the bottom alive and well for three weeks would be challenge enough.

For an operation whose essential mission was to make a convincing case for saturation diving and sea dwelling—and to avoid major mishaps and Keller-style catastrophes—Sealab seemed to be tempting fate. It would be placed at the northern tip of what had become known as the Bermuda Triangle. This 500,000-square-mile expanse of ocean between Bermuda, Miami, and Puerto Rico had also earned the epithet of the Devil's Triangle after a century of mysterious disappearances of sailing ships and, more recently, aircraft. In December 1945 six Navy planes inexplicably vanished in good weather during a routine training mission out of the naval air station at Fort Lauderdale. The fully equipped rescue plane sent out after them also vanished without a trace. One member of the Navy panel that investigated said: "They vanished as completely as if they had flown to Mars." A year before Sealab arrived, a pair of four-engine jets disappeared on a refueling mission in the area. Probable debris turned up about 260 miles southwest of Bermuda—along the route to Sealab's destination—but that only deepened the mystery of the planes' true fate.

Yet if God intended man to have Dominion over the Seas, why fear diving in the Devil's Triangle? Ed Link had also set his sights on Bermuda for an ambitious attempt at a saturation dive to four hundred feet. Link wrapped up his work with the Deep Submergence Systems Review Group at the beginning of 1964 and sailed *Sea Diver* from the icy Navy Yard on to Key West and warmer climes. In anticipation of this next, much deeper saturation dive, he had designed a bare-bones little habitat he called SPID, for Submersible Portable Inflatable Dwelling. Like the cylinder, the SPID was more pup tent than house, not only in size but in design. Link's creation had a thick, sausage-shaped rubber skin about four feet around and eight feet long that was mounted horizontally on a metallic frame. A hatch in the floor could be kept open to the sea once the interior was pressurized and inflated like a balloon. There was also the steel cylinder from Villefranche, originally envisioned as an elevator for Cousteau's first undersea house. Link would use the cylinder with his inflatable dwelling.

Link still paid many expenses out of his own pocket but now had backing from the National Geographic Society and again had the assistance of a Navy ship, the *Nahant*, this time near Bermuda. The *Nahant* used its echo sounder to scope out the sea floor but couldn't find a suitably level site at four hundred feet. The islands of Bermuda are at the top of an extinct volcano and the surrounding bottom of the Atlantic Ocean is irregular, often dropping steeply from coral plateaus straight down to depths of more than twelve thousand feet. Link decided instead to set up his SPID on an undersea plateau much closer to Florida, in the western tip of the Triangle.

The site was about 150 miles east of Miami at the northeastern end of the Great Bahama Bank. On June 30, the seas were calm as Link prepared to lower his inflatable dwelling to the ocean floor, just ahead of Sealab. That timing appeared to be more happenstance than the product of an intentional race to the bottom, although the friendly rivalry may have been a contributing factor. Some pesky helium leaks had to be plugged that morning before the rubber tent was put in the water. The two divers rode down inside Link's cylinder. In addition to Robert Sténuit, the Belgian diver who had spent a day in the cylinder, Link had recruited a celebrity diver of his own—or at least a diver with a celebrated name. He was Jon Lindbergh, the eldest son of aviator Charles Lindbergh. A former Navy diver, Lindbergh had gone into commercial diving and was known for his daring deep-diving exploits on West Coast offshore oil rigs. He bore a noticeable resemblance to his famed father and was affable but considerably more taciturn than his outgoing Belgian counterpart. By early afternoon, Lindbergh and Sténuit were inside Link's inflatable dwelling, 432 feet down. As Link had announced in London at the Second World Congress, they intended to stay for several days, roaming in and out of their pup tent in the sea.

A thousand miles away, the Devil's Triangle was up to its old tricks. On June 29, two U.S. Air Force jets had flown out of Bermuda's Kindley Air Force Base and collided in midair. This time there were plenty of witnesses and little mystery as to what had happened. Seven airmen survived the collision but seventeen were killed. The Sealab crew got an unexpected assignment: searching for wreckage and bodies.

Since most of the jet wreckage landed on the floor of the Atlantic at depths down to 250 feet—similar to the planned depth for Sealab—those in charge considered putting Sealab down near the crash site. Bond found the idea tempting indeed. They could do a significant operational job and pointedly demonstrate what he had been preaching for seven years—that saturation diving and undersea living would be the best way to accomplish all kinds of useful underwater work. The accolades that might come from handling an important recovery job would be good for Sealab and generate some favorable headlines. On the other hand, the physiological studies would have to be cut. They wouldn't be able to run a controlled experiment and focus on their recovery work at the same time. The irregular depths that compelled Link to head for the Bahamas would undoubtedly complicate the process of decompression and also the mooring of the support ship. By adding more variables and unknowns to the operation they could wind up in a situation worse than the flooding at Panama City. Captain Bond pondered the options over a few pipefuls of Sir Walter Raleigh, his preferred tobacco, but in the end, he and several other commanders recommended that they stick with their original plan.

But the Air Force still wanted the divers' help. So the Sealab support barge sailed out to St. David's Head, at the opposite end of the hook-shaped island chain from the Navy base, and divers from the Sealab crew, including Bond, Mazzone, and Cyril Tuckfield, Bond's partner on the 300-foot submarine escape, went looking for wreckage. Charlie Aquadro had arrived in Bermuda just in time to dive for the dead, though officially he was there only as an observer and emissary from Jacques Cousteau, whom Aquadro had begun working for the previous summer, during Conshelf Two. Ever since his days as a scuba-diving medic with the Underwater

Demolition Teams, he always seemed to pop up where the action was, such as on Link's early expeditions to Jamaica and Israel. In mid-1961, Aquadro was part of a group that made perilous dives to three hundred feet, fraught with narcosis, to reach the wreck of the *Lusitania* off the coast of Ireland for a television documentary. Indeed, to glance at Charlie Aquadro's peripatetic résumé was to understand why some would call this period a golden age for diving.

Aquadro, like Scott Carpenter, had been an admirer of Jacques Cousteau and found his way onto the Frenchman's team by introducing himself at a meeting in Miami. Aquadro apparently made a favorable impression. Before the second Conshelf got under way in the Red Sea, Cousteau invited Aquadro to lunch in Washington and surprised him with an offer to join his team. Delighted, Aquadro arranged to leave the Navy. He took a reserve commission and happily accepted the low pay that had been a disincentive for Scott Carpenter. The other problem for Carpenter, Cousteau had said, was that he didn't speak French. Neither did Aquadro, but because Aquadro was an experienced diver and a medical doctor to boot, Cousteau may have considered him more valuable than the former astronaut. Now, through an amicable arrangement with the Office of Naval Research, Cousteau sent Aquadro, along with a longtime leader of the Cousteau team, Jean Alinat, to observe during Sealab I. No sooner had Aquadro arrived in Bermuda than the jet crash put him close to the action again.

Over the next few days the Navy divers made conventional bounce dives from the surface using scuba and ordinary compressed air instead of a helium mix. Down to almost 250 feet the visibility was good. The jet remains were widely scattered, shredded into small pieces. Sharks made frequent appearances. Crabs and other sea life were quick to move in on the human remains. George Bond found this type of underwater work especially unhappy and unrewarding. Bond was a doctor, not prone to squeamishness, and he and the others had brought up dead bodies before. But Bond's feeling was that except in cases of a simple drowning in shallow water, when a body can be recovered almost immediately, it is kinder to consign human remains to the ocean depths, marking the end with appropriate ceremony, preserving dignity in death for survivors and the deceased alike.

One of Tuckfield's finds pretty well illustrated Bond's philosophy. Tuck came up with a foot that had been severed at the ankle. While making a decompression stop, fuzzy with narcosis, hanging on to a shot rope from the surface, the good-humored sailor felt nauseated by the body part he was holding. Tuck tried to pass it off but no one would take it. Intact bodies proved difficult to find, but Charlie Aquadro had managed to extricate one, then grabbed hold of the dead airman as best he could, wrapping an arm around him and tugging on tattered clothing. Aquadro used his free hand to shinny up a line while kicking like mad. Decomposing limbs could tear off, and bowels could burst from a waterlogged belly. Carpenter, who had just rejoined the Sealab crew in Bermuda, didn't make any dives but was with the crew aboard the YFNB-12. He was especially impressed with Aquadro, who seemed the least fazed by the depth, the cold, the narcotic haze, and the horrific nature of the work.

While the Navy divers were carrying out their macabre search and recovery mission, Jon Lindbergh and Robert Sténuit were doing their best to avoid becoming corpses themselves. Four hundred and thirty-two feet down on a fringe of Great Bahama Bank, they were well beyond the reach of scuba divers. They rode down in Link's pressurized cylinder, which was tethered near the Submersible Portable Inflatable Dwelling as a backup safe haven in case something went wrong with the SPID. Swimming to the surface, even just part of the way there, would be suicide. Their saturated organs would fizz like the contents of a soda pop bottle shaken before opening.

They soon needed their safe haven. During the first hour there was a problem with the scrubber for purging exhaled carbon dioxide and the two divers were panting like winded dogs. They took refuge in the nearby cylinder and spent several hours crammed into the aluminum tube, waiting for a functioning filter to be lowered from the surface so they could fix the scrubber. Once resettled inside the SPID their heater failed. They put on several sweaters and shivered through the first night, the heat-sapping, helium-rich atmosphere abetting their persistent chills. A lightbulb imploded from the whopping fourteen atmospheres, cracking like a gunshot. Helium speech rendered them virtually incomprehensible, their falsettos all the more garbled by the pressure. At times they communicated with *Sea Diver* and the topside crew by Morse code. Rather than struggling to decipher each other's helium gibberish, the divers scribbled notes back and forth, sometimes using the inside wall of their undersea pup tent as a blackboard.

Their prototype shelter had its shortcomings, but it was working and the divers could simply drop through the open hatch in the floor and into the sea. They were limited to a distance of fifty feet, the length of the umbilicals that fed breathing gas to their specialized dive gear.

Aqualungs would have been little use because under the pressure of fourteen atmospheres, scuba tanks would have emptied within minutes. The two men aimed to dive for several hours a day. The water temperature was about seventy-two degrees Fahrenheit—not bad, considering the depth—but combined with the heat-sapping effect of helium Sténuit nearly froze, especially after tearing his custom-made wet suit. He took pictures of Lindbergh working on their dwelling, hopeful that his would be the deepest photographs ever shot by a diver. Lindbergh was able to get a heater working, and on the second night they might have slept better if not for the unnerving thuds against their rubber tent made by some grouper fish.

Forty-nine hours after Lindbergh and Sténuit first entered the rubber dwelling, Link called down to congratulate them. They had made the longest deep dive and reached their goal. Having gotten things under control, the divers thought they might as well spend the rest of the week down there, extending the duration of their deep dive. But Link thought they had been fortunate, both in terms of weather conditions and the prototype equipment. They had proven a point about the potential of saturation diving and, as he told them, there was nothing more to be gained by extending their stay on the bottom. It was true that in theory, once the divers were saturated, their decompression would be the same whether they stayed another day, a week, or even longer. But Link was not going to put the divers at risk to prove the theory. His decision would cause the Sealab crew to say that the SPID experiment had been little more than a stunt—neatly timed to beat them to the bottom.

Upon Link's order, Lindbergh and Sténuit packed up and swam out of their inflatable dwelling and into the cylinder. They closed its lower hatch, sealing the pressure inside, and rode it like a slow-moving elevator back to the surface. The Sea Diver crew maneuvered the cylinder onto the deck and shortly after three in the afternoon on July 2, they affixed the hatch end of the cylinder to a separate decompression chamber, as planned. The saturated SPID divers then made an airtight transfer, crawling from cramped cylinder into the more spacious and comfortable chamber, and waited out their three-day decompression schedule as Sea Diver sailed back to Miami.

Supervising the decompression were Link's three medical specialists from the University of Pennsylvania, with whom Link had been collaborating in advance of the four-hundred-foot sea trial. They also had the decompression schedule that Dr. Workman had devised while Sea Diver was still berthed at the Navy Yard. Workman later followed up with a similar chamber test to four hundred feet and passed the results along to Link.

As Lindbergh and Sténuit were nearing the end of their decompression, Sténuit felt a burning sensation on the inside of his leg. It was a sure sign of the bends, just like that ache in his wrist two years earlier at Villefranche. To err on the side of safety the two divers' decompression schedule was extended by a day to almost four days, nearly twice as long as they had spent on the bottom. Lindbergh felt fine but would have to wait for Sténuit's symptoms to subside before they could both leave the chamber. It was another reminder that decompression schedules were not a one-size-fits-all proposition, despite the deft calculations of a maestro like Dr. Workman and Link's Pennsylvania doctors. Under the record-breaking circumstances, it was not surprising that one diver would be symptom-free while the other got bent, but why? It might be a common case of differences in the divers' physiological makeup, or perhaps the decompression schedule needed adjustment, or both. Or had living in a helium-rich artificial atmosphere at two hundred pounds per square inch of pressure for two days taken them to the limit that most human bodies would tolerate? Had they hit some kind of barrier?

Neither Link nor Cousteau was producing the physiological data to answer questions like these, a fault George Bond considered a scientific sin. To push further down the continental shelf safely, and ultimately achieve Dominion over the Seas, they were going to need more of the kind of data Bond compiled during Genesis and was now seeking during Sealab I. And as the world well knew, a similarly detailed process of discovery was taking place in outer space. A year earlier, in the sixth and final Project Mercury mission, Gordon Cooper successfully completed more than thirty-four hours in orbit so that NASA could continue to assess a man's ability to endure the long trip to the moon—about four days of space travel each way. In addition to issues of physiology, the well-being of an astronaut was of course inextricably tied to the proper functioning of equipment. Diving was no different.

Lindbergh and Sténuit were coming out of their long decompression when the end of a high-pressure air tank blew off, causing enough of an explosion to rock *Sea Diver*. Shrapnel badly bloodied the legs of Annie Sténuit, Robert's wife, and one of the lead crew members was struck in the ribs. The blast dented *Sea Diver*'s steel deck. Ed Link, who prided himself on precision, was upset but also relieved that the accident wasn't worse. The injured were patched up at a Miami hospital, and Link later traced the problem to human error, an omnipresent wild card more likely to turn up under experimental conditions. Someone had incorrectly hooked up the air tank.

Sténuit and Lindbergh had emerged from their prolonged decompression on July 6, six days after they had begun living under pressure and just in time to witness the explosion. When they reached Miami, the Belgian diver still had some odd sensation in his shoulders and a kind of paralysis around his ankles. When he tried to lift his feet and stand on his heels, nothing happened. Usually, such physical aftershocks from the bends could be expected to subside, and they eventually did—not that the world was holding its breath for Robert Sténuit.

"BREATHE!"

From the Navy base at Bermuda, Sealab I was towed twenty-seven miles to Plantagenet Bank, a seamount chosen for its level topography and the presence of a Navy observation platform called Argus Island. Standing alone out in the Atlantic, its two-story building and construction crane perched some six stories above the waterline atop a handful of steel pillars, Argus Island looked like a misplaced section of a pier.

As the Sealab team went through the tricky process of lowering the lab, the sea kicked up a steady roller coaster of swells. The line handlers were trying to keep the habitat from slamming into the barge when Sealab suddenly vanished. Captain Bond's heart sank with the lab. It was like Panama City all over again. Despite new precautions, Sealab had somehow flooded and sunk to a depth of about sixty feet before they stopped it. Had the line snapped, the lab would have dropped two hundred feet down to the bottom, and the Bermuda Triangle could claim another victim. Fortunately, they were able to raise the flooded lab and reel it in to the barge, then return to Bermuda for another round of cleaning and drying. Considering the negative reaction from Washington after the flooding at Panama City, Bond had good reason to worry that this latest accident might end the experiment.

Back at the base, Bond and Mazzone were called into a conference with others working on the project, including Captain Melson and his top engineer, Al O'Neal. Bond listened to complaints about the monstrous handling difficulties and the near loss of the flooded lab. Some on-the-surface support crew seemed to be saying, in so many words, that the job was hopeless, at least with the materials and methods their limited budget afforded. Bond had heard this kind of pessimism before. It was a source of tension aboard Your Friendly Navy Barge, but so, too, was Captain Bond's attitude that he was the one running the show. As the senior medical officer and principal investigator, Bond had his share of responsibility, including the well-being of the aquanauts, but he was a medical officer, not a line officer, and in the Sealab chain of command he was outranked by other officers, namely Roy Lanphear, Melson's on-scene commander, who had been summoned to Washington for a chastening after the first flooding. There was also William "Red" Hollingsworth, the officer-in-charge of the YFNB-12. Both men were sailors and divers with impressive credentials. Lanphear complained to Melson that Bond ignored his commands. Even Mazzone, who was usually accepting of Bond's unorthodox, sometimes Walter Mitty ways, reminded Bond, none too gently, of the basic hierarchical fact: *A medical officer cannot accede to command!*

As Melson became aware of this uneasy tug-of-war between Bond and Lanphear, his worst suspicions about Bond seemed to be confirmed. Tampering with the chain of command! Such insubordination was worthy of court-martial, Melson believed, but he did not pursue it. He and Bond were, after all, accidental allies in this quest to live in the sea. Melson did have a serious talk with Bond, and Bond was politic enough to avoid getting sidelined.

During the otherwise downbeat meeting after the flooding at Argus Island, O'Neal suggested what sounded to Bond like an ingenious approach: They could use very large buoys to lower and raise Sealab on chains with jacks. Bond had always liked the engineer for his can-do attitude. But not everyone agreed with O'Neal, and someone suggested relocating the test to a sheltered spot in the bay at Bermuda, not far from the Navy base in the Great Sound. The depth there, however, would be just seventy feet, no deeper than Cousteau's Deep Cabin experiment.

Melson listened to the various difficulties and possible remedies and gruffly noted that those in the room had apparently spent more time figuring out why the job could *not* be done than how it could. It was about then that Bond and Mazzone exchanged hopeful glances. Melson said the best way to lower Sealab would be to use the big crane on Argus Island. It had been off-limits to the Sealab project, but he would see that they got access to it. He left it to O'Neal and the others to work out the details.

After seven years of butting heads throughout the Navy hierarchy, Bond was relieved to have someone in a position of authority stand up for

the cause of undersea living, even if he and Melson didn't always see eye
to eye. Unfortunately, no sooner had Sealab been given another reprieve
than they got some bad news about Scott Carpenter. A blazing front-
page headline in the *Bermuda Sun* pretty well summed it up: "Astronaut
Comes to Grief Here." It added, cheekily: "He may be a whizzo in space
but he was let down by a bicycle."

On July 15, Scott Carpenter had talked Charlie Aquadro into renting the little single-horsepower motorbikes popular with tourists. The
newly acquainted colleagues rode the ten miles into the colonial capital of
Hamilton and as the night of socializing wore on, Aquadro headed back
to the base. Sometime after midnight, he was awakened and called to the
dispensary. Captain Bond was summoned, too. It was Carpenter. As soon
as they saw him, the two doctors could tell that the thirty-nine-year-old
spaceman would be in no condition to take up residence on the ocean
floor. On the ride back from Hamilton, about halfway to the base on the
gravelly road, Carpenter had dozed off at the handlebars and crashed into
a coral embankment. He suffered a compound fracture in his left arm and
a pulverized joint near his big toe. The accident left him looking badly
bruised and battered. It wasn't the first time Carpenter had encountered
hell on wheels. He had totaled half a dozen cars in his youth. At twenty-
one, on another drowsy late-night drive, he had nearly killed himself in his
beloved '34 Ford coupe along a winding mountain road outside Boulder,
Colorado, his hometown.

Before Carpenter was taken by helicopter up the island chain to
Kindley Air Force Base, and from there to Houston for further treatment,
Bond had to tell him that he was off the roster for Sealab I. Carpenter
seemed more pained about getting scratched from Sealab than about
his scrapes and injuries. Bond, too, was sorely disappointed. He expected
that Carpenter's presence on the bottom would have drawn welcome
attention to the new undersea program.

There was no shortage of divers who would gladly have volunteered
to be among the first to live in the sea—*three weeks at two hundred feet!
Unbelievable!*—but Bond stuck with the four-man crew he had lined up
prior to Carpenter's arrival, including Bob Barth and Tiger Manning, the
hospital corpsman who was Barth's chamber mate for two of the three
human Genesis experiments. Bond also wanted a doctor on the Sealab
crew and chose Robert Thompson, a younger submarine medical officer
he had known at New London. Dr. Thompson, a lieutenant commander,
had graduated from the University of Southern California medical school

and had additional graduate degrees in nuclear science and radiobiology. The fourth diver was Lester "Andy" Anderson, who had spent much of his Navy career diving and was working at the Experimental Diving Unit during Genesis D. Unlike some of his more skeptical fellow divers, Anderson had been enthusiastic about the week-long hundred-foot test and told Captain Bond that he would gladly volunteer for any future saturation diving ventures.

Anderson was thirty-one, a couple of years younger than Barth, and the two hit it off from their first encounters at the Unit. Anderson had swarthy features, a slight paunch, and an impish grin. Above all, Anderson was an inveterate prankster whose antics had transcended into legend. Barth himself was no stranger to pranks, which was undoubtedly one of the reasons the two became fast friends. During training that spring, for example, Barth and Anderson engaged in some mischievous "midnight requisitioning"—their way of helping to keep Sealab I's costs down. Under cover of darkness, they drove a jeep over to the Mine Defense Lab's salvage yard, where Bond had first shown Barth the surplus floats, and grabbed any old hardware that looked like it could be put to use—steel cables, winches, weights, whatever they could get their hands on.

Like Barth, Anderson had joined the Navy as a teenager. Andy Anderson was just fifteen when he left his rural hometown of Weldon, Illinois, to enlist after World War II. He worked as a gunner's mate for a few years but after going through the dive school in the early 1950s, with hardhat baptisms in the Anacostia River, Anderson scarcely returned to gunnery and was able to spend much of his time diving. All four divers chosen for Sealab I were in their thirties and all were fathers. Barth had two kids, Manning had four, Thompson had six, and Anderson five daughters.

The U.S. Navy's bureaucratic structure with respect to its divers could be complicated, as Anderson knew as well as anyone. No one joined the Navy as a diver or was ever exclusively a diver, at least on paper. There were also several different types of diver training, including conventional hardhat diving using air, helium hat diving for the deepest dives, and, by the early 1950s, the Aqualung and some other forms of scuba. But diving was considered a means of performing jobs that happened to be underwater, and not a job in itself. Not every sailor who qualified as a Navy diver necessarily got in the water each day. In most cases, those qualified as divers continued to do their primary jobs—gunner's mate, boatswain's mate, engineman, mineman, and so on—and were called on to dive as needed. Some job assignments had more need for divers than others. Serving as a quartermaster on submarines, Barth managed to find ample opportunities to get wet. At New London, while an instructor at the escape training tank, he ran scuba training sessions and would take part in the kind of bleak recovery work that had just been called for at Bermuda.

Anderson's helium hat training at a specialized six-month Navy course got him assigned to submarine rescue ships, the only Navy vessels equipped for helium-oxygen diving. Anderson may have done a lot of diving in those days, but in the eyes of the Navy bureaucracy he was still primarily a gunner's mate.

Anderson joined Captain Bond on the tug for the day-long ride back to Argus Island. Sealab I, freshly cleaned up after being flooded, tagged along on its towline. Some devilish weather had passed, and the seas were blessedly calm, sparkling like an endless sheet of gently rippled glass as Sealab arrived after the journey from Bermuda. They floated the lab away from the barge and into the skeletal shadow of Argus Island. With the aid of support divers, Sealab was secured by a bridle of steel cables and fastened to the island's crane, as Captain Melson had suggested, in the hope of executing a more controlled descent. Even in the absence of storms, swells, and strong currents, lowering a large, buoyant structure to the bottom of the sea was never easy. Part of the problem was that a stationary sea floor habitat didn't have the engine power or maneuverability of a submarine. There was also the issue of gradually raising the lab's internal atmospheric pressure to keep the sea from rushing in.

By now, after having flooded Sealab a couple of times, they had gained some valuable experience in how to lower the lab without mishap. Support divers put ballast into the open troughs of the lab's pontoonlike feet, the "ballast bins," in the form of train axles. The axles were another product of the bargain hunting to stretch the program's tight budget. Although somewhat unwieldy, the old train axles could be had at a much lower cost than more traditional ballast, like steel pellets, which Ed Link had used to add weight to his inflatable dwelling. Some of the ballasting could be done before Sealab was towed out to the site, but dozens of train axles, one or two at a time, were lowered into the water and support divers, including Tuckfield, wrangled them into the ballast bins. Concrete blocks, each one about half the size of a hay bale, were also used to help distribute weight evenly and prevent the lab from listing.

The day after Sealab's arrival at Argus, it hung sixty feet below the surface on the crane cable. Barth and Dr. Thompson made a dive to hook up the hefty umbilical lifeline for the breathing gas, fresh water, power, and communications. They worked inside one end of the cigar-shaped lab in a compartment about the size and shape of an airplane cockpit that was sealed off by a bulkhead from the main living quarters and had its own open hole in the floor where the sea formed a pool just below the brim. They called this section the "air space" because it was pressurized with ordinary compressed air instead of the helium-rich mixture pumped into the adjacent main living quarters. As a safety precaution, the Mine Defense Lab engineers decided to isolate all the umbilical hookups, transformers, and electronic circuits in the air space. Their concern was that some systems might malfunction in the habitat's artificial atmosphere.

Following further inspections, the habitat continued its controlled descent. Shortly before two o'clock that afternoon, July 19, Sealab I landed safely on Plantagenet Bank, 193 feet down and under nearly seven atmospheres of pressure. At that point Bond and Mazzone made a final inspection of the lab before the aquanauts were sent down. They rode to the bottom inside the Submersible Decompression Chamber, a stout cylinder about ten feet tall and four feet around that looked something like Hannes Keller's Atlantis bell. The SDC, like Link's cylinder, would serve as a pressurized elevator to take aquanauts to and from the habitat.

Essential to the operation, the SDC was another hand-me-down piece of equipment, this time one that they remade from an old single-lock pressure chamber that Walt Mazzone had to pick up in Virginia on short notice. Mazzone, now promoted to captain, arrived with it on a flatbed truck. An officer of his rank was rarely seen behind the wheel on a delivery job, but Mazzone could be counted on for many things. After they modified the chamber, he gave it a trial run, riding it down to survey the wreckage of the jets at Bermuda.

When the Submersible Decompression Chamber reached a depth of about 165 feet, Bond and Mazzone each took a full breath, as if preparing for a plunge into the escape training tank, dropped through the hatch in the floor, and swam the twenty yards or so to Sealab. They swam the distance wearing only masks, fins, weight belts, and swim trunks. It occurred to Bond, Mitty-like, that he might be making the deepest breath-holding dive in history—although that distinction likely belonged to Lindbergh and Sténuit, who had to make a similar swim from Link's cylinder over to the inflatable dwelling. The late-afternoon sun shone down, and though filtered through many feet of cobalt seawater, it was still strong enough to light up the white coral sand on the sea floor.

Bond swam into the fenced area around the entry hatch—the shark cage. In a pinch, the aquanauts could duck into the cage and slam the gate on unwelcome predators. Like a chain link fence, the cage created a little backyard around the two entry trunks protruding from the lab's underbelly—one leading up into the main living quarters, and one leading into the cockpit-sized air space. Once inside the shark cage Bond crawled up into the main entry trunk, like a sewer worker going up a manhole. He climbed a few rungs, popped through the liquid looking glass in the floor of the lab, and rolled onto his back in the dry shelter. Mazzone followed him in. The artificial atmosphere felt good—no alarming smells or traces of impurity, although that didn't necessarily tell them anything. The presence of deadly gases could be imperceptible—until it was too late. The atmosphere was to be maintained at nearly seven times ordinary surface pressure with a mix, ideally, of about 79 percent helium, 17 percent nitrogen, 4 percent oxygen, and less than .5 percent carbon dioxide. A potentially great chapter in human achievement was about to begin, Bond sensed, even if Sealab had had a choppy start.

Bond and Mazzone popped in to the area around the entry hatch that was a kind of foyer for donning and doffing dive gear. The rest of the interior was a cross between a camper and submarine. A compact shower and toilet were situated near the entry. A simple table with folding chairs was along one side of the living quarters; along the other side was a countertop, cabinets, refrigerator, and electric stovetop. Two portholes the size of dinner plates were built into each side. Four bunks were in the tapered end of the living quarters, at the tip of the lab opposite the entry area. The entry hatch itself was next to the steel bulkhead that separated the living space from the adjacent air space. Bond and Mazzone took a breath and dropped through the liquid looking glass in the floor, down through the short trunk and into the sea. A couple of feet away they crawled up a similar trunk to check out the air space before taking a final breath to make the swim out of the shark cage and over to the SDC, which was hanging about an atmosphere up. Just as they had when they practiced escapes in the training tank, they exhaled purposefully along the way to avoid giving themselves an embolism.

Within a half-hour the four designated aquanauts followed the same sequence to get to Sealab. The SDC was cramped with four divers so perhaps it was just as well that Scott Carpenter wasn't with them. Minutes passed as Bond and Mazzone awaited word from the first man in the habitat. They listened to the intercom inside their control room, a converted trailer that made it look like someone set up a campsite on the barge's deck.

They knew each diver would have to make the breath-holding swim from the SDC into the habitat. As Bond and Mazzone had seen for themselves, the waters around the Argus tower teemed with marine life—schools of barracuda, amberjacks, tuna, groupers, countless tropical fish. Some large groupers—one must have weighed almost two hundred pounds—seemed to enjoy congregating inside the shark cage. A four-foot moray eel glided by.

At last came the sound of a Chipmunk—singing! It was 5:35 P.M., July 20, 1964, and the Navy's first undersea dweller had arrived, singing a helium-spiked version of "O Sole Mio" as he entered the lab. Bond got on the intercom and said: "Sealab I, this is Sealab Control. I hear you loud and clear—and I believe that is Robert A. Barth, QMC, am I correct, over?" Actually it was Lester Anderson. The familiar Italian folk song was a favorite of Andy's and the irony of crooning this ode to the sun he would not be seeing for three weeks must have amused him.

"Well, Lester Anderson, congratulations," Bond said. "Were you able to hold your breath all the way or did you have to breathe some water?"

The Chipmunk just chuckled and said, "You know how things are."

Bond could hear some falsetto muttering over the intercom and Anderson called out the names of the others as they appeared, one by one, through the liquid looking glass. Next to enter was Dr. Thompson, then Tiger Manning, and finally Bob Barth.

"As I take it now all four of you are together," Bond said.

"All four are together, that's right," came the Chipmunk reply.

Bond then quipped: "If anyone else comes in, kick him out!"

It was about seventy-four Fahrenheit in the lab, warm by ordinary room temperature standards but chilly in the heat-sapping helium atmosphere. The aquanauts turned on a couple of electric heaters to boost the temperature into the mid-eighties. They also set up the cameras that would allow the topside controllers to watch them, Big Brother style, via black-and-white monitors no bigger than license plates.

"Sealab I, this is Sealab Control. We have your picture," Bond said.

"Smile! You're on *Candid Camera*," Mazzone interjected.

Anderson shouted back a greeting, then said something that sounded like giddy gibberish in helium speech. Anderson tried again to make himself understood, squeaking as distinctly as possible: "I said: We're not going to call you 'Sealab Control.' We're going to call you 'topside.' *Is* . . . *that* . . . O . . . K?"

"Sealab Control" perhaps had more of a space age ring to it, but "topside" was standard sailor jargon.

"Well, I guess that's okay," Bond said. "I don't know; I kind of like the title of 'Sealab Control' but if I haven't got control, go on and call me 'top-side.'" Before long they were calling Bond "Papa Topside," a nickname he warmly embraced. Papa Topside watched the monitors as the aquanauts— "*my* aquanauts"—began running through a variety of setup procedures, checking life support systems like the carbon dioxide scrubbers, checking circuits, calibrating instruments, organizing the equipment and supplies carefully stowed prior to Sealab's descent, including a stockpile of boxed and canned foods to last the three weeks.

Barth and Anderson had done most of the shopping for Sealab and they had packed a sparkling Mateus rosé, inspired by Cousteau and his oceanauts, who had indulged in wine. The Mateus was of course not as fine as anything poured in Starfish House and something was wrong with the thermoelectric refrigerator, so the aquanauts would have to drink their rosé warm. (It would also be flat, as would a soda pop when opened, because the pressurized atmosphere prevents carbonated drinks from fizzing.) Fortunately they still had the use of their electrowriter, an electronic notepad that enabled the aquanauts to transmit facsimiles of handwritten messages to a similar device in the topside control trailer. The electrowriter could only send messages, not receive them, but it was a godsend when helium speech became exasperating. Tiger Manning gave the electrowriter a try, scribbling laconically: "HELP!" He was only kidding.

The aquanauts found it convenient and somewhat of a thrill to swim out with just a lungful of air. It was like diving at the training tank, but a single breath got them from inside Sealab to the bottom of the Atlantic. Experienced tank instructors had no problem holding their breath long enough to pick up supplies lowered from the surface—or just for the fun of it. Sometimes they'd go out on their own instead of with a buddy, which would irritate topside commanders, who viewed solo swims as a breach of buddy diving protocols.

Once outside the shark cage the aquanauts could see for a hundred feet or more in crystalline conditions worthy of a Cousteau film, even if the scenery was less dynamic than in a place like the Red Sea. The immediate area was flat like a beach, covered in white sand with scattered patches of marine shrubs. Cables associated with Argus Island operations crisscrossed the sandy sea floor like errant strands of giant spaghetti. The island's steel legs were encrusted with sea life. The water was about seventy degrees— warm considering the depth, but they would nonetheless find themselves fending off chills. Being saturated with heat-sapping helium was one problem, but water wicks heat away quickly on its own.

For their working dives, the aquanauts had a choice of several kinds of gear, including scuba and a "hookah," a type of breathing apparatus that drew its gas directly from the lab's atmosphere and delivered it to a diver through an umbilical. It wasn't bad for diving near the habitat but the umbilical could get tangled and also the hookah's compressor chugged like a lawn mower, killing conversation for those still inside the lab. There was no diver-to-lab communication, and any diver-to-diver dialogue in the water had to be carried on the old-fashioned way, with hand signals. Mostly they used a newly developed variety of Navy scuba called the Mark VI. The name implied that it was a replacement for the venerable Mark V hardhat, but the old-style hardhat system was still very much in use and not soon to be retired.

To the untrained eye the new Mark VI looked pretty much like ordinary scuba—a pair of tanks worn on the back, with a hose from over each shoulder that connected to a mouthpiece. The most noticeable difference was a pair of breathing bags that hung like a Mae West life vest over the diver's shoulders and fastened around his chest. A canister about the size of a thermos was wedged between the tanks. Bags and canister were the key components of a built-in filtering system that enabled the Mark VI to recycle a good portion of exhaled helium and any unused oxygen. With ordinary scuba, all the exhaled gas—carbon dioxide, nitrogen, and unused oxygen—is released after each breath as a hail of bubbles in the water. The Mark VI burped just a few bubbles from an exhaust valve near the diver's left shoulder about every third breath. The bulk of the gas, mainly helium, was routed back through the filtering system. With fresh oxygen injected, the proper gas mixture could then be recirculated and rebreathed. This system conserved the expensive helium and also made the gas supply last longer. A diver could get an hour or more out of a Mark VI at two hundred feet, about twice as much time as ordinary scuba.

The Mark VI rig was not designed with Sealab in mind, but it afforded longer dive times and was thus a logical choice for the project. Its one drawback was that it could be much more temperamental than ordinary scuba. Although less likely to kill a man than its prototype predecessor, the Mark VI could still dupe its user into thinking his breathing gas was flowing properly when it wasn't. Everything might seem fine—until it was too late. There were no warning lights or alarms. A diver had to develop a sixth sense of sorts to recognize telltale danger signs. A lack of burping exhaust bubbles was one, feeling light-headed was another, but both were subtle enough that a diver might miss them, especially if preoccupied with work.

In an emergency on a conventional, short-duration dive, a direct ascent to the surface was still preferable to drowning—provided you remembered not to hold your breath, out of nervousness or some other misbegotten impulse, so as not to give yourself a paralyzing, possibly fatal embolism. You'd likely get bent and run the risk of injury from surfacing without making any decompression stops, but you at least stood a chance of surviving, especially if you could get to a chamber quickly and undergo recompression. But for the saturated Sealab I aquanauts, any swim for the surface was out of the question. They were allowed to rise about one atmosphere above the habitat, but any higher and they would run the risk of "explosive decompression," a virulent case of the bends. The resulting bubbles on the brain would cause a saturated diver to black out en route to the surface—a merciful thing, really, since he would be oblivious to the excruciating pain from the bubbles fizzing throughout his innards. He would be dead before he hit the surface, and the direct ascent from seven atmospheres to one would turn him into a puffy corpse, his body ravaged first by the initial explosion of gas bubbles and then their continued expansion, per Boyle's law, as the enveloping water pressure diminished.

During the first week the aquanauts came and went from the lab as they pleased. In the water, they tackled some specific tasks. Barth inspected Argus Island's support pillars, and a few technical difficulties initially needed attention—a troubled fresh water hose, a valve installed backward on a helium line, a fickle TV monitor cable—but the aquanauts were left with ample time to swim along Plantagenet Bank. They typically stayed within a two-hundred-foot radius of Sealab, being careful not to drift up more than thirty or so feet. Barth and the others didn't care much for their canned sardines so Barth made a ritual of feeding them to the neighborhood fish. The amberjacks and groupers would snatch them greedily from Barth's fingertips. Two groupers that became regulars around the habitat were soon named George and Wally, in tribute to Bond and Mazzone. Less welcome were the two sharks that appeared on the third day, causing a stampede of frantic fish. They circled the habitat for five minutes, like curious envoys from the abyss. They were impressive in size, about ten feet long. With any luck the pair would not be hanging around enough to merit nicknames.

The aquanauts occasionally engaged in behavior that irritated the topside commanders. They covered the lens of the closed-circuit TV camera inside the habitat, scribbled derisive messages on the electrowriter, adjusted their atmosphere without proper approval, or made sorties from the lab without reporting their movements as required. Dr. Thompson upset some topside sensibilities when a camera outside the habitat caught him making what might have been the world's deepest skinny dip.

Bond attributed such digressions to what he called the "aquanaut breakaway phenomenon"—a show of independence, in effect, that came from trying to live autonomously on the ocean floor while being constantly watched and hounded with instructions. Falco and Wesly had much the same feeling during Conshelf One. Consciously or not, the aquanauts seemed to resent the intrusions and found little ways to rebel. The most important thing, however, was that they were all *surviving*, without dramatic mishaps or program-killing catastrophes. In many ways, they were just four guys having the time of their lives. *Silly skeptics! Look at us! We're living in the sea!*

Dr. Thompson was not as experienced a diver as the other three, but Bond had wanted a physician on the team and Thompson had studied marine biology in college. He spent time in the water collecting sea life from around the lab, which made for an informal demonstration of the kind of scientific research that Bond hoped saturation divers could carry out someday from sea floor bases. The only downside to Thompson's effort was the fishy aroma that permeated the lab.

The aquanauts' blood and urine samples were sent to the surface for analysis. Support divers shuttled supplies and equipment between the barge and the bottom, including replenished scuba and Mark VI rigs, but unlike the Conshelf habitats, which received a steady stream of visitors, including Cousteau, Bond wanted to minimize the experimental variables, so Sealab was off-limits to everyone but the four divers.

The habitat was kept in the mid-eighties Fahrenheit, a comfortable temperature in the pressurized, helium-rich atmosphere. Still, a variety of physical discomforts was noted, some with more concern than others. On the first day everyone's joints ached, which Genesis had shown they could expect as their bodies acclimated to the pressure. The pain in Barth's left shoulder lingered longer than he would have liked. Dr. Thompson and Manning had headaches one day, a possible sign that something was wrong with the artificial atmosphere, but the various gauge readings appeared to be normal. Thompson and Anderson spiked fevers and remained feverish for a couple of days, but neither one showed symptoms and both said they felt fine. Thompson had periods where he felt he didn't write or think clearly, but his vital signs seemed in order. One afternoon they all complained about rapid breathing and traced the problem to an elevated carbon dioxide level, the serious problem that had forced Lindbergh and Sténuit to evacuate their inflatable dwelling. They put in fresh canisters of carbon dioxide absorbent and injected another three hundred liters of oxygen into their artificial atmosphere for good measure.

Living in the sea was bound to have its share of surprises, as Dr. Thompson was harshly reminded on the very first morning. He nearly burned his lips when he sipped a cup of instant coffee, even though it showed none of the telltale signs of bubbling or steaming. Water boils and bubbles at a lower temperature at high altitudes because of the lower atmospheric pressure, as any mountaineer knows. Under the pressure of seven atmospheres it's just the opposite. A pot of water on the Sears hot plate didn't reach the boiling point until it was more than three hundred degrees Fahrenheit. To prevent future scaldings, the Sealab crew checked liquid temperatures with a thermometer.

Settling in for their three-week stay, the aquanauts added a few homey touches to their humble habitat. Barth taped a couple of *Playboy* pinups to the wall—tactfully placed beyond the omnipresent eye of the camera. One showed a nude blonde sitting poolside, a pensive look on her face as she extended a hand toward the water. The gesture—reach for the deep— seemed vaguely symbolic. A half-dozen books were propped up on the lone table, an odd assortment that included Rachel Carson's *Silent Spring*, though not her earlier work, *The Sea Around Us*, and a guide to Caribbean seashells, although they were miles from the Caribbean.

In the evenings, Captain Bond, ever the raconteur, recited prose and poetry to his captive aquanauts, perhaps a few passages from *Beowulf* or Chaucer he had committed to memory or "The Cremation of Sam McGee," a Bond favorite. He knew all eight hundred words of the galloping Robert Service ballad by heart. On the first Sunday, Bond read a bit of verse he penned for the occasion. He called it the "Sealab Prayer" and it began with a paraphrase of the lines that had inspired him in the Book of Genesis:

"Almighty God, who declared through Holy Scripture that mankind would one day acquire dominion over the seas, and the creatures therein, grant that this day fulfillment of Thy word is at hand. To the brave and dedicated men who have committed themselves to this project, grant Thine unending watch and safeguarding care in all the many hours of their life under the sea. Give unusual wisdom to each of us topside who might somehow control their work and safety as they perform their duties below. And when their work and Thy will together be done, grant us all a safe and worthy respite from our labors for a time to come. We ask all this in the name of Jesus Christ our Lord, Amen."

The Lord may have been watching over them, but the aquanauts could be fairly certain that their countrymen were paying little attention. They had hooked up a ham radio and when they reached several ham operators in the United States, they got only perplexed responses when they tried to explain, in their Chipmunk falsettos, that they were living in a Navy capsule on the ocean floor.

Fish feeding, sightseeing, skinny-dipping, specimen collecting, and ham radio chats—to Captain Melson it looked like undersea summer camp. He was not pleased. An attempt to perform the sort of useful work Melson found lacking came at the end of the first week with the arrival of *Star I*, a one-man submarine. The teardrop-shaped mini-sub, brought out to the support barge for an open-sea test, was a prototype designed to reach a sunken submarine under its own power. *Star I* could then attach itself to an escape hatch, like the old McCann Rescue Chamber, without the need for cables or tethers to the surface. Barth and Dr. Thompson placed a dummy hatch near Sealab. The aquanauts were then supposed to photograph the mission and come to the sub's aid if it ran into any trouble. The mini-sub pilot, Albert "Smoky" Stover, was a retired Navy diver the aquanauts had known at New London. Stover was to practice "landing" on the dummy hatch, as if for a real rescue. The aquanauts also set up a TV camera outside Sealab so Captain Bond and the others in the command trailer could witness the little sub's maneuvers. The scene began to resemble the artist's rendition of an undersea future that had appeared in *The American Weekly*, with mini-subs ferrying divers around a colony on the sea floor. Here was Bond's vision as vérité.

Anderson had a slight fever that day and asked Manning to take his place on the dive and shoot motion pictures while Barth and Thompson snapped away with waterproof still cameras. Anderson would stay on watch inside the lab. The three men in the water could soon hear the approaching whir of electric motors. The sub wasn't much bigger than a Chevy van. Barth hadn't seen Smoky Stover in a year or two but recognized him in the clear water behind the sub's hubcap-sized front window. They waved at each other and exchanged hand signals. At one point Stover found himself unable to go forward. Barth had grabbed on to a piece of the sub and the propeller whirred in place like an eggbeater. Stover haplessly gunned the sub's electric motors as Barth and Manning giggled hard enough to flood their face masks.

When Manning ran out of film—each magazine held only fifty feet— he swam back to the lab to reload, making his way through the shark cage and up the entry trunk. He gave the camera to Anderson, who then handed the reloaded camera back. As Manning swam the twenty yards or so back to the demonstration site and resumed filming, no one noticed that bubbles had ceased to burp from the shoulder of his Mark VI rig, a very bad sign. Within minutes Manning suddenly swam for the lab. On the TV monitor, Bond could see him scurry off the screen, but Manning appeared to be going to reload his camera again, so no cause for alarm. The divers had no voice communication, so neither Bond nor anyone else could know that a crisis had begun to unfold.

As Manning swam for the safety of Sealab, he pulled the ring that hung near his tailbone to activate the bypass, a fail-safe mechanism that was supposed to inject a fresh burst of gas into the Mark VI system. Nothing. A former training tank instructor and skilled breath-holding diver, Manning was not prone to panic. As he made it into the shark cage, Anderson heard a clang and put down his cup of instant coffee. He figured it must be Manning back for more film, and walked over to the pool of seawater in the floor. Anderson looked down and there was Tiger Manning, curled up on the sea floor as if taking a nap. Anderson jumped through the looking glass and lifted up Manning's deadweight, managing to get Manning's head high enough to reach above the waterline. When he pulled out Manning's mouthpiece Manning bit his tongue, so Andy stuck his index finger in his mouth to clear it. He and Manning were wedged awkwardly inside the trunk. Manning was the shortest of the aquanauts by a good measure, but stocky and muscular, and he had more than a hundred pounds of diving gear strapped to his body. As Andy struggled to loosen Manning's Mark VI, he shouted for help and rapped on the steel trunk to send an SOS to the other divers. Those watching the topside TV monitors still assumed Manning had gone for film.

Tiger Manning began to wheeze.

"Breathe! Breathe deep!" Anderson shrieked, in Chipmunk falsetto. "Take all the air you can get!"

Manning's entire face, lips, too, were as blanched as the sandy sea bottom. Dr. Thompson was closest and swam over when he heard the SOS. He helped Anderson pull off Manning's Mark VI and get him into the dry lab. Barth followed and Manning finally flopped onto the Sealab floor. By then, mercifully, he started breathing. He was weak, speaking but not making much sense. He insisted nothing was wrong with him. Captain Bond and the topside crew, who by then had been informed of the accident, awaited the results of Dr. Thompson's checkup of Manning.

Thompson soon reported that Manning's vital signs were back to normal. Manning even beat Anderson at cribbage that night. The only trace of the accident was in the whites of Manning's eyes, which had turned a devilish bright red. Manning had suffered a case of "face mask squeeze." Once out of breath, he was unable to equalize the pressure inside his face mask through his nostrils. The blood vessels in his eyes then swelled as if locked into a suction cup. "I've seen eyes like those before," Captain Bond would say, "but mostly they were on dead men."

That evening the aquanauts received a special supper from above. Steak and potatoes were sent down in airtight containers—a welcome change from the canned fare. But also on the menu that night was a good talking-to from Captain Mazzone. They needed stricter discipline down there, he told them, including tracking of gauge and equipment functions and reporting when anyone came or went from Sealab. And no more solo swims. They had to stick to the buddy system.

Within a couple of days of Manning's mishap, hurricane conditions were picked up on radar seven hundred miles to the southeast and forecasters predicted that high winds and high seas would pass uncomfortably close to Argus Island within twenty-four hours. A tropical tempest was not much of a problem for Sealab. On the seabed, nearly two hundred feet down, a major storm would pass virtually unnoticed. At the surface, however, it could tear the support barge from its four-point mooring, or slam it into nearby Argus Island, pulling out Sealab's umbilical in the process.

The engineers had anticipated that a storm might force Sealab to cut its ties to the surface and tried to make the lab at least temporarily self-sufficient. They built in tanks of oxygen and helium, packed extra carbon dioxide absorbent and a modest reservoir of fresh water. Electrical power could be drawn through an auxiliary cable from Argus Island. The barge could sail back to Bermuda and return after the storm had passed, and in the meantime, the aquanauts could live out their twenty-one days as planned.

The aquanauts were ready, even eager, to hunker down and take their chances with the backup systems, but Bond and Melson and the topside crew agreed that it was best to raise Sealab before the storm hit. Bond was satisfied that after ten days they would have adequate physiological data to advance the concept of saturation diving, and ample experience with the habitat and its equipment. But how to bring the lab and the aquanauts safely back to the surface became another source of tension between Bond and the other commanders.

They settled on a plan to begin two-plus days of decompression inside Sealab as the Argus crane gradually raised the thirty-ton lab to the surface. One concern was that surging storm swells could strain the crane to the breaking point, sending everything crashing to the bottom, aquanauts included. When Sealab began its slow ascent at midnight on July 29, the sea was dead calm and a three-quarter moon glowed. The ascent rate was to be three feet per hour, ideally in increments of eight to twelve inches, with corresponding adjustments made in the breathing gas mixture. The cable connecting the top of the crane to Sealab was about the length of a football field, and the crane operators were up on Argus Island. Bond and Mazzone controlled the gas changes in the artificial atmosphere from the barge and had to coordinate their adjustments with the crane operators. It was a little like having one driver steer and another operate the foot pedals.

By late in the afternoon the next day, they had brought Sealab to a depth of 112 feet. At that point, with enough oxygen in the mix to light a match, Anderson and Barth promptly swam into the adjacent air space for a smoke. As the night wore on, swells of up to fifteen feet began rolling through. The topside crew opted for the first of many holds so the lab wouldn't yo-yo too hard on the crane. Between holds, Mazzone coaxed Sealab I up to ninety feet below the surface. When Bond awakened in his trailer bunk, the support barge was bucking like a bronco. He could hardly brush his teeth. By seven-thirty in the morning Sealab had been inched up another ten feet. The growing seas and increasing winds were turning the lab into a noisy, slow-motion roller coaster. The bridle of cables slung around Sealab scraped and drooped and then went taut as the lab rose and fell. The water in the lab's open floor hatch sloshed and gurgled.

Bond had hoped the aquanauts could complete their decompression inside Sealab, for safety and comfort. But when dynamometer readings showed that the crane was reaching the breaking point, Bond had to concede that it was time for his aquanauts to move to the Submersible Decompression Chamber. Support divers struggled in the roiling sea to remove the train axle ballast, lightening the load as Sealab rose, and attach Sealab to four large mine buoys so that the crane could be freed up to handle the SDC.

When the SDC was nearby, hanging vertically like a big tin can on a string, Anderson, Barth, Manning, and Thompson all took a last breath of the Sealab atmosphere, dropped through the looking glass, swam over, and crawled up through the open hatch. Once inside, the aquanauts sealed the hatch in the floor to maintain the chamber's high pressure as they were lifted onto Argus Island. Anderson avoided looking out the porthole as the SDC dangled like a wrecking ball on the crane cable, more than seventy feet above the waterline. He had no qualms about living on the sea floor, but he did not like the sensation of swinging in the stiff breeze, crammed inside that confounded chamber.

The chamber was minutes from being lowered onto the Argus platform when a cable caught on an exterior fitting and snapped it off. The aquanauts heard a *pop*, followed by the eerie sound of their high-pressure breathing gas whistling out of the chamber. Dr. Thompson was standing near the leak and stuck his thumb over the hole. The blood blister on his thumb was minor compared to what a sudden drop in pressure could do to their saturated bodies—explosive decompression for one and all. As soon as the SDC landed on Argus the support crew hurriedly plugged the leak and dodged disaster.

The cylindrical SDC had been fitted with a box-shaped frame so it could be turned on its side and double as a decompression chamber. The SDC, like many early prototypes, lacked amenities that would come later, but the little World War II–era chamber at least had a medical lock, an airlock about the size of a suitcase, for moving supplies into and out of the pressurized chamber. Coffee, served in mayonnaise jars, and air mattresses were locked in. The men lay like fish in a barrel; if one moved on his mattress, they all moved—and sloshed in several inches of residual seawater. The mayonnaise jars, once emptied of hot coffee, served as makeshift urinals. One jar fell off a ledge and broke. By the last night they might as well have been locked in a cesspool.

On August 1, shortly after midnight, adjustments were made to finish the decompression process by about half past seven that morning. A dozen or so reporters were due to arrive from Bermuda to see the aquanauts emerge after nearly twelve days under pressure. No divers had ever stayed down so deep for so long, and for that the Sealab four could arguably be called the first men to live in the sea. The boat ferrying reporters and photographers to Argus hadn't yet arrived when the long decompression was complete, but Captain Bond wouldn't let his aquanauts out until the press was there. The reaction inside the chamber was enough to blow the hatch off. Barth, for one, was holding in more than just frustration. A proper toilet and a moment of privacy were just a few steps away and there were limits to what Barth and the others could or should have to do in a mayonnaise jar. Finally, at about eight-thirty that morning, reporters made it to Argus and watched as the American aquanauts, grimy and stubble-faced, emerged from their decompression of almost fifty-five hours.

An informal press conference was held in a small dining area and a couple of microphones were thrust at the aquanauts, who sat on one side of a table. Manning wore sunglasses, hiding his eyes, which were still discolored from his close call. Anderson smoked a Camel and sipped coffee. Thompson was shirtless. Barth was annoyed. The reporters wouldn't know the difference between jock itch and air embolism. Barth was never a big fan of the press, but for Captain Bond, he would squelch his smirk and cooperate as best he could.

After a few days of rest, they would have a more formal press conference, this one inside a cavernous hangar at the base in Bermuda. Every word said over the PA system ricocheted off the brick walls, as did the triumphal refrains of "Under the Double Eagle" coming from a record player. The aquanauts sat on a temporary stage, this time in service dress khakis or whites, along with Bond, Mazzone, and some Navy brass, including the chief of the Office of Naval Research. Manning again wore sunglasses. There were more men onstage than in the front row, which was reserved for the press. A couple of dozen people, mostly Navy personnel, were scattered around the room in the remaining folding chairs.

Bond felt some adrenaline when it was Lester Anderson's turn to speak. Anderson looked sharp in his crisp dress whites, tattoos adorning his powerful forearms—including the one of a lady in a champagne glass, which his wife detested. Bond was concerned that Anderson might lapse into his trademark salty syntax, but as Anderson sat down, Bond breathed a sigh of relief. He figured old Andy had just made his longest printable statement.

As Sealab I ended its run, *The New York Times*, like many papers across the land, was filled with coverage of the spectacular lunar photographs that had just been sent from the spacecraft *Ranger 7*. A front-page *Times* story got one of those triple-decker banner headlines most often seen during wartime. The *Times* did finally produce more than a squib about Sealab I, but the story wasn't front-page news and didn't appear until a week after the press conference. The paper ran sixteen paragraphs by one of its more renowned reporters, Hanson Baldwin, a Pulitzer Prize–winning military affairs correspondent. Under the headline "Navy Men Set Up 'House' Under Atlantic and Find Biggest Problem Is Communication," it reported that the dive set a record for depth and duration, albeit with considerable understatement.

Ed Link was taken aback when he read the *Times* report. The article made no mention of his own record dives. Not the recent 432-footer in the Bahamas, or even the day-long two-hundred-footer at Villefranche—a true first. He didn't begrudge Bond and the Navy for getting the paper's attention, but the only other aquanaut the *Times* mentioned was Jacques Cousteau and his Conshelf exhibition. Link, who kept an apartment in Manhattan, fired off a letter the next day to a top *Times* editor: "As a reader of *The New York Times*, I am somewhat shaken at the lack of knowledge on the part of your editors and the apparent need for more data in your files on important scientific events which are totally ignored in the before mentioned article."

A few of Link's friends, including a retired Navy captain who lived on Park Avenue, contacted Hanson Baldwin directly. One friend extended an offer to Baldwin to arrange a luncheon, "so that whatever Mr. Link can contribute to your knowledge of this growing field could be made available." Baldwin responded with a cordial note and expressed interest in a meeting. But the *Times* editor who received Link's missive was terse and unmoved: "Simply because everyone you have mentioned was not mentioned in Mr. Baldwin's story is no reflection whatsoever on the knowledge of our editors, and our files are quite adequate on the subject." The officer-in-charge of the Experimental Diving Unit lamented that coverage in the Washington newspapers was no better than in the *Times*. He told Link that the papers had failed completely to recognize the significance of the four-hundred-foot saturation dive.

The single *Times* story and a smattering of magazine articles were not enough to make household names out of Sealab I, saturation diving, or Captain Bond and his aquanauts. The proof was in the invitation that Bob Barth soon received to appear on *To Tell the Truth*. Each week on the popular TV game show a virtually unknown man or woman who had achieved or created something remarkable sat alongside two impostors. All three answered questions from the celebrity panelists whose job was to guess which of the three was the notable person. For this installment of the game, they had to figure out which one of the anonymous contestants seated before them had actually lived in the sea.

10

THE TILTIN' HILTON

In January 1965, just a few months after the press conference in Bermuda, planning began in earnest for Sealab II, a structure similar to its cylindrical predecessor but with a number of improvements. It would be the centerpiece of a more ambitious experiment in which three teams of both Navy and civilian divers would live and work out of their base at 205 feet for forty-five days. The budget was close to $2 million, ten times that of Sealab I. Although the target depth was only slightly greater than before, this second foray would be more challenging by virtue of its location: the edge of an undersea canyon about a mile offshore from La Jolla, California. This was just south of the location of the abortive Keller dive, with undersea conditions that would make the Devil's Triangle look like a paradise. The water was both murkier than at Argus Island and markedly colder—around fifty degrees Fahrenheit—and therefore more like most of the world's ocean depths. If the aquanauts could survive in this "hostile environment," they could succeed anywhere on the continental shelf.

One of the Navy's principal scientific partners for Sealab II was the famed Scripps Institution of Oceanography in La Jolla. The renowned institute, part of the University of California at San Diego, was perched along rocky bluffs overlooking a serene, two-mile-long crescent of sandy beach bisected by the utilitarian Scripps pier. It was to serve as the mainland support base and a half-dozen Scripps scientists were among the civilians picked to live on Sealab II. Twenty-eight volunteer aquanauts were divided into three teams, each team taking turns on the bottom for fifteen straight days. Scott Carpenter had recovered from his Bermudan motorbike injuries and was scheduled to spend a longer block of time on the ocean bottom than anyone else. As the designated leader of the first two teams, the former astronaut would live and work out of Sealab II for thirty consecutive days.

The first of the three Sealab II teams would do most of the setting up of the undersea house, and all three teams would have their share of experiments in oceanography and marine biology that Scripps researchers devised, along with projects on the Navy agenda, including a performance test of a specially trained porpoise. The Navy was happy to foster scientific interests and possible civilian spin-offs, but its main concern was in developing promising methods for military operations, like submarine rescue and especially salvage. Among the Team 3 aquanauts were some highly experienced Navy divers whose useful work would include salvaging the fuselage of a downed jet plane, like those that had crashed at Bermuda. Medical monitoring of all the aquanauts would continue throughout the experiment.

In the end, the most important mission of all, as with Sealab I, was survival. The serious injury or, God forbid, death of an aquanaut could derail the program. But from the looks of things at the christening ceremony for Sealab II, the Navy was showing a heady confidence, even pride, in this latest quest. The ceremony, held on Friday, July 23, 1965, was itself a telling contrast to the first Sealab's sideshow status at the Mine Defense Lab's open house the previous year. A number of Navy dignitaries were at the christening, including Dr. Robert W. Morse, who had replaced Bond's friend and supporter Jim Wakelin as assistant secretary for research and development. Morse and a dozen others were seated onstage for the outdoor ceremony at the Long Beach Naval Shipyard, just south of Los Angeles. Sealab II, painted a gleaming white instead of international orange, was festooned with billowing red, white, and blue ribbons.

With the habitat as an imposing backdrop, Secretary Morse gave a keynote address that often sounded as though George Bond had been his speechwriter. Morse hailed the Navy's emerging efforts at exploration and exploitation of the continental shelf—not since the days of the Louisiana Purchase, he proclaimed, had the United States stood at the frontier of a territory so rich with promise and yet so completely unknown. The secretary paid rare public homage to Bond, Walter Mazzone, and Robert Workman, who were seated nearby, and called Sealab I "successful beyond expectation," particularly in view of its "very limited budget." Morse got on quite a rhetorical roll, foretelling a day when excursions to deep-sea habitations, six hundred to a thousand feet below the surface, could be "as commonplace as jet air travel is today at thirty-five thousand feet." The secretary of the navy, Paul Nitze, did not attend the christening but dispatched his daughter, Heidi, to do the traditional honors of smashing a champagne bottle on Sealab II's hull. An ocean breeze blew softly and a Navy band launched into "Anchors Aweigh." A chaplain read a benediction, blessed the Sealab crew, and prayed to Almighty God, King of Kings, Lord of Lords. Little wonder that as the ceremony came to a close, George Bond wept, unashamedly.

All the aquanauts at the ceremony had heard their names read aloud in triumphant fashion. Scott Carpenter, Bob Barth, Cyril Tuckfield—Navy divers made up most of the crew for Sealab II, and Bond had favored divers he knew personally. Many more had applied to be aquanauts than could be accepted. Some, like Tuck, had served as support divers for Sealab I. Bond's pal Charlie Aquadro might have made the team but he was still in Monaco with the Cousteau group, which was busy preparing for a third Conshelf, this one scheduled to overlap with Sealab II. The nearly dozen civilians who filled out the Sealab roster had been picked for their familiarity with diving and their scientific specialties.

The only aquanaut missing at the ceremony was Lester Anderson, who had disappeared the day before, following a spat with Scott Carpenter. Commander Carpenter was in charge of physical training—PT—which included regular exercises like calisthenics and running. Soon after the aquanauts arrived in Long Beach from their training in Panama City, Anderson flatly refused to do PT. He could be stubborn sometimes, as everyone knew, but he was blatantly disobeying an order. Carpenter didn't know what to make of it. He and Andy had seemed to get along fine during the preparations for Sealab I. Andy had even asked the world-famous astronaut to autograph some pictures for his five daughters. Billie Coffman, one of the new aquanauts, had sensed a souring of his Navy buddy's attitude since earlier that year, when he and Anderson had witnessed a horrific accident at the Experimental Diving Unit.

Anderson and Coffman were on watch at the Unit on February 16, 1965, when their friends Fred Jackson and John Youmans were locked inside the chamber to test a decompression schedule for a two-hour dive to 250 feet. Anderson and Coffman had switched with Jackson and Youmans that morning so they would be available to meet with Captain Bond about their upcoming roles for Sealab II. That simple schedule change had fateful consequences. As the divers in the chamber approached the end of the prescribed decompression, additional oxygen was pumped into the atmospheric mix, a routine procedure.

Coffman was looking through one of the little portholes and talking over the intercom with the divers inside when Youmans said, "We have a fire in here." Coffman saw a flame shoot out of the carbon dioxide scrubber, the same cauldron-sized filtering device used in the chamber during Genesis. In the next instant it was as though Coffman were staring into a blast furnace. His friends vanished in a superheated flash of a thousand degrees—a "thermal explosion." Within a hellish minute, during which Coffman and Anderson joined a dozen others in a futile rescue attempt, the chamber was blackened like the inside of an old barbecue pit.

By their nature experimental dives carried an added risk, which is why not everyone clamored for billets at the Unit. While the risks were known, everyone was stunned and distraught that day, including Dr. Workman, who always took a cautious approach to the many chamber dives he supervised. Anderson seemed to take the accident especially hard. Some of the guys took him over to the Green Derby, a favorite neighborhood bar across the Anacostia River from the Unit. The Derby's owner, Al Vogel, was well known to divers from the Navy Yard for his generous spirit. Depending on what a diver needed at any given moment, Vogel could be banker, teacher, tailor. At a time as bleak as that Tuesday morning it would have been just like Al to announce that drinks were on the house. Anderson wasn't alone in needing a stiff one, and for some time the normally upbeat prankster was down in the dumps.

In the weeks that followed, Captain Bond noticed a rebelliousness in Anderson but figured that Andy would come around, out of loyalty to the program if nothing else. But then Bond heard that Andy was refusing to do PT and had goaded Carpenter by saying that he couldn't kick him off the team, that Captain Bond would overrule him if he did. Bond went looking for Anderson but couldn't find him that night, and then Anderson didn't show up for the christening. Bob Barth, like other friends, hoped that Andy would be allowed to stay, despite his recent transgression. They would miss him as a diver and they would miss his inimitable lighter side and nonpareil pranks.

The day after the christening Bond caught up with Anderson and tried to find the reasoning behind his insubordination but heard nothing that made any sense. Perhaps the EDU tragedy was to blame. Or maybe he couldn't accept Carpenter, a less experienced diver, as a leader—spaceman status be damned. Or perhaps it was some combination. Without giving specifics, Andy frequently complained to his wife about Carpenter, saying, "He's going to kill somebody, Ma. He's going to kill somebody on this trip!"

Bond, Mazzone, and Carpenter agreed that Anderson would have to go. Bond still believed Anderson to be a valuable diver—it was his quick thinking, after all, that had saved Tiger Manning, and possibly the entire program—but Andy would have to learn to cope without staging revolts against authority. Bond met privately with Anderson to let him know he was being taken off the Sealab roster. He humbly accepted his fate.

Now that Sealab was less of a shoestring operation it was becoming more regimented. Captain Melson and the Office of Naval Research again took the lead in running Sealab, now formalized as the Man-in-the-Sea program and set up under the auspices of a new Deep Submergence Systems Project, an outgrowth of the group that had reviewed the Navy's shortcomings in the wake of the *Thresher* loss. A busy agenda was designed to keep the aquanauts fully occupied, as Captain Melson thought the foursome on Sealab I should have been. Bond would reprise the Sealab I role of principal investigator, and they would still fondly call him Papa Topside, though his formal title was "deputy for medical affairs" in the organizational chart.

After the christening festivities in Long Beach, Sealab II was towed eighty miles south to its ultimate destination, within sight of the shoreline at La Jolla. The lab had been built from scratch over the past six months at the Hunters Point Division of the San Francisco Bay Naval Shipyard. The shipyard's builders had a handful of mentors from the Navy's Mine Defense Lab and the lessons learned from the engineering of Sealab I, but they had been given none of the usual clear-cut, detailed specifications for a Navy vessel.

They followed Sealab I's basic layout, only the new habitat would be nearly twenty feet longer and have greater girth—a diameter of twelve feet and an overall length of fifty-seven feet. Instead of resembling a cigar with tapered ends, Sealab II looked like an extra-large tank car plucked from a freight train. It had nearly a dozen portholes along its sides and in the center on top was a stout conning tower painted with the new Sealab insignia. It looked like the Navy's version of the NASA "meatball" logo— that familiar blue orb with a red swoosh and speckling of stars. The Sealab insignia was an upright rectangle with blue and white horizontal stripes symbolizing the ocean depths. Overlaying the stripes was a broad red arrow in which a fine white line etched the silhouette of a manly profile.

The red arrow pointed down, the direction in which the program hoped to go, atmosphere by menacing atmosphere.

A pair of barges, two boxy vessels with decks the size of basketball courts that had previously been used for test-launching Polaris missiles, comprised the support ship and mission control for Sealab II. In a bit of cost-conscious recycling, they were being modified and joined in such a way that they formed a U-shaped staging platform. A decompression chamber was installed in what had been a missile bay. The clutter of equipment, struts, beams, cables, crane, tarps, and color-coded tanks of compressed gas scattered around the deck made it all look like a floating construction site. Many of the crew on deck wore protective hardhats. A flag with the Sealab insignia fluttered in the breeze. The support ship was dubbed the SS *Berkone*, a moniker Mazzone coined one day by fusing his own name with that of Joe Berkich, a top civilian technician from the Naval Ordnance Test Station with whom Mazzone worked closely to set up the barges.

By August the *Berkone* was moored almost a mile offshore from La Jolla, over an area known as Scripps Submarine Canyon that the nearby institution's oceanographers had mapped as thoroughly as any comparable swath of ocean floor in the world. Below a rim two hundred feet deep a canyon plunged to depths of more than six hundred feet. Finding a suitable site had proved to be a challenge, even with the assistance of a Navy ship that scoured the sea floor with underwater cameras to assess a number of possible locations. On one exploratory dive to get a firsthand look at a possible site, a scuba failure left Walt Mazzone lying unconscious on the ocean floor, more than two hundred feet down. In that moment, the whole program might as well have been lying down there, too. Mazzone's alert buddy diver, an aquanaut-to-be named Bill Bunton, came to his rescue, but in the confusion Bunton had to abandon his bazooka-sized camera.

Once Bunton and Mazzone made it back to the surface, Bob Barth and Scott Carpenter went down to retrieve the camera, a valuable piece of equipment that Bunton, a former Army paratrooper turned civilian specialist in underwater photography, would need while living in Sealab II. Carpenter spotted it and swam at top speed for the bottom, leaving a great trail of bubbles in his wake. Barth tried to slow him down, knowing the dangers of breathing hard when in the realm of the narcotic haze, but couldn't stop Carpenter before he got to the camera. As the two paused for a decompression stop, holding on to the line they followed between the surface and the bottom, a dizzying array of synaptic bells and whistles went off in Carpenter's head. Disoriented, he hung upside down on the line like a drunken trapeze artist, eager for his disconcerting first bout with nitrogen narcosis to pass.

The fact that such near disasters could arise on these brief and largely conventional scuba dives put the risks lying ahead for Sealab II in sobering perspective. The job of a Sealab II aquanaut wasn't for everybody. Each of those who volunteered had his own reasons for embracing the risks of undersea living. Berry Cannon, for example, one of the dozen civilian aquanauts, had joined the Navy after high school. During four years of service he became a mineman second class and developed a keen interest in electronics. He earned a degree in electronic engineering from the University of Florida, back in his home state, and was working as a civilian electronics specialist at the Mine Defense Lab in Panama City when Sealab I was pieced together on Alligator Bayou. When word got around about Sealab II, Cannon made sure to volunteer, and was among the several hundred others in the running for a spot.

Cannon was a young father, thirty years old, lean and muscular, with dark hair and chiseled features reminiscent of a Roman statue. He had always admired the independence and self-reliance of the Old West, read a lot of Zane Grey, and liked to imagine himself as a Western pioneer. He and the other aquanauts shared a genuine sense that their sea floor sojourn could be the start of something historic, a pioneering exodus into the sea that mirrored the moon shot. It was an extraordinary opportunity that ordinary guys like Cannon didn't want to pass up, despite the risks.

On August 26, 1965, Sealab II touched down on the sea floor, following a scare from helium leaks, a tangled umbilical, and a few other technical difficulties. The lab had to be held just below the surface for a day, but once the problems were fixed the lowering went relatively smoothly. The *Berkone* had been fitted with a counterweight and an innovative system of pulleys and winches for easing the two-hundred-ton habitat to the bottom. Sealab II, which weighed about seven times as much as Sealab I, also had an improved ballasting system. Instead of relying on old train axles, the team could control the lab's buoyancy by taking in and releasing water from built-in tanks, much as a submarine operates. That system, along with the counterweight, made it possible to lower the lab within a couple of hours—and with considerably less drama than had been the case with Sealab I. No yo-yoing, no major flooding.

A sizable press corps witnessed the event, which pleased Captain

Bond. It was a welcome contrast to the scant attention in Bermuda, and might bode well for coverage over the forty-five days to come. Two dozen photographers snapped away as the first team of ten aquanauts posed for pictures, and later that afternoon they donned wet suits and scuba, moving about awkwardly on the *Berkone* as they prepared to swim down to the lab in pairs. Captain Bond and numerous crew members crowded onto a lower deck at the waterline to see the divers off. Someone handed Scott Carpenter a cigarette and he took a last puff. Carpenter, who was forty, and Wilbur Eaton were first to drop into the water. Eaton, who was like Bob Barth's alter ego, was a thirty-nine-year-old gunner's mate and an experienced diver who had been a support diver during Sealab I.

The lab had landed about fifty yards to the southeast of Scripps Canyon, atop a silt-blanketed bluff that would have afforded a spectacular view if not for the opaque saltwater setting. Carpenter entered first, and Eaton followed him up the stepladder and through the four-foot liquid looking glass in the floor at one end of the lab. No one was singing "O Sole Mio" but Carpenter and Eaton exchanged a few words and found themselves struck by fits of laughter. For some reason the pitch of their Chipmunk voices sounded more preposterous than usual. Maybe it was a release of pent-up tensions. The project had been run on a tight schedule. For seven months they were focused on preparations and training. Now, just a few days behind schedule, they had suddenly arrived. The plan was to have Carpenter lead the first two teams, for fifteen days each, so he could spend thirty consecutive days living and working on the ocean floor. Cousteau's Starfish House oceanauts had lived for as long a stretch in the Red Sea, but that was a much less hostile environment, with balmy, crystalline water, and they were just two atmospheres down, not seven.

Carpenter and Eaton were followed by Berry Cannon and Cyril Tuckfield. Tuck, a chief engineman, and Cannon, the electronics specialist, had a few preparatory tasks to do around the lab before entering it—adjusting some valves, hooking up the freshwater lines, and checking out the Personnel Transfer Capsule. The PTC was a pressurized elevator similar to the Submersible Decompression Chamber used for Sealab I, but specially fabricated for Sealab II and large enough to hold the ten aquanauts in each team. This new capsule, painted international orange and shaped like an oil drum the size of an elevator, was lowered by the *Berkone* crane to the sea floor. It stood upright atop a four-legged steel base and was placed ten yards or so from Sealab II's entry hatch. The first aquanauts had swum down to the habitat but the little capsule would take them back to the surface and be attached, under pressure, to the decompression chamber installed in the former missile bay on the *Berkone*.

Cannon and Tuck were working in the water for perhaps twenty minutes before they swam into the shark cage, the protective fencing around the entry hatch similar to the setup on Sealab I. A lot of small fish had already congregated, attracted to this alien fixture that landed in their midst. Cannon and Tuck climbed the short ladder, then pulled themselves up through the liquid looking glass. Cannon immediately sensed the warmth of the lab's interior and heard the Chipmunk chatter. Out of one porthole he could see the light he just turned on inside the PTC. The light reassured him, like a beacon in thick fog. If Sealab II became uninhabitable for any reason, the PTC would be a safe haven.

Once inside the lab Cannon and Tuck doffed their Mark VI gear and peeled off their wet suits. The suits' porous latex material became noticeably thinner and fit more snugly, like a second skin, under more than a hundred pounds of pressure per square inch. Cannon, like Wilbur Eaton, looked like a lab rat. They were both sporting electrodes on their scalps, with wires that ran down their necks, over their shoulders, and onto their chests. For the first four days the electrodes and wires would detect any irregularities in brain and heart function.

Dive gear was stowed around the entry hatch, in a kind of foyer. From the entry, Cannon took his first walk to the bunks at the opposite end of the lab, about twenty paces along a path of rough carpet running straight up the middle of the lab. As on Sealab I, these living quarters, with a galley and workspace, had the austere comforts of a camper. But the entire lab was on a slant. It had come to rest on a canyon slope and Cannon was among the first to experience the gentle uphill grade en route to the bunks. The lab also listed to the port side. The tilt increased slightly during the first hour, which made Cannon a little nervous. No one wanted Sealab sliding toward the edge of Scripps Canyon. To prevent any further slippage, an anchor with a six-inch-thick line was dropped from the *Berkone* onto a ridge about two hundred feet from Sealab. From there the aquanauts would have to mule-haul the line across the sea floor and tie it to the lab.

Within the next few hours the rest of the first team arrived in pairs, including the team doctor, twenty-seven-year-old Lieutenant Commander Robert Sonnenburg, a burly six-foot-four and 240 pounds. His beefy hands had been the catalyst for some crass commentary when the aspiring aquanauts gathered in Panama City for physicals, including a routine prostate exam—a "digital rectal examination," which the guys just called "the finger wave." Lester Anderson had used the exam to perpetrate a characteristically outrageous prank. Anderson flummoxed Dr. Sonnenburg—and greatly entertained the divers in the adjoining waiting room—with an exaggerated, operatic wail during the beefy finger wave. He followed up by giving the earnest young medical officer a big kiss—right on the mouth—then threw in a crude, mock proposition. After that encounter, Dr. Sonnenburg may have been relieved when Anderson was bumped from the program.

Sonnenburg was the only other aquanaut besides Scott Carpenter scheduled to live a full month on the bottom, but instead of duplicating Carpenter's consecutive days, Sonnenburg would leave with Team 1 and come back for the final fifteen days with Team 3 to shed light on the possible effects of going in and out of prolonged saturation under real working conditions.

Dr. Sonnenburg swam down with Billie Coffman, who had witnessed the EDU fire with Anderson. Coffman was a torpedoman who had earned the Bronze Star during the Korean War for his efforts to rescue fellow sailors aboard their stricken destroyer. Like Cannon and Eaton, he was sporting electrodes. Carpenter, Eaton, Cannon, Tuck, Dr. Sonnenburg, and Coffman still awaited four more aquanauts, including Tom Clarke, a lanky and bespectacled Scripps graduate student in marine biology who liked to keep goldfish in the claw-foot bathtub of his La Jolla apartment. At twenty-five, Clarke was the youngest of the aquanauts—their average age was thirty-six—and one of the few who wasn't married with children. On Team 1 alone were the fathers of eight boys and fourteen girls. Clarke swam down with Jay Skidmore, a thirty-seven-year-old Navy photographer who was involved with the record deep dive of the bathyscaphe *Trieste*. Even before all ten were in, the full spectrum of aquanaut personalities and their varying degrees of experience was plain to see.

Sealab II made an ideal setting for a study of "groups under stress," so from an office onshore at Scripps, psychologists planned to observe the captive aquanauts over the closed-circuit television system. Each team member had to fill out extensive daily questionnaires, such as a "mood checklist." The living conditions alone were enough to get on a person's nerves—a crowded tank, stuffy and humid, suffused with the sweaty stench of a locker room in which everyone sounded like Chipmunks. Many of the aquanauts had known each other only a short time. The smokers couldn't smoke, of course—not enough oxygen in the pressurized atmosphere to sustain fire—and outside was a very hostile environment.

Captain Mazzone's wartime submarine patrols had taught him to appreciate the value of decent food to a captive crew's morale and he took a personal interest in creating the Sealab menu. Some foods could not be prepared inside the artificial atmosphere—frying was verboten, as was cooking meat, for fear of spewing potentially dangerous hydrocarbons and other elements. The lab was loaded with more than two thousand cans, many containing heat-and-eat meals of the sort for which Chef Boyardee was well known—spaghetti and meatballs, franks and beans, ravioli, beef stew. All of these the aquanauts could safely heat on their electric stovetop. They had also stocked peanuts, pickles, raisins, crackers, assorted condiments, and four varieties of freeze-dried omelets—cheese, ham, mushroom, and Western.

For the first team life inside Sealab II was initially like a dormitory on move-in day, with lots of unpacking and housekeeping to be done in a space not much bigger than a city bus. The lab's tilt didn't make things any easier. Pots and dishes had a tendency to slide. Someone came up with a suitable nickname for their sea floor accommodations: the Tiltin' Hilton.

Lying in the narrow bunk that first night, electrodes protruding from his scalp, Cannon listened as Scott Carpenter struggled to make himself understood for a radio hookup with his pal Gordon Cooper, who was then whizzing around the globe, a hundred miles up, aboard *Gemini 5* with astronaut Pete Conrad. Carpenter sat next to the Sealab II console in his bathrobe, wearing headphones with a jutting microphone, like the one sportscasters wear, patiently if futilely repeating himself. This "sea-to-sky" voice link had struck someone as a fruitful way to siphon some of the enthusiasm for outer space to inner space. Unfortunately, the NASA radio operators seemed unable to comprehend Carpenter's otherworldly helium speech. Captain Bond, the erstwhile philologist, stepped in via telephone from the *Berkone* to help and the call was patched through about eleven-thirty that night, Pacific time, as *Gemini 5* made its 117th orbit. Carpenter might have kidded his fellow spaceman and delighted reporters with a favorite diver maxim of the day: "The ocean's bottom is more interesting than the moon's behind." Or he could give Gordo a little ribbing about being just one atmosphere up while he and his Sealab crew were more than seven atmospheres down. But Carpenter wasn't that sort of a josher and he and Cooper were barely able to greet each other before the historic communications link fizzled.

The newspapers devoted scant attention to the sea-to-sky stunt but were filled with stories about *Gemini 5* seeking to break the Soviet Union's duration record by orbiting the earth for eight days. Still, there was ample time for Sealab II's pioneering quest, including its own unprecedented duration, to make the papers. A month earlier Captain Bond had appeared on an NBC current events program called *Survey '65*, but the timing was not ideal. His interview aired on a Saturday night in mid-August, shortly before Sealab II was towed from Long Beach to La Jolla—and just as the Watts neighborhood in Los Angeles exploded in riots. Thirty-four people would soon be dead, more than a thousand injured, hundreds arrested, and dozens of buildings burned and destroyed. A couple of weeks later, as the aquanauts entered Sealab, those on board the *Berkone* could practically hear the ecstatic shrieking from across the water as the Beatles put on one of their final American shows of the year in San Diego—another reminder that in the mid-1960s Sealab had ample competition for attention.

The aquanauts on Team 1 followed the splash-down phase of the Gemini space mission on a little television that had been sealed inside a pressure-proof housing. Captain Bond liked the idea that his aquanauts felt a kinship with the spaceship. He was also relieved that they were safely on the bottom at last, completing their move-in tasks. When they asked that some music be piped in, Papa Topside gladly obliged. He also pulled out his harmonica. It soothed his nerves to play. He serenaded the men below over the intercom and relished the wild howls that came back in response. It had been eight years since Bond first began to lobby for a grand undersea quest. His thoughts, as recorded in a daily logbook, reflected his dreamy jubilation: "What a life in the deep! Commercial TV, unlimited fish-watching, FM music, gourmet-grade food, and hundreds of willing servants! Beats space exploration any day." The first full working day on the bottom, August 29, was a Sunday. Bond knew his aquanauts were busy but called for a halt in the action below to draw some inspiration from above. Over the intercom, in his soothing pulpit brogue, Bond recited the Sealab prayer he wrote for Sealab I. "Almighty God, who declared through Holy Scripture that mankind would one day acquire dominion over the seas, and the creatures therein, grant that this day fulfillment of Thy word is at hand . . ."

11

LESSONS IN SURVIVAL

A key to the aquanauts' survival, apart from their own wits, was the efficacy of their equipment, from the hulking Tiltin' Hilton itself on down to the delicate needles in their depth gauges. When they went swimming outside the habitat, nothing was more important than the breathing gear on their backs. For much of their work in the water, the aquanauts would rely on the Mark VI. Despite its peculiarities and the diver's need to maintain a sixth sense for subtle signs of malfunction, the gas-recycling rig was still the best of its kind for dives out of a habitat at seven atmospheres.

In addition to the Mark VI, the aquanauts could use standard scuba and would try out a new prototype, called the Arawak—named for a seafaring group of South American and Caribbean Indians. The Arawak was umbilical-fed, drawing breathing gas from the artificial atmosphere like the noisy "hookah" used on Sealab I. But the Arawak was designed to conserve gas, like the Mark VI, and would return the diver's exhalations to the lab through a secondary umbilical. The dual Arawak umbilical—about a hundred feet long and as thick as a pair of garden hoses—was attached to breathing bags worn as a vest in a manner similar to the Mark VI. As with any umbilical-fed rig, the Arawak offered an indefinite gas supply, but also limited range. It was useful for work right around the outside of the habitat, but the growing web of lines and cables could complicate its use, as Tuck and Dr. Sonnenburg found out. Their Arawak umbilicals got so tangled up that it took Berry Cannon and Billie Coffman a half-hour to free them. The visibility was poor that day, as it often was, and it was almost six o'clock in the evening, so the minimal sunlight was beginning its fade to black.

Scott Carpenter gave the Arawak a try early on and what happened was enough to make him wish he were back in orbit. He was out with Billie Coffman, working on something at the opposite end of the cylindrical habitat, twenty yards or so from the entry hatch, his umbilical trailing behind him. It was dark and cold, but otherwise everything seemed to be all right—until Carpenter had difficulty inhaling. He sucked harder on the mouthpiece but the blockage just got worse, until it was like sucking on a rock. The inability to breathe is an immediate, terrifying threat to survival, especially when you don't know what is going on. Carpenter swam for the shark cage and the entry hatch inside. It seemed a world away. At moments like this, somewhere in the diver's adrenaline-drenched mind, lurks the possibility that his trailing umbilical could get caught on something, yanking him to a sudden halt like a dog on a leash. *Then what?*

The aquanauts still had no voice communication in the water, just hand signals. So no one inside the lab nor anyone topside was aware of Carpenter's plight any more than they had known when Tiger Manning was swimming for his life. They only realized there was trouble when Carpenter burst through the liquid looking glass. He pulled out his mouthpiece and speaking . . . *between* . . . *gas-gobbling* . . . *gasps*, he said that Coffman might be having the same problem. Someone had to get in the water and go after him. No sooner had Carpenter spit out the words than Coffman appeared next to him in the open hatch. Coffman had seen his team leader take off in a hurry and he followed, guessing that something was wrong.

If Carpenter hadn't made it back, the national press would undoubtedly have noticed. ("Scott Carpenter, the second American astronaut to orbit the earth, became the nation's first aquanaut to die on the ocean floor . . .") That would not have been good for the program, but Carpenter lived and his close call went unreported, like so much else about Sealab. A bad kink in Carpenter's umbilical was to blame, a simple but potentially deadly malfunction. After straightening out their hoses, Carpenter and Coffman disappeared down the hatch once more into the dark, cold water to get on with their work.

The inability to talk while in the water remained an obvious threat to survival, so Berry Cannon and Wilbur Eaton, Barth's good friend and alter ego, prepared to try out an Aquasonic, the space age name of a promising prototype for diver voice communication. If it worked, the aquanauts wouldn't have to be incommunicado in emergencies and diver-to-diver dialogue would be especially helpful on complex jobs. Now, when hand signals didn't suffice, the aquanauts had to stop what they were doing and swim back into Sealab so they could pull out their mouthpieces and talk things over, in their helium falsettos. Even old-style hardhat divers had had telephone systems in their bulbous helmets for half a century, since about 1915. But such systems were more easily devised for the dry interior of a helmet than for a mouthpiece and face mask.

The Aquasonic looked like something a fighter pilot might wear, a type of mask fitted around the mouth that was supposed to allow for both breathing and talking underwater. Cannon, whose expertise included underwater communications, hooked up the Aquasonic system with a Mark VI and Wilbur Eaton took it out for a test dive. Back in the habitat Cannon received Eaton's signal but his helium speech was incomprehensible. Worse, the mask leaked badly. Cannon's verdict: Old-fashioned hand signals would have to do. A piece of equipment as significant as the Aquasonic that wasn't thoroughly tested and safe to use was another indication that the Man-in-the-Sea program, despite a surge in funding and enthusiasm, was still in its infancy. As Scott Carpenter summed it up, they were "working with mail-order equipment in marginal conditions."

Besides the special projects the aquanauts had to do more regular kinds of jobs—routine wouldn't be the right word—but jobs to keep their undersea house in order. Tending to the "pots" was one. These were airtight containers the size of trash cans that ran like dumbwaiters on cables, often several times a day, to ferry supplies and materials between the *Berkone* and a spot just outside the shark cage. Arriving pots had to be met, hauled through the water, and lifted up through the hatch into the lab, and pots such as those filled with aquanaut blood samples had to be sent back. Open baskets were set up on a dumbwaiter system similar to the one for the pots to carry Mark VI and scuba rigs to the surface for gas refilling. The lab was not yet equipped for the aquanauts to refill their gear themselves.

When a number of larger pieces of equipment were lowered from the surface, the first order of business was often just to find them in the obsidian gloom. Then they had to be moved into position, and moving anything around on the ocean floor could be exhausting with water being about eight hundred times as dense as air. Try walking waist-deep in a swimming pool or just clapping your hands underwater and—even without the burden of the cold and the bulk of diving gear—you will experience the added effort needed for even the simplest moves a diver makes. But depth and the greater pressure that comes with it do nothing to create additional resistance to movement. As Pascal's law explains, pressure applied to a fluid is transmitted equally in all directions, so there is no net effect.

The *Berkone* crew had to be careful not to drop anything on the lab and could only be so precise about where auxiliary equipment might land after drifting and dangling on a cable through two hundred feet of seawater. A few items did get lost on the way down, even fairly large ones like the "way station"—a hollow steel box about the size of a phone booth sandwiched by a smaller pair of adjoining boxes for ballast. Two or three way stations were to be set up as emergency safe havens along frequently traveled routes, but one tumbled into Scripps Canyon early on and disappeared. Several "psychomotor" machines were lowered and put into place outside the habitat, including one resembling a pinball machine. These devices comprised a kind of underwater arcade for testing and recording the aquanauts' strength and dexterity.

Another piece of equipment lowered from the surface was a weather station that looked something like a teepee frame. It was part of a Scripps project conceived to demonstrate the kind of scientific research that might be done more usefully from an undersea base than from the surface. The weather station hit bottom just twenty-five feet or so from the shark cage, on the port side. No one inside Sealab could see that far through the portholes—too dark and murky—but Carpenter and Eaton eventually found the contraption shrouded in the muck. They kicked up dirty clouds of the silt that blanketed the bottom as they dragged the weather station to a sandy slope near the rim of the submarine canyon, about fifty yards away, which the Scripps scientists had decided was a better data-gathering site than in the sheltered valley around the Tiltin' Hilton.

Cannon was carrying a fifteen-pound instrument out to the weather station for hookup one day and soon realized that he was too heavy to swim. He had to walk the distance, working against both gravity and water density each step of the way. Being encased in a full-body, tight-fitting wet suit added another layer of resistance. Cannon was huffing and puffing by the time he reached the station. In situations like this it could be hard for an aquanaut to tell whether his heavy breathing was due to the physical exertion or whether he might actually be experiencing a telltale sign of trouble with the Mark VI—Is my exhaust valve burping properly? Am I feeling dangerously light-headed? Am I about to pass out, like Tiger did? Will anyone even notice if I collapse out here in the liquid void? Cannon would survive this taxing trek to the weather station, but anytime any aquanaut's fate was in question, so, too, was the fate of the program. Things might go relatively smoothly for hours, even days at a time, with the entire team surviving, if not always thriving. Headaches, fatigue, ear infections, and a few other ailments dogged the aquanauts.

Within a few days of Cannon's arrival on the bottom, he and Jay Skidmore, the Navy photographer on Team 1, donned the Arawak and swam over to the PTC so Cannon could hook up a battery charger. They climbed into the short trunk and up through the liquid looking glass into the dry capsule. They immediately sensed that something was fishy—literally. Anchovies! Schools of the little silver fish apparently had become enraptured by the glow emanating from the open hatch, and hundreds of them, now lifeless and putrefying, had swum over the circular rim of the hatch and met with a dry fate. Cannon swam back to Sealab to report that their safe haven and elevator to the surface stank—and not just a little. The effect was nauseating. It was intolerable for minutes, much less hours. There was no immediate crisis, but if Sealab II were to lose pressure, or spring a leak, or if the atmosphere turned toxic—the very problem that forced Lindbergh and Sténuit to evacuate during their first hours in Link's SPID—the ten aquanauts would have nowhere to go for shelter. It took Cannon, Tuck, and Dr. Sonnenburg a day to rid the PTC of the decaying fish and purge its putrid atmosphere. Despite their efforts, the capsule had to be raised to the *Berkone* for a more complete flushing and ventilating. A makeshift screen was installed to prevent further anchovy invasions, and this additional work meant the aquanauts would have to go for another night without their backup safe haven.

As it turned out, anchovies were not their only problem. The shark cage surrounding the Sealab II hatch was meant to give the aquanauts a quick getaway from large predators. And while encounters with a fearsome creature like the great white shark would be rare in these waters, the shark cage did nothing to keep out smaller fish. One species was gathering in ever larger numbers, not just inside the cage and near the entry hatch but all around Sealab. They were *Scorpaena guttata*, also known as sculpin or scorpion fish. They're brownish and speckled, with a stout body that can grow to be about the size of a football. Unaggressive and slow-moving, they often wriggled into the silt-blanketed bottom. But along the backs of scorpion fish is a spiky dorsal fin. They look like little dragons, and their caudal fins are similarly spiked. When they sense danger, they respond like porcupines, sticking up their protective spikes. Those spikes, sheathed with tiny venom sacs, are sharp enough and strong enough to skewer human flesh, even through a wet suit. The sting can cause excruciating pain, but not death, not usually, although multiple stings could be serious enough to require hospitalization. Soon the spiky creatures were everywhere, about one every square yard or so. It was like a minefield of scorpion fish, which made the hostile environment even more hostile than anticipated.

The lighting outside the white habitat created a dusky scene in chiaroscuro. The oasis of light usually dissolved into darkness within twenty yards or so, long before the aquanauts arrived at job sites like the weather station. Some found it unnerving to lose sight of Sealab, enveloped in murk with no points of reference and only the sound of their own breathing as reassurance. Many aquanauts made a practice of carrying tethers with them to avoid getting lost or thrown off course by currents. Mazzone noted that the added sense of isolation from living with limited visibility might be another area ripe for physiological research. There was still much to learn.

As a practical matter, almost everything seemed to take longer than expected, a source of frustration for those on the bottom, but also just another fact of underwater work. Donning a snug-fitting wet suit and strapping on breathing gear could take an hour (greasing themselves with liquid soap helped speed the process of squeezing into wet suits). Once in the water, every movement unfolded in belabored slow motion. Fatigue set in more quickly, and along with the biting cold and limited visibility a job could get complicated. Things floated away. Lines got tangled. Manual dexterity was further diminished for those who wore wet suit gloves to fend off the cold. All the while they were dodging the ubiquitous scorpion fish and keeping a sixth sense trained on the functioning of their diving rigs, especially the Mark VI. Berry Cannon found that the floodlights around the outside of Sealab, for example, frequently burned out and could take an hour to replace. Other maintenance and repair issues, both the expected and the unexpected, added to an already full agenda of useful work.

More than a week into the experiment, neither the scorpion fish minefield or the anchovy invasion or even Carpenter's Arawak scare had made the daily papers. A few stunts managed to produce a smattering of innocuous notices, though, as when Captains Bond and Mazzone put on scuba rigs and made their way down into Sealab to witness Billie Coffman's reenlistment ceremony, led by Scott Carpenter. "This will certainly be a first in naval annals," Bond mused. Usually, as with Sealab I, no one other than the aquanauts was allowed inside the habitat, but Bond had made an exception for his brief ceremonial visit. Reenlistment aside, the visit might have made good advertising for the advantages of a saturation dive over a conventional bounce dive. For their brief thirteen minutes on the bottom, Bond and Mazzone had to spend more than an hour decompressing.

A few days later Dr. Sonnenburg's wife came aboard the *Berkone* with a devil's food cake topped with ten homemade aquanaut figurines. The three-layer cake had crumbled, but was put in one of the airtight transfer pots and sent down to Sealab for her husband's twenty-eighth birthday, on September 6. The couple exchanged pleasantries over the intercom and from the command van Captain Bond led the singing of "Happy Birthday." This cheerful episode, too, produced some bemused notices in the papers.

Bond's prospects for better coverage brightened when he was asked to sit for an interview with a reporter from *The New York Times*. The *Times* didn't send its esteemed military affairs writer Hanson Baldwin, whom Ed Link and his friends had sought out, and as far as Bond could tell, the reporter who came to interview him on the *Berkone* had never even heard of Sealab I. When the article appeared the next day, it was two dozen paragraphs consigned to page thirty. In it, the reporter wrote that the aqua-nauts "were suffering considerable hardships." He noted the "inadequacy" of the wet suits, the "paralyzing coldness," the "painful ear infections," and the "invasion of stinging scorpion fish." This was all true. But the *Times* made no mention of saturation diving, or how this newfound ability to live on the seabed was breaking century-old diving barriers—despite the unavoidable hardships that came with the hostile territory. Papa Topside was always sensitive about what the skeptical Washington bosses might think, and the story did cause some unwelcome consternation back east. Like Ed Link, who had fired off a letter to the *Times*, Bond tried to lodge an official protest but mainly he blew off steam by blue-penciling the of-fending article and adding a few earthy comments of his own. He posted the marked-up clipping on the *Berkone*'s event blackboard for the edifica-tion of other visiting news reporters.

A few days after the *Times* story appeared, the first team was getting ready to wrap up its fifteen-day boot camp on the bottom. Commander Carpenter had just returned from a final, midday inspection of the PTC. As Carpenter came up through the liquid looking glass, dragging the Arawak umbilical behind him, someone dropped a wet suit glove into the open hatch. Carpenter instinctively stuck his bare hand in the water to grab the glove before it sank. Then came the pain. Instead of grabbing the glove he grazed his right index finger on a sculpin's spiky fin. The pain soon spread from finger to hand and arm. It was an excruciating, debilitating ache, like having your hand slammed in a car door. Carpenter got out of the water, undressed, and went to lie down on his bunk. His nose and sinuses filled with fluid, forcing him to breathe through his mouth.

Bond sent down an assortment of drugs for Dr. Sonnenburg to administer. Bond also had Carpenter soak his hand in ice water, but that seemed only to exacerbate the pain. Woozy from the drugs, Carpenter slept for a few hours. Sonnenburg went back and forth on the intercom with Bond, who was prepared to dive down himself and help if necessary. Cannon, Tuck, Coffman, Eaton, and the rest of Team 1 postponed their pressurized PTC ride back to the *Berkone*, where thirty-plus hours in the decompression chamber awaited them. Carpenter was supposed to stay down another fifteen days and lead the second team, but if his condition didn't improve soon he would have to catch the PTC back to the surface with the others. A motorbike spill had kept him off Sealab I, and now a venomous fish threatened to cut short his record thirty-day stay on Sealab II.

It's safe to say that *Sculpaena guttata* had never made such headlines. The papers seemed to find it irresistible to point out that a former astronaut, who orbited the earth three times without a scratch, had fallen prey to . . . *a fish!* Not a dreaded great white shark, but a spiky little menace no bigger than a football. Carpenter was not even the first aquanaut to have been stung. About a week into the experiment, a scorpion fish punctured Billie Coffman's wet suit and zapped him on the leg. That incident never made the papers. As far as Sealab's place in the popular psyche went, hardly a word was printed to mark a genuine milestone of the sort often heralded in headlines, as with the recent *Gemini 5* flight: A few days before Carpenter's sting, on September 7, Sealab II passed the ten-day duration mark set by Sealab I.

Within a few hours the doctors were satisfied that Carpenter's condition had stabilized enough to leave him on the bottom. The rest of the first team swam over to the PTC and rode it back to the surface that afternoon. Three members of Team 2, including Bob Barth, came down to begin the transition. While Commander Carpenter was still laid up, Captain Bond appointed Barth as the team leader's acting deputy.

Like the other aquanauts waiting to take their turn living in Sealab, Barth had spent his days working topside. There wasn't room for everyone to live on the *Berkone* so he and the others would ride a boat to and from the beach at Scripps each day. Making the ten-minute trip at night, Barth was struck by the sight of the *Berkone*, its myriad lights twinkling in the darkness that fused sea and sky. The hum of its machinery echoed across the water. Barth had been looking forward to living and working in the sea again, but he was not happy to learn that his latest effort was going to cost him money. Some stingy bureaucrat had decided that the aquanauts, while working their fifteen days in Sealab, were not entitled to the standard $16 per diem for travel expenses, so Barth and the other Navy divers would be out two hundred and forty bucks. But Barth had made sacrifices for the program before. He wanted Sealab and his friend George Bond to succeed, and right now Scott Carpenter could use his help, per diem be damned.

Throughout his days as team leader, Carpenter spoke frequently over the intercom with Bond and Mazzone. Papa Topside, while delighted to have the former spaceman on the team, had to admit that he was getting his fill of Carpenter's persistent questions. Carpenter seemed to raise questions about everything, even something as straightforward as the proper functioning of a pneumofathometer, a common depth gauge affixed to the end of a hose. The "pneumo" had been a standard and reliable piece of diving equipment for years. So it was perhaps a sign that Carpenter was back to his old self after the scorpion sting when the questions started coming over the intercom again. Bond fielded them like a patient father.

A thinly veiled response to Carpenter's queries could be found in one of the first homespun Sunday sermons Captain Bond planned to give throughout the experiment. The topic: faith. "Our Lord gave us the admonition that if man has faith so small as a grain of mustard seed then he has faith enough to accomplish any of the ends which he might desire in this life," Bond preached over the intercom from inside the topside command van. His distinctive baritone drawl filled the Sealab II habitat as if piped in straight from the Church of the Transfiguration. He picked up a preacher's rhetorical steam as he warned about those who would accept nothing on faith and instead *question everything presented to them.* Of course neither Carpenter nor anyone else would have been down there in the first place if they didn't have more than a mustard seed worth of faith—faith in Captain Bond, faith in Captain Mazzone, in the concept of saturation diving, in so-called mail-order equipment and the many prototypes on which their survival depended.

One prototype was an electrically heated wet suit developed to improve upon a standard wet suit's limited ability to fend off the water's bone-biting chill. With an ordinary wet suit of rubbery black neoprene, a thin layer of water is trapped between the skintight suit and the diver's skin. The body warms the water and the suit provides insulation from the cold outside, although more water typically seeps in past the elastic around the wrists, ankles, head or neck, a sensation akin to getting an ice cube shoved down your collar. Breathing a gas rich in heat-sapping helium didn't help, and because the external pressure compressed and thinned the wet suits' porous neoprene, the suits were more like bedsheets than thick, cozy blankets. Berry Cannon could get chilled within a half-hour or so, even before his breathing gas had run out. Others felt shivers coming on in just ten minutes or so. It wasn't unusual to see an aquanaut pop up through the liquid looking glass, his arms shaking uncontrollably and teeth chattering like castanets. There were a couple of showers in the foyer around the open hatch and after a dive the sensation of that fresh, hot water piped down from above was "unbelievably delicious," as Carpenter put it—or "better than sex," according to Barth. They would stand there, stripping off their wet suits and gear, uttering absurd Chipmunk moans of relief.

Eight aquanauts, a few from each team, had been fitted to try out the electrically heated wet suit. It looked like an ordinary wet suit, but wires ran like copper blood vessels throughout the jacket, hood, trousers, boots, and mitts. An outer suit was worn like a coverall to protect the underlying rubber suit from scuffs, tears, or punctures. This "snag suit" was light-colored and slightly rumpled, like a space suit, and used the novel fastener Velcro instead of zippers. The electric suit was powered by a half-dozen brick-shaped batteries that hung around a belt, like extra ammunition.

Two aquanauts on Team 2, Wally Jenkins and Bill Tolbert, were among the first to wear the new wet suit on a dive. Both were civilian oceanographers from the Mine Defense Lab, and Jenkins, who was thirty, was a Navy veteran. He had a youthful face and a bald head. A few years earlier he'd wrapped up a four-year stint at the nearby Miramar Naval Air Station during which he often drove out to the coast at La Jolla to dive. Tolbert, who at thirty-nine was among the older aquanauts, was especially grateful to have been picked for a Sealab team. His leg had been crushed by an ice truck in a boyhood accident, and it left him with a slight hitch in his gait—and prevented him from going out for the football and basketball teams. To be part of something as extraordinary as Sealab II was a welcome consolation for the sporting opportunities missed in his youth. For Tolbert even a menial task could be a joy. As one of his first jobs he was handed a bag of trash to send up to the surface. He swam out of the shark cage, let the buoyant bag go, and gleefully watched as it vanished overhead. He couldn't help but marvel: This was garbage detail like no other.

While the first team was often focused on setting up the Sealab II encampment, Team 2 would have more time to devote to undersea research projects, such as those Tolbert planned with George Dowling, another oceanographer from the Mine Defense Lab. One was to release scores of detectors resembling mechanical portobello mushrooms around Sealab. As the hours passed, they would observe how these partially buoyant objects drifted over time, which could yield insights into how to track fragments scattered on the sea floor, one of the problems that had arisen during the search for the *Thresher*. The two scientists also planned to set up a bioluminescence meter—as soon as Dowling got over a mysterious and rather severe case of rash and swelling. Like astronomers training their sights on twinkling stars, the oceanographers wanted to measure the light produced by plankton and other phosphorescent undersea creatures. Tolbert said that seeing the dark water lit up, as if by fireflies, was the most beautiful sight he had ever seen: "You are completely enclosed in a layer of sparks. You can wave your hand through the water and the movement makes it bright enough to see for ten feet."

Before that project got under way, Tolbert swam out with Wally Jenkins to put up four "targets," which looked like little road signs. The targets were eye charts of sorts, with different shapes and colors designed to test the aquanauts' ability to see in their surroundings. Jenkins donned a standard wet suit but Tolbert put on an electrically heated prototype. The visibility test had to be set up beyond the reach of Sealab's lights. The two were swimming along the seabed when Tolbert realized he was drifting upward. The battery packs around his waist weighed thirty-five pounds and were doubling as a weight belt, but the belt had somehow slipped toward his knees.

He cocked his legs to keep it from falling completely off, but up he went, as if swept up in a vertical rip tide. The bulky battery belt had got caught on the bypass switch near Tolbert's tailbone. Normally the bypass was a fail-safe mechanism the diver activated by yanking a ring to inject extra bursts of fresh gas into the Mark VI breathing bags. But in Tolbert's case, the belt was holding the bypass switch wide open and the gas kept flowing, overfilling the vest so that it inflated like a helium balloon. It popped up around his neck and made him rise faster still. He had no voice communication, so he couldn't shout for help. He tried to reach around and shut off the bypass but couldn't dislodge it from the battery belt. He swam for the bottom as hard as he could but made scant progress, like a man walking down an ascending escalator.

Wally Jenkins saw that his buddy was in deep trouble and swam over to him. Jenkins cranked the exhaust valve wide open on Tolbert's ballooning vest. With the valve open, Jenkins literally gave Tolbert a big hug to squeeze the excess gas out of the overinflated vest. At the same time Jenkins kicked furiously, adding some propulsive power to Tolbert's own effort to stay near the bottom. It crossed Jenkins's mind that if the battery belt slipped any further or fell off, Tolbert could become buoyant enough to take them both on a deadly rise to the surface.

Tolbert managed to hang on to the belt and they were within twenty feet or so of the lab when another aquanaut, Ken Conda, saw Jenkins and Tolbert doing their emergency tango. Conda helped get Tolbert to where he could grab hold of a tether leading back to the shark cage. Tolbert then followed the line, using it like a handrail, until he finally popped up through the liquid looking glass. As an oceanographer, Tolbert had spent hours diving in the sea, but nothing had ever scared him quite like this. It was as close as they had come to losing an aquanaut since Tiger Manning collapsed under Sealab I.

Emergency situations like this were what the Navy had in mind when Tuffy was brought into the Sealab fold. Tuffy, an Atlantic bottlenose porpoise, was part of a cadre of marine mammals the Navy rounded up to see how they might be useful in underwater work. Tuffy was no astronaut or celebrity diver, but he might also draw some favorable attention to Sealab. Porpoises and their dolphin cousins had lately leapt into the public imagination through Flipper, the dolphin star of a hit TV series.

In recent weeks, a few designated Sealab aquanauts joined Navy trainers to focus on teaching the porpoise some tasks applicable to Sealab II or a similar sea-based operation. One was to act as a courier for fast delivery of tools or other small objects, either from the surface to Sealab or between divers. But most important, Tuffy was trained to act as an undersea St. Bernard by responding to the sound of a buzzer and racing to the aid of a lost or ailing aquanaut, or one in unexpected danger, as Tolbert had been. A web of tethers had been set up around the lab and the aquanauts usually took a come-home line with them on a dive. But if anyone dropped a line, didn't have one, or somehow got lost in the liquid shadows, he could press a buzzer and the 270-pound porpoise would speed to the rescue—in about a minute from his holding pen at the surface. The aquanaut could then hook on to Tuffy's harness for a tow back to Sealab.

That was the idea, anyway, but even with a buzzer in hand, Tuffy's aquanaut trainers on Team 2 couldn't always get him to respond properly during several days of simulated rescue and delivery exercises. Tuffy seemed to be wary of the droning *Berkone* generators, the web of heaving lines and transfer pot cables, and the floodlights around Sealab. But he had enough success to show that porpoises had the potential to work with future Sealabs, especially as habitats were placed at greater depths. When news of the porpoise's mixed results made it into some papers, it was given a light touch, as if aquanaut rescue was of no real concern. "Porpoise Tuffy Chickens Out on Sealab Test," a mirthful headline in the *Los Angeles Herald-Examiner* said.

Carpenter, Barth, and the rest of Team 2 were scheduled to make a series of deeper dives from Sealab that shared the pioneering spirit of Ed White's space walk earlier that summer. Tethered by a twenty-five-foot golden umbilical, White had become the first American astronaut to crawl out of his orbiting capsule. He drifted effortlessly into the vacuum a hundred miles up, and considering the way he floated, this historic half-hour could have rightly been called a space dive, but space walk entered the lexicon. The eight-hundred-seat auditorium in the new Manned Space Center in Houston wasn't big enough to hold all the newspaper, magazine, radio, and television reporters and cameramen. NASA had to lease an additional building for the news horde.

The Sealab aquanauts' dives had so far been limited to the same depth as Sealab itself, about two hundred feet, give or take an atmosphere. Going too far up toward the surface was definitely out—lest they risk drifting into the dreaded realm of explosive decompression. A plan to make deeper dives—"excursion dives," as Walt Mazzone called them—constituted an important step in learning to live in the sea. They would test the premise that someone saturated at a particular depth, living and breathing in a helium-rich artificial atmosphere, could spend some time working at a deeper depth and then swim directly back to his undersea base.

The goal was to make the excursion a no-decompression dive, like one that typically begins and ends at the surface. In recent months, Bond and Mazzone had been captivated by the potential of deep excursions and had run a series of experimental dives in their new chamber at New London and at the Experimental Diving Unit in Washington, with animals and with Navy divers. Dr. Workman wielded his slide rule to formulate schedules for allowable depths and durations, and the results were promising enough to give the researchers faith that a saturated aquanaut could survive a deep no-decompression excursion. Still, like Ed White on his space walk, the aquanauts would be taking a risk. No one had ever made excursion dives at such depth, and the aquanauts would be making these inaugural dives at sea, a more dangerous setting than the controlled confines of a chamber. Cousteau's Conshelf Two oceanauts had made some excursions, but with shallower starting depths and different breathing gas mixtures. Even if the celebrity diver had produced substantive physiological data—Mazzone never saw any—the Cousteau team's findings would have been based on such different circumstances that they wouldn't have been readily applicable to the Sealab excursions.

The preliminary Navy tests indicated that from a habitat at, say, 430 feet, the aquanauts could work at depths down to six hundred feet for more than an hour without having to decompress on the way back to their habitat. Even if there was some decompression penalty, making deep excursions could have an advantage over placing the habitat at the deeper depth, in part because a deeper habitat is much more expensive to operate. There's the additional breathing gas needed to maintain a higher-pressure artificial atmosphere, and final decompression would also take longer and thus cost more.

Bond and Mazzone could foresee a time, possibly within a few years, when an aquanaut living at six hundred feet could journey down to seventeen hundred feet, nearly a third of a mile below the surface, do a job and return to his base at six hundred feet. Perhaps, as Secretary Morse said at the christening, living and working at such depths might become as commonplace as jet travel. Also, as Bond had envisioned in his "Proposal for Underwater Research," sea floor shelters could be placed like way stations at depths corresponding to appropriate decompression stops. An aquanaut could carry on with undersea work, gradually moving from deeper stations to shallower ones, instead of returning to the surface and whiling away days in a decompression chamber.

The aquanauts were instructed to take their maiden excursions cautiously. Bond and Mazzone set an initial excursion depth limit of 266 feet—conservative, but that still added another two atmospheres of pressure. It was also deep enough to allow the aquanauts to explore the canyon rim, a bold step that Captain Bond, in his poetic way, believed would yield greater insights into an aquanaut's psyche than fiddling with test gadgets in the psychomotor arcade or filling out mood checklists.

Scott Carpenter and Barth took a first excursion but theirs and several others were cut short, in part because of concerns—and more queries from Carpenter—about the accuracy of their depth gauge readings. Time was nearly up for the second team, including Carpenter, who was about to complete his record thirty-day stay. The next day Carpenter set out with Wally Jenkins, the Navy veteran who came to Tolbert's rescue. The two swam along the bottom for what seemed to be about fifty yards—distance could be deceptive in the saltwater fog. The lights of Sealab dissolved into darkness as they followed a gentle downward slope in the ocean floor, swimming along until they reached the rim. There, the sea floor dropped out from under them. It reminded Jenkins of standing at the edge of the Grand Canyon, except that the view was less grand. The beams from their diving lights revealed lonely outcroppings of soft gorgonian coral, waving like barren trees along the rocky canyon rim. A few drab solitary fish hovered around, including the ubiquitous sculpin.

At 253 feet, Carpenter noticed a discrepancy in depth gauge readings and feared they would overshoot the limit of 266 feet. The two divers stopped where they were, suspended in darkness, then retraced their strokes, swimming up the canyon wall, over the rim, and back to the habitat. They had been out for about a half-hour. After reviewing the gauges it became clear that they had reached their depth goal, and they had done so in open water, not in some laboratory wet pot. It was good news for the concept of excursion dives, but not a milestone that brought out the droves of reporters who covered the space walk.

On the other side of the blue planet, Jacques Cousteau had just lowered Conshelf Three deep into the Mediterranean near Villefranche, where Ed Link first tested his cylinder. Shortly after midnight on September 22, Cousteau's six oceanauts reached a depth of almost 330 feet, more than a hundred feet deeper than Sealab and many atmospheres beyond Cousteau's first undersea houses. Accordingly, this third Conshelf habitat's design was altogether different, but like Starfish House it was an eye-catching bit of undersea architecture. Conshelf Three was a steel sphere, eighteen feet in diameter, its exterior painted in a checkered pattern of yellow and black. The living quarters were like those of a modest yacht, with a laboratory, lounge, and dining area in the upper sphere, and bunks, bathrooms, and a spacious dive station down spiral stairs. The checkered sphere sat atop a rectangular base, like a globe on a shoe box. The box contained food storage, ballast tanks, and nine tons of fresh water,

one of several features that made the Conshelf Three habitat more autonomous than its predecessors and even Sealab II.

The habitat's power and communication lines came from shore, but
Cousteau had sought to cut a few umbilical cords and minimize the habitat's reliance on surface support. Like Bond, he wanted to move toward
designs that would make future habitats as autonomous as possible once
placed on the seabed. In addition to its own supply of fresh water, Conshelf Three carried its own breathing gas for added self-sufficiency. The
habitat was also designed to rise like a hot air balloon so that the oceanauts
could float it back to the surface on their own, without the aide of cranes.
The sphere would also double as their decompression chamber.

The six-man Conshelf crew was far smaller than Sealab's, so during
its planned two-week stay on the bottom it would collectively spend less
time in the water and accumulate less physiological data than the three
Sealab teams. But these rival oceanauts faced many of the same hostile
conditions, darkness and the bite of helium-enhanced cold among them,
and they were doing it at greater depths, with the help of Bond's friend
Charlie Aquadro, who was entering his third year as a medical adviser.
Cousteau's younger son, Philippe, twenty-four, was in charge of shooting
a film while living with the five others in Conshelf, but a big part of the
oceanauts' mission was to demonstrate that undersea living was more
than just a boon to cinematographers or oceanographers. One of Conshelf
Three's prime sponsors was the French government's petroleum research
office, which had also underwritten Conshelf Two's Red Sea sojourn. So
while Sealab II focused on work designed to be useful to the military,
the oceanauts of Conshelf Three had an industrial raison d'être. Several
of Cousteau's men had been trained to work on a vertical stack of pipes
and valves known in the oil business as a "Christmas tree," the wellhead
structure for seabed drilling, which looked like an elaborate thirty-foot
spire. Surface divers making short-duration dives the old-fashioned way
already played an important role in the growing offshore oil industry, and
Cousteau's team planned to lower a mock Christmas tree to demonstrate
that the oceanauts could make repairs on it at a depth of 370 feet, an
atmosphere deeper than their spherical habitat. Oil engineers would be
watching via closed-circuit television as the oceanauts showed off the
value of saturation diving for putting in long hours and working at the
greater depths the oil industry hoped to reach.

Back in the Pacific, on Sunday, September 26, at seven-thirty in the
morning, Scott Carpenter took a last breath from Sealab II and made

the short swim to the Personnel Transfer Capsule. Fifteen minutes later, after the others had followed, Bob Barth was the last to leave Sealab. He crawled up into the PTC and closed the hatch, sealing in the pressure of seven atmospheres. All ten aquanauts squeezed into the capsule, taking seats on either the upper or lower bench around the inside wall. Those seated up top dangled their legs in front of the guys on the bottom. In case of severe jostling, especially once the PTC broke the surface, everyone put on football helmets—the state of the art in personnel transfer safety.

The big crane on the *Berkone* began gradually reeling in the PTC so it could be mated to the deck decompression chamber, a process that would take a couple of hours. Once out of the water Barth could see out a little porthole as the thirteen-ton capsule started swinging like a wrecking ball over the *Berkone*. It looked like the Keystone Kops out there. *Berkone* crew members ran every which way as they struggled to unbolt and remove the capsule's four-legged base, and manhandled the swinging hulk into position over the hatch of the deck decompression chamber. Each passing swell rocked the *Berkone* and threatened to send the PTC into an unwieldy orbit over the deck. Barth could hear the shouting and cussing outside and he might have found the scene funny except that his life was literally on the line. The return to the surface exposed the aquanauts to as much danger as anything on the bottom, much like the dramatic reentry into the atmosphere for an orbiting spacecraft. The drama here revolved around getting a solid, airtight seal between the PTC hatch and the chamber hatch. As always, the seal was essential to prevent a lethal loss of pressure. So was avoiding accidental damage to the swinging PTC, like the hole in the capsule that Dr. Thompson plugged with his thumb during the return from Sealab I. Any real gusher of a leak and—within seconds—they'd have a capsule full of puffy corpses, casualties of explosive decompression. The aquanauts all knew that no matter how well things went on the bottom, they still faced the inherent dangers of surfacing at the very end.

Once the PTC was safely in place, the aquanauts inside opened the hatch in the floor and climbed into the chamber below as if going down a manhole. The chamber climate had been adjusted so that it was less steamy than when the first team had had to sweat out its decompression. But for the next thirty-four hours, as they eased back to the surface at a rate of about six feet an hour, they would still feel as if they were crammed into a tiny locker room. The chamber was just over twenty feet long and ten feet in diameter. It did have a toilet, a big improvement over mayonnaise jar urinals.

Scott Carpenter broke up the monotony of decompression by playing standards like "Goodnight, Irene" on his ukulele. His singing sounded exactly like a crooning Donald Duck. Carpenter took quite a few congratulatory calls, including one from President Lyndon Johnson, who was at his ranch outside Austin that weekend. A call from the commander-in-chief could raise Sealab's profile and should have been less complicated to put through than the fast-fizzling sea-to-sky linkup with Carpenter's orbiting astronaut pal had been. Alas, it was not.

Captain Bond, from his post on the *Berkone*, brought Carpenter into a party line with a telephone company operator and a presidential operator. Carpenter's garbled voice befuddled both operators, and the presidential operator sounded mortified by the idea of patching the president of the United States through to someone speaking an incomprehensible Chipmunk gibberish. She apparently thought a poor phone connection was to blame. Carpenter tried to explain his situation, enunciating as best he could, but hearing the operator's obvious confusion Bond came back on the line. He gave the operator soothing but authoritative assurances that it was just helium speech and the president would understand. She should put the call through.

The operator seemed to have no idea what Bond was talking about and Carpenter was prepared to give up on the call. But Bond persisted and got through to a presidential secretary who said: "Astronaut Cooper, are you able to understand me, over?"

Carpenter replied: "Uh, yes ma'am. This is astronaut Carpenter, not astronaut Cooper, however. Could you understand that?"

"I'm sorry. I'm not able to understand you."

The secretary apparently had astronaut Gordon Cooper on the cortex—not too surprising, since it had only been a couple of weeks since the big headlines about Cooper and the longest-duration space flight he made with Pete Conrad in *Gemini 5*. Carpenter was at that moment decompressing from the longest, deepest undersea stay—thirty days living and working on the seabed—but the secretary had no idea. Captain Bond again broke in to clear up the confusion and talked his way through to another secretary, who asked them to hold for a minute. Then their receivers resonated with a familiar and expansive Texas twang.

"Scott, do you read me all right?"

"Yes, sir, Mr. President. I read you loud and clear. How me?"

"Fine, Scott, mighty glad to hear from you. You've convinced me and all the nation that whether you're going up or down you have the courage and the skill to do a fine job." Carpenter responded, "It's a great crew out here . . . an honor to be a part of the program . . . everyone is happy and we are running on schedule."

President Johnson didn't seem to understand Carpenter any better than the telephone operators, but he forged ahead with perfunctory praise that sounded as though scripted with Carpenter's personal status as an astronaut hero more in mind than the program of which he was a part.

"I know that being two hundred and five feet under water for thirty days and making the excursion dive that you did has been very valuable to us and has advanced our knowledge of how humans can perform in these conditions, and I want you to know that the nation is very proud of you. You are very brave and skillful, and I am grateful that you have successfully completed this experiment."

Carpenter put in a word for the rest of the Sealab crew, thanked the president for his interest, and apologized for the effect of the helium atmosphere on his voice.

Chuckling, Bond told Carpenter, "I'm not sure he got much of it, but at least it sounded like a good conversation!"

The day after Carpenter, Barth, Jenkins, Tolbert, Conda, and the rest of the second team emerged from the decompression chamber, they came ashore for a more formal press conference than had been the case for the first team, whose members had sat willy-nilly around a big table, as if attending a picnic, on the deck of the *Berkone*. The members of Team 2 took their places at a long, cloth-draped table on the stage of an auditorium at Scripps. Each aquanaut had a nameplate in front of him and microphones were placed along the table. The red, white, and blue Sealab insignia, with its downward-pointing arrow, hung on a large banner just behind the seated aquanauts. They wore their green Sealab working uniforms with an insignia patch on the left shoulder. Down in front of the stage was a scale model of the habitat, about ten feet long. Captain Bond and several others spoke from a lectern at stage right. Carpenter sat front and center at the table. It all looked as official as a NASA briefing, although not nearly as well attended.

Captain Bond opened the proceeding with some triumphant statements, highlighting the many man-hours spent on the bottom. "If we had been seeking records there would have been no shortage in that department," he proclaimed. Bond then introduced his aquanauts, and with every introduction tossed in the sort of kind-hearted ribbing that showed why they called him Papa Topside. Each aquanaut took a moment to sum up his experience, accentuating the positive rather than dwelling on the threats they had encountered—tangled Arawaks, useless Aquasonics, anchovy invasions, the sculpin scourge, or Bill Tolbert's close call. Mostly they just expressed gratitude for the opportunity, and gleefully echoed the sentiment that it had all been, in a word, "thrilling."

Bond left a glowing introduction of Scott Carpenter for last. The former astronaut, no stranger to press inquiries, did his best to describe some of the "hardships"—the ocean's biting cold, the chronic fatigue, scorpion fish stings, and the difficulty sleeping in the humid Sealab atmosphere. The Navy pilot and veteran of rigorous Project Mercury training said he had never worked harder in his life than during his thirty days on the bottom. Bond couldn't have asked for a better pitchman.

Not surprisingly, most of the press questions were directed at Scott Carpenter. He had just lived for an unprecedented month on the hostile ocean floor, seven atmospheres below the surface, but one television reporter's earnest question was an indication that the Man-in-the-Sea program either remained low on the national radar or was not well understood, or maybe both.

"Commander Carpenter, is undersea habitation possible and how soon?"

"It's going on at this moment," Carpenter replied, sounding mystified.

Also going on at that moment was Cousteau's Conshelf Three, but no one mentioned it. The French were not the Russians, after all.

Walt Mazzone sat quietly throughout most of the proceeding. He had no use for the limelight but when ignorance seemed to cloud a room, he could be moved to clear the air. At one point in the press conference there was some talk about the habitat's structural shortcomings, such as the lack of space for donning and doffing gear just inside the entry hatch, the lab's foyer. Mazzone didn't disagree, but no one seemed to see the big, pioneering picture.

"This is the second, really, manned undersea dwelling we have," Walt Mazzone said. "Now unlike the space program, we've been forced to redesign on the basis of trial and error. In the space program, at least according to papers, before the first man was put into a rocket to go up, they had to guarantee to the president of the United States his absolute safety. There had to be precautionary ejection mechanisms developed and tested and tried over a period of time. Funds were no problem. When you have to build one of these for the first time and in many cases are actually taking dollars out of your own pocket to buy little things for it, then the empirical becomes the best way to build one."

It was an honest, no-nonsense assessment, from a top Sealab officer no less, but no one followed up on it. Toward the end of the press conference, which had drifted along for a couple of hours, Mazzone spoke up again, sounding frustrated as he tried to underscore, once and for all, the point that didn't seem to be registering. "We are really in the dawn of defining a whole new era of diving," Mazzone said. "We can keep people on the bottom, divers on the bottom, working and available for work, and pay the sacrifice for decompression at one time." Of course the experiment wasn't over yet. There was still one team down, fifteen days and a very long decompression to go.

12

THE THIRD TEAM

Team 3 was in many ways the grand finale for Sealab II. The third team's aquanauts would take part in the biggest showcase of useful work involving undersea jobs specifically tailored to meet Navy needs. Scott Carpenter's thirty days as team leader were up, but if the papers were looking for a substitute celebrity, or if President Johnson wanted a worthy recipient for another congratulatory phone call, they needed to look no further than the astronaut's successor as team leader, Robert Carlton Sheats. Like the other members of Team 3, Sheats had been working for the past month on board the *Berkone*. Captain Bond unabashedly called him "the finest man I have ever known."

Bob Sheats was a war hero, widely admired for his die-hard exploits during World War II after being taken prisoner in the Philippines. At forty-nine he was also the oldest of the aquanauts, compact and muscular, his face a study in angles and the creases from a lifetime at sea. He looked like a gracefully aging superhero. Sheats was a master chief torpedoman but had been diving since the 1930s and attained the elite status of master diver. He had recently served as the topside diving supervisor for Sealab I, heading such novel and delicate operations as ballasting the lab with train axles. He also took part in the macabre recovery dives after the jets crashed at Bermuda, fending off the heavy narcotic haze to reach the wreckage.

Sheats's wealth of underwater experience included the dives he was forced to make by his Japanese captors, who were after tons of silver pesos that had been dumped into Manila Bay. Filipino divers were literally killing themselves for lack of decompression schedules, so the Japanese

rounded up a few imprisoned American divers, including Sheats, to lead
their treasure hunt. Like many prisoners of war, Sheats endured hellish
conditions and suffered from dysentery, disease, near starvation. He and
a few other Americans figured they might be better off diving—and have
more opportunities for escape. They made their dives using hardhat-style
gear and played some ballsy tricks to undermine the enemy's silver-
salvaging efforts. If anyone was unlikely to be bothered by Sealab's hostile
environment, it was Bob Sheats.

George Bond had first met Sheats in 1954, when Bond was newly stationed at Pearl Harbor. Sheats found Bond's brand of dedication to diving
infectious and the two quickly became friends and diving buddies. Both
men also enjoyed good, deep conversation on almost any subject. The two
remained friends and Bond was especially glad to have Sheats leading the
third team, whose performance would in many ways be the most critical
to bolstering Navy support for Man-in-the-Sea. Team 3 aquanauts would
continue to tend to equipment like the weather station and carry on some
other planned Scripps research projects, but a primary function over the
final fifteen days would be to show what sea-dwelling aquanauts could do
in the arena of salvage and rescue. Navy divers like Sheats were steeped in
submarine rescue techniques and the often grueling business of salvaging
ships and other objects, large and small, from the sea floor.

Among Bob Sheats's many admired qualities as a master diver was his
obsession with safety. Before his team went down he ordered five pairs of
soled wet suit boots for better protection from the ubiquitous scorpion
spikes. Sheats believed Commander Carpenter had made a great leader
for the first two teams, but also felt there had been some lapses in safety
procedures, at least by Sheats's standards. Also on Sheats's mind were some
unresolved problems with the gas-recycling Mark VI, beyond the breathing gear's usual peculiarities. Divers were finding that a rig set up and
calibrated to last, say, an hour and seven minutes, might run out of gas after
only three-quarters of an hour. This forced the aquanauts into an unwelcome game of rig roulette. When they started a dive with the Mark VI they
couldn't be sure how long their gas supply would actually last. They were
also having problems with leaks. With water in the breathing circuit the rig
would lose its ability to filter out carbon dioxide. The Mark VI still had no
warning lights or alarms, so these problems meant the diver had to be that
much more attentive to any of the subtle signs of trouble, like feeling light-
headed. Sheats couldn't help thinking that where safety was concerned, the
Mark VI had a long way to go. Yet for some reason—perhaps his friendship

with Captain Bond, perhaps the allure of living in the sea before his impending retirement—neither the dubious behavior of the Mark VI nor the program's trial-and-error approach were enough to dissuade even a stickler for safety from taking his place with the pioneers of undersea living.

Soon after Sheats and the third team settled into Sealab, the gutted fuselage of an F-86 jet, minus its wings, was lowered to the ocean floor for a first major test of a method for salvaging lost planes. Sheats was going to run a first test with Bill Meeks, a chipper thirty-four-year-old boatswain's mate with a long face and deep-set eyes. Sheats had known Meeks since the two served together on a Navy submarine rescue vessel a decade earlier. Meeks was then fresh out of diving school and was paired with Sheats during a training mission off Lahaina on Maui. They were doing breath-holding dives—the kind practiced at the two escape training tanks—to identify mines before their ship recovered them. On a lungful of air, they would swim down a hundred feet to where the practice mines were tethered. When Meeks injured his lung lining from excessive hyperventilating, a medic was dispatched from Pearl Harbor on an old DC-3. That's how Meeks was introduced to Dr. George Bond.

In the ensuing decade, Meeks garnered considerable experience as a working diver, including a stint in Alaska waters so numbingly cold they gave him nightmares for years. He later got transferred to the Navy's Ship Repair Facility in Yokosuka to do salvage and repair work, mostly using hardhat gear. When Bond was looking for Navy divers with salvage experience, he put in a call to Meeks, who had just arrived for duty in Bermuda. Meeks really hadn't heard much about Sealab—might have read a blurb in the *Navy Times*—but when Bond brought it up, sure, he was interested. Ten days later Meeks was on his way back to the States, ready to accept Bond's tempting offer to join Sealab's third team.

Given their backgrounds, Sheats and Meeks were logical choices to be first to try out a novel technique called "foam in salvage," or FIS, one of several high-priority tests. The procedure was straightforward enough: Pump the jet carcass full of a frothing goo designed to force water out. The goo then morphs into a buoyant and spongy foam. Once there was enough foam the plane would float to the surface. That was the idea, anyway. This method had been tried before, but only in water no more than thirty feet deep. Laboratory tests indicated that the foam would work at greater depths and in colder water, but Team 3 would put it to a realistic test on the aircraft hulk. If successful, the foam could be used to raise other large objects—missiles, submarines, even a space capsule.

The day of the debut foam-in-salvage test Sheats and Meeks went through the tedious ritual of donning their skintight wet suits. They also took the necessary steps to prepare the Mark VI—checking regulator assembly, checking valves and hoses, gas flow rate, and always making sure the canister was topped off with dry, sandy granules of the carbon-dioxide-absorbing compound called Baralyme. A high-pressure line enabled the Team 3 aquanauts to refill the gas tanks on their diving rigs themselves rather than sending them to the *Berkone* in dumbwaiter baskets, which gave the divers an added measure of autonomy.

Sheats and Meeks swam a hundred feet over to the sunken aircraft and got the FIS gun, which looked like a cross between a rifle and a spray-painting gun. Dangled down from the *Berkone* it was attached to a pair of hoses carrying the frothing goo, a mixture of Freon and other chemicals. Meeks picked up the goo gun to take the first shots while Sheats assisted. As usual they had no voice communication, only hand signals. The procedure was to shoot the goo into a series of holes around the fuselage. Each one was supposed to take between two and eighteen minutes to fill. They started at the tail, then worked their way to the holes in the side. As Meeks fired away, some of the foam escaped through cracks in the fuselage, creating near whiteout conditions and coating the aquanauts with the sticky goo. They were perhaps twenty minutes into the project when Meeks noticed he was having trouble staying in place. He was becoming overly buoyant. As the silent alarms went off in his head, Meeks quickly recognized that foam was clogging the exhaust valve on the shoulder of his breathing bag and causing it to balloon around his neck. The blockage could also disrupt the proper gas flow, another threat to his survival. Meeks and Sheats signaled to each other, dropped the goo gun, and made a brisk swim through the darkness toward the shark cage.

Back in the lab Sheats let it be known that the FIS gun was unfortunately not compatible with the Mark VI. Like the unready Aquasonic, this was another by-product of the trial-and-error approach to Man-in-the-Sea that Mazzone had talked about at the recent press conference. The escaped foam had also coated watches and depth gauges, making them impossible to read, another unforeseen difficulty. As for the incompatibility of the Mark VI rig and the errant foam, the aquanauts would switch to standard scuba rigs for the rest of the testing. The gas supply would last about half as long, even with two large tanks, but there would be no breathing bags to clog and overinflate.

Once the aquanauts got the jet hulk to rise and hover about twenty feet off the bottom, they watched to see how well the foam maintained its buoyancy. After a few hours, the jet settled back on the bottom. The next day an aquanaut team shot up the jet with more foam. It rose and sank again. There was either a problem with the foam, or the aquanauts may not have injected enough, or both. Still, Sheats found the work gratifying, and believed it was an important step forward for the program. The aquanauts had better luck with several fifty-five-gallon oil drums that also got the foam treatment. They rose like balloons on strings and showed no signs of sinking.

In another test Sheats and Meeks blasted studs into steel as thick as a ship's hull with a stud driver. Each time the aquanaut fired it, he had to keep his head out of the line of sight to avoid a painful shock loading on his eardrums, but Meeks preferred it to the messy foam. They successfully patched a mock section of a submarine hull and dragged a cumbersome prototype rubber pontoon to the repaired hull, kicking up silty clouds as they shuffled through the scorpion fish minefield. Once they wrestled the pontoon into position they inflated it to raise the mock sub section to the surface.

Sheats and the third team were down less than a week when a telephone call was put through from Jacques Cousteau's Conshelf Three, then on its tenth day on the bottom of the Mediterranean. The French oceanauts spoke some English, and one of the Sealab aquanauts, Scripps graduate student Rick Grigg, tried out his French. The result was mostly helium-spiked bilingual confusion that produced little more than comic relief. Bond blithely summed up the exchange this way: The oceanauts and aquanauts were doing well; the oceanauts congratulated the aquanauts; the aquanauts congratulated the oceanauts; the water in the Pacific was dark and cold; the water in the Mediterranean was dark and cold; no one understood each other very well.

The dark and the cold were constants, but the Sealab II aquanauts contended with many variables as well. Both Meeks and Grigg were stung by scorpion fish. The Mark VI continued to present technical difficulties, including unnerving occurrences of rig roulette. Sheats worried that they were going to lose someone to the rig's inconsistent breathing times. The whole team got a scare when gauge readings showed elevated carbon monoxide levels inside Sealab II. The CO may have been responsible for some of the aquanauts' headaches. Captain Bond sent down Hopcalite, a chemical used to oxidize carbon monoxide aboard nuclear submarines, which the aquanauts mixed into the lab's atmosphere-scrubbing system to eradicate the colorless, odorless, silent killer that had mysteriously crept into their midst.

The Team 3 aquanauts were doing most of their own cooking, although one of the wives was allowed to send down a cake decorated with the Sealab insignia for Sheats's fiftieth birthday. To supplement the canned fare, Sheats soon had his team eating sashimi, slicing up cod or octopi or whatever else they caught around the habitat, even scorpion fish. The old master diver speared perch and rockfish and sent some up to Captain Mazzone and the night crew.

As part of a Scripps research project, grad student Grigg had been netting hordes of plankton from the water, like dust from the wind. Sheats decided to serve some in a soup. They were not to everyone's taste, but Meeks found the little critters pleasantly nutty. The next morning Sheats had a bowl of raw plankton for breakfast. Not bad with cereal. Plankton could make a nutritious food on the deep frontier, he thought—perfect for future sea dwellers. Captain Bond was delighted with his friend's show of self-sufficiency. A true aquanaut ought to have a woodsman's knack for living off his environment, Bond reckoned, and who better than Bob Sheats to show the way? In that same spirit, the Team 3 aquanauts did some impromptu training of a curious sea lion that regularly popped up in the entry hatch. Compact, agile, and cooperative, the precocious little pinniped, which they called Sam, rivaled Tuffy the porpoise in its ability to respond to the porpoise's buzzer. Sam's performance indicated that his kind might do well as couriers for future undersea missions, especially since a sea lion could bring deliveries right to Sealab's doorstep.

All the Team 3 aquanauts, starting with Sheats, made more excursion dives, this time over the canyon rim and down to three hundred feet, adding another atmosphere to the maximum depth attained by Carpenter and the second team—another first for Sealab II, as proud Papa Topside would point out, and another step downward. At that depth they were also just an atmosphere shy of Conshelf Three's depth in the Mediterranean, though that seemed to be coincidence more than a product of conscious competition.

While the first two teams had each managed to rack up totals of around one hundred hours diving outside the lab, the Team 3 aquanauts, with their heavier work schedule and the benefit of the foundation laid by the previous teams, would spend almost 150 hours in the water. On one especially productive day, the Team 3 aquanauts amassed almost eight hundred minutes in the water—more than thirteen hours of dive time.

Captain Bond got to thinking about a recent plane crash off Long Island, and how much more efficient the recovery operations might have been using a Sealab approach instead of conventional surface dives.

Sheats and the third team spent their final hours essentially breaking camp—packing up, disconnecting cables, dismantling equipment. Bond gave his okay to keep their television circuit on through the last full day of their run so they could catch the third game of the World Series between the Dodgers and the Twins. Los Angeles, the favorite, had lost the first two games. Shortly after nine o'clock the next morning, Sunday, October 10, the day after the Dodgers won that third game, Captain Bond preached a bittersweet farewell of a sermon, inspired by *The Road to Bithynia*—one of his favorite religious novels. Following Bond's customary recitation of the Sealab prayer, Bob Sheats led the aquanauts on the final swim to the PTC. They took their places on the benches around the inside of the cramped capsule and put on their football helmets. They had a smooth, uneventful ride to the surface and by eleven o'clock the PTC was successfully mated to its chamber hatch. All they had to do now was wait out their decompression time of approximately thirty hours. Team 1 was finished in thirty-one hours, but Team 2 had taken thirty-five. Where decompression was concerned, patience was always a virtue, and this was a small price to pay for two weeks at two hundred feet.

Sheats had found his time in Sealab beautiful, adventurous, rewarding. But he was also frank enough to admit that fifteen days on the bottom had been enough, even for him, a master diver who had endured thirty-nine months as a prisoner of war. Toward the end of his stay in Sealab, Sheats felt as though he had to force himself into the chilling water. The others seemed ready, even eager, to get back to the surface, too. Some of Sheats's fellow aquanauts had disappointed him in their performance, but others had performed superlatively. He planned to rate them all in his report.

For the first twenty-four hours as the atmospheric gases were reconstituted and the chamber pressure systematically dropped, from seven to two atmospheres, Sheats passed the time reading and played some chess with Rick Grigg. They all listened to the Dodgers even up the series. Shortly after eleven the next morning, Sheats began to notice a telltale pain in his right knee. He did not want to be the one to ruin Dr. Bond's perfect batting average for Sealab decompressions, but he knew what the pain might mean. A couple of hours later, when they were just twenty-three feet and a few hours from being released from the chamber, Sheats's nagging knee ache had spread to his right hip, along the sciatic nerve. It was the same sensation he had the year before, back home in Washington state, when he got bent while helping a stricken civilian diver in a chamber. Before that, he had been hit just twice in twenty-seven years of diving. It was a reminder that no matter how experienced the diver or how precise the decompression schedule, the bends were always a threat.

Sheats grudgingly reported his condition to Bond at one o'clock that afternoon. Bond locked into the chamber to verify his friend's diagnosis, but knew Sheats was probably right. Indeed Sheats had the bends, but his case appeared to be the less serious variety—just a pain in the leg, not bubbles in the central nervous system that could do serious damage. Bond decided to have the nine others move into the chamber's outer lock, a smaller section where they could finish off their six remaining hours of decompression before stepping out to breathe the cool, salty Pacific air. Sheats would stay behind in the main chamber, where the pressure would be raised to about three atmospheres to quash the bubbles fizzing in his leg. A hospital corpsman locked in, trading places with Bond to monitor Sheats's condition. He would sit for the next twelve hours with Sheats, watching for any symptoms that might have to be treated with a hyperbaric roller-coaster ride, like the hair-raising one Bond once took with ailing Charlie Aquadro back in Pearl Harbor.

While Sheats waited out his extended decompression, the other Team 3 aquanauts finished theirs in the outer lock. They soon went ashore for the final press conference, taking their places onstage while the aquanauts from the first and second teams assembled alongside reporters and guests in the Scripps auditorium. The tenor of this last conference was not much different from the previous one, although there was some added pomp as the Navy aquanauts received the Navy Unit Commendation. Captain Melson announced that it was "among the proudest ribbons that can be given by the secretary of the navy to military personnel." The rear admiral in charge of the Office of Naval Research touted the oceans as "the next frontier." Captain Bond offered fond introductions of his aquanauts, and each of the Team 3 members said a few words, mostly accentuating the positive—"The deep dive we made is something we'll never forget," as Bill Meeks said. No one dwelt much on close calls or chronic threats, like the need to play rig roulette with the finicky Mark VI.

In Bob Sheats's absence, Captain Bond recounted a heartfelt story about a sage piece of diving advice he once got from Sheats—a psychological tip about breath holding and maintaining one's composure in underwater emergencies. Bond believed that advice once saved his life. Bond now returned the favor. His prescription for treating Sheats's case of bends worked. By the time of the press conference the old master diver was safely out of the decompression chamber, but was held on the *Berkone* another day for observation.

The two-hour press conference was not packed with a NASA-style horde, although *The New York Times* dispatched a reporter to the conference and in the next day's paper heralded the Navy experiment as a success. Sealab II was indeed a success, and so was Cousteau's Conshelf Three, though it went unmentioned at the final Sealab press conference, despite the recent habitat-to-habitat phone call and the simultaneity of the experiments. The French sphere surfaced the following day, having remained on the sea floor for an extra week—three weeks in all—to make up for some unexpected troubles, including stormy weather and helium leaks. Cousteau would later describe his Conshelf Three venture in another big article for *National Geographic*, and in a TV special shot by his son Philippe and narrated by Orson Welles.

Sealab II, like Conshelf, had its share of troubles, but they paled in comparison to the overriding success of the mission. Never had so many people lived and worked for so long at such depth. *Lived.* That was the key. Captain Bond pointed out that the Sealab II aquanauts had logged a grand total of three and a half man-years living on the bottom, and when a reporter asked Bond to comment on the program's achievements, he didn't hesitate. The greatest achievement, he said, was sending twenty-eight men down and getting twenty-eight back. No one had been lost to the hostile environment, to spiny fish, finicky rigs, mail-order prototypes, poisonous gases, explosive decompression, or any other looming threats. With the mission of aquanaut survival accomplished, the Man-in-the-Sea program itself could hope to survive.

Summing up, Bond sounded homey and proud. He proclaimed that the deep excursion dives, especially those made by the third team to three hundred feet, were a major breakthrough in saturation diving technique. Papa Topside also predicted that within a few years there would be aquanauts living at six hundred feet and commuting from there down to job sites at seventeen hundred feet—a bold prediction, to be sure, though perhaps not as eye-popping as the coming of Homo aquaticus. Bond then gave what amounted to a spontaneous mini-sermon on the Man-in-the-Sea program, and how far it had come despite an uncertain start. He told of the program's genesis in a "sub-rosa fashion" eight years before, in the dark corners of a laboratory. Work was done on weekends, sometimes without official sanction, and in the face of cries from officials high up in Navy bureaus that this quest to live in the sea was "Buck Rogers–ish," "madness," and had no "social significance." It was deemed a waste of his time and the Navy's time, Bond said. "We have seen a program grow, and we have seen it grow into a healthy child." Now, flush with triumph, they could nurture a deepening quest to live and work in the sea, one that might grow healthy enough to rival the moon shot, as George Bond always believed it should.

13

THE DAMN HATCH

Within a few months of the successful conclusion of Sealab II, the Navy transformed its Deep Submergence Systems Project into a higher-prestige, stand-alone organization. With that hierarchical change, Sealab and Man-in-the-Sea were elevated from their experimental, R&D status to full-fledged Navy programs. That meant Sealab's budding methods and equipment—decompression schedules, diving rigs, personnel transfer capsules, possibly even Tuffy the porpoise and an amiable sea lion named Gimpy—were all to be refined and certified for practical, "operational" use. A half-century earlier the Navy had embraced hardhat diving, followed by helium-oxygen diving and scuba. Now saturation diving was to become operational—a fully developed addition to the undersea toolbox. Saturation diving could be developed for a variety of underwater Navy operations—oceanography, search, salvage, construction, and the rare but potentially urgent need for submarine rescue.

It looked like George Bond's dream of an undersea future was coming true. Sealab II was shipped back to the Hunters Point shipyard in San Francisco, where it had been built, to be expanded and remodeled into the roomier, more sophisticated Sealab III, a habitat designed for a depth of at least four hundred feet. Also at the shipyard, the *Elk River*, originally built during World War II for amphibious landings, was being remade into a state-of-the-art support ship and mission control. It would be outfitted with the latest saturation diving systems, including two decompression chambers, two Personnel Transfer Capsules, and a sturdy gantry crane designed to move the PTCs less precariously—no more need for football helmets—and to lower and raise them through a protected opening in the hull, called a moon pool, amidships.

Some $10 million, about five times as much as was spent for Sealab II, was being pumped into Sealab III. Close to sixty aquanauts and support divers, both military and civilian, were being selected to train for a two-month stay on the sea floor near the northeastern shoreline of Southern California's San Clemente Island, a craggy twenty-mile-long slice of dusty chaparral run by the Naval Undersea Warfare Center. The island served as a testing range for submarine-launched missiles, among other things, and would give Sealab III an additional base of operation close to the *Elk River*, which would be moored a little less than two miles from San Clemente. To the northeast, the Long Beach shipyard was a half-hour away by plane; directly to the east it was about the same distance to San Diego and neighboring La Jolla, the site of Sealab II. About twenty-five miles to the north, halfway between Sealab III's planned location and Long Beach, was Catalina Island, where six years earlier Hannes Keller got tangled up in his Swiss and American flags, his partner perished, and a support diver vanished.

Along with Sealab III's operational stature and fresh infusion of money came a burgeoning bureaucracy and an influx of new people, most notably Commander Jackson M. Tomsky, the line officer chosen to lead this new organization. Captain Melson and the Office of Naval Research oversaw R&D, and they were now out of the picture. Jack Tomsky had enlisted thirty years back, at age nineteen, in his hometown of San Francisco and was trained as a hardhat diver. He worked on submarine rescue and salvage ships for a decade before he became an officer and a ship commander. Tomsky was a mustang, an enlisted man who attains the rank of an officer, and had recently commanded a submarine rescue vessel. Just prior to joining DSSP he was an officer-in-charge of the deep-sea diving school. Tomsky was tall and imposing, with dark hair swept back, hawklike, over a broad head. His hard-nosed manner and booming quarterdeck voice earned him the nickname "Black Jack."

When the prospect of running the Sealab program came his way, Tomsky had to do some talking to persuade those in charge of the Deep Submergence Systems Project that he was the right man for the job. Some in DSSP understood that George Bond had been the de facto leader of the previous two Sealab experiments, even though Bond was a medical officer and not a line officer, an issue that had caused some headaches for Captain Melson. With the entire program moving from experimental curiosity to

operational capability, DSSP heads were looking for a proper commanding officer to infuse Man-in-the-Sea with "authority, responsibility, and
accountability," as John Craven, who was then director of DSSP, liked to
put it. Tomsky, if chosen, would serve as Sealab III's on-scene commander,
a role that fit with his background, but he would also have to assume a
bureaucratic role that was new to him, that of project manager of DSSP's
Ocean Engineering Branch, run from DSSP headquarters near Washington,
D.C. Tomsky would be responsible for running several major projects related to submarine rescue and salvage operations in addition to Sealab III.

Tomsky impressed Craven and convinced him that he could wear both
organizational hats. By 1966, Tomsky had the job and would oversee the
design and testing of new diving systems, the outfitting of the *Elk River*, and
the construction of Sealab III, while also managing his branch's other projects. The rise of a commander known as Black Jack over Papa Topside was
itself indicative of how Sealab was maturing from its anomalously familial
beginning, with George Bond as benevolent patriarch. Tomsky and Bond
got along well enough. Bond was an easy man to like, and in the new management structure he was less involved with day-to-day operations. Bond
retained his favored title of principal investigator and was to preside over
Sealab III personnel health and safety, and conduct a continuing program
of biomedical research. But like a peripatetic professor emeritus, Bond also
had ample time to travel, lecture, and spread his gospel of undersea living,
which was fine with Tomsky.

Within the expanded DSSP and Sealab ranks there were now a number of medical officers besides Bond, including one of Tomsky's key lieutenants, Bob Bornmann, the Navy doctor who worked with Ed Link and
his cylinder on that first saturation dive in the south of France. Bornmann
was one of many new faces among the doctors, corpsmen, master divers,
engineers, civilians, scientists, and Seabees, members of the naval construction forces. Many of the aquanauts, too, were new to the expanded
program. Just ten of fifty-four aquanauts and surface support divers had
previous Sealab experience, among them Bob Barth, Berry Cannon, Barth's
pal Wilbur Eaton, Cyril Tuckfield, and Wally Jenkins. Lester Anderson did
not return. Bob Sheats, the seasoned master diver and key participant in
Sealab I and II, had concerns about the new DSSP management structure,
how he would fit in and what authority he would have to ensure that
the operation adhered to his strict standards for safety. Sheats ultimately
decided not to participate, despite Bond's encouragement that he do so.
Bond's old friend seemed to lack faith in Sealab III.

Two British divers, two Canadians, and one Australian were on the aquanaut roster. All were volunteers from navies that used similar equipment and cooperated in diving training. Philippe Cousteau, the famous diver's dashing twenty-eight-year-old younger son and collaborator who shot the footage for a TV special about Conshelf Three, was now signed up to turn his camera on Sealab III. The arrangement greatly pleased Captain Bond, who had some ideas of his own for opening scenes that featured him and Jacques Cousteau seated by a rock fireplace in a guest house on San Clemente Island, smoking their pipes and pondering the potential of undersea living.

Scott Carpenter might have been preparing for another stint as an aquanaut, but X-rays revealed lesions in his femur, a sign of bone necrosis. Carpenter was asymptomatic, but Bond and other doctors were worried that his condition might worsen with another long stay under pressure. So Commander Carpenter became Tomsky's deputy. The number three man was Captain Mazzone. Everyone understood that there would be no Sealab III without Walt Mazzone. Tomsky, like many others, had no personal experience with saturation diving prior to taking his post at DSSP. Mazzone was named the diving operations officer and would oversee much of the diving activity, including decompression cycles and atmospheric monitoring. He was also put in charge of what was called the Deep Submergence Systems Project Technical Office, an all-purpose West Coast hub for aquanaut training and equipment testing that was set up at the naval station at Point Loma in San Diego, close to the planned Sealab III site at San Clemente Island.

Although the bigger budget for Sealab helped, the project was not being lavished with NASA-caliber funding, which might not have mattered except that there never seemed to be enough money to pay for adequate personnel or to buy the time needed to polish the rough edges and avoid the kinds of mistakes that cause delays. The project was originally scheduled to get under way by 1967, within a couple of years of Sealab II, but a variety of design problems and equipment failures, some related to the ever-challenging containment of helium, caused postponements that pushed the project back a year. A flood-out of a PTC during an unmanned test in late 1968 was typical of the setbacks the program experienced and it contributed to yet another delay. A further challenge came from a decision made, with little explanation, to extend Sealab III's initial target depth of 430 feet to six hundred feet, a substantial increase of five atmospheres and fully three times the depth of the two previous

Sealabs. In addition, excursion dives were to be made to 750 feet, possibly even deeper.

Just about everyone, aquanauts and the *Elk River* crew alike, were growing weary of delays. It had begun to look like man would land on the moon, leaving footsteps in the ashen dust of the Sea of Tranquillity, before he was ever able to make his mark with Sealab III deep in the cold primordial soup of inner space. A new target date for getting Sealab III on the bottom was set for mid-February 1969, and all indications were that this deadline, unlike the earlier ones, would be met. Some worried about having to cut corners in the final push to get the operation started, but the February deadline also proved to be a morale booster for those tired of delays. It was for Bob Barth, and the rest of his team felt the same way. Barth was to lead the first of five teams scheduled to occupy the habitat for twelve days each, giving mankind an unprecedented sixty-day presence on the continental shelf, by far the longest, deepest stay ever attempted and a giant leap toward George Bond's undersea future.

In the early evening of February 15, a rain-swept Saturday, a barge crane lowered the three-hundred-ton habitat to its place on the bottom, 610 feet below the surface. The process took three hours but was largely uneventful, a credit to the experience gleaned from previous trials and errors. Sealab III was painted a traffic-light yellow, with the name U.S. NAVY SEALAB on either side and on each end of the tank. Two symmetrical sections, each twelve feet square and almost eight feet high, were added to the underbelly at each end of the cylindrical habitat, creating an arch-shaped, two-story structure nearly forty feet tall. One of the new sections would serve as a marine observation room; the other was the diving station with an entry hatch. The station was the foyer where the aquanauts would have ample space for donning and doffing wet suits and diving gear. On the outside, in between the two new additions that formed the base of this golden arch, a block of ballast tanks was suspended as part of a novel anchoring system. The idea was to lower the tanks all the way to the bottom while the lab maintained just enough buoyancy to float a few feet above. Regardless of any unevenness on the ocean floor, the lab could be made level, so no more Tiltin' Hilton.

Not long after the successful lowering, the topside commanders could tell they had a problem. The lab was leaking badly. Its helium-rich artificial atmosphere was bubbling out at an alarming rate. More and more gas had to be pumped in to compensate for the leaks and maintain the lab's internal pressure—a finger-in-the-dike fix to prevent a disastrous flood. By the next morning the leaks had become steadily worse. The lab's high-pressure atmosphere was now escaping so briskly that within a day the gas tanks would be depleted on board the *Elk River*. If the gas ran out, the lab would surely flood.

One possible solution was to raise the habitat, fix the leaks while it was back on the surface, and then start over. But after all the delays of the past few years, and with Sealab III already on the bottom, the commanders decided it was best to hurry up and get Barth and his team down there to deal with the problem in situ. Captain Bond, in his more strictly defined role as chief medical adviser, approved an expedited compression schedule—four hours instead of fifteen—to take the divers from the benign ambient pressure of slightly less than fifteen pounds per square inch at sea level to the full pressure they would experience at six hundred feet, a formidable 280 pounds per square inch. This rapid pressurization, in a chamber on board the *Elk River*, could bring on some additional soreness, but that initial discomfort should subside as the aquanauts adapted to their depth.

At six hundred feet, the habitat was well beyond the reach of conventional bounce divers from the surface, so a mini-sub called *Deep Star*, with Scott Carpenter on its three-man crew, took a dive down to the habitat to try to pinpoint the source of the leaks. No one on board could see much except the chaotic streams of bubbles rising from the habitat as if it were a dissolving Alka-Seltzer tablet. The worst leaking appeared to be around the electrical stuffing tubes, where power, communication, and other lines passed through the midsection of the lab's steel hull.

On February 16, shortly after five in the afternoon, the *Elk River* gantry crane plucked a white Personnel Transfer Capsule from its perch atop the hatch of a decompression chamber, where the capsule sat like a golf ball in a tee. The decompression chamber with the aquanauts was below the main deck, alongside a second, matching chamber in a cramped control room. The PTC, a steel pod seven feet in diameter and ringed by ten scubalike tanks, looked like a sci-fi transport for space aliens. With the twelve-ton pod in its grip, the crane rolled aft along the deck of the *Elk River*, then stopped over the moon pool amidships to begin the lowering to six hundred feet.

Inside the pod were Bob Barth, now thirty-eight years old, who had been promoted to warrant officer, and Berry Cannon, the able thirty-three-year-old civilian electronics engineer from the first Sealab II team. They were the ones chosen to get inside the habitat first, and find and fix the leaks. Once inside, they would be joined by the two others riding with them in the PTC, Richard "Blackie" Blackburn and John Reaves. Blackburn was new to Sealab, but not to Navy diving. Blackie was a twenty-nine-year-old aviation ordnance man first class. He had spent six years in explosive ordnance disposal, EOD, the inherently dangerous and delicate business of diving to locate and disarm mines and bombs. John Reaves, a last-minute replacement for this first PTC ride to the lab, was a decade older, a Navy photographer, and a former Sealab II aquanaut. He had been a principal trainer of Tuffy the porpoise, who was scheduled to get another try during this deeper venture.

The lowering took almost an hour, and long before the PTC reached its destination, about twenty feet from the bottom, the foursome inside had been chilled to the bone. Their capsule was acting as much like a refrigerator as an elevator. Its electric heater was out of order, which might have been bearable, especially to a band of divers accustomed to inhospitable conditions. But a deflector was also missing on the carbon dioxide scrubber and it blew what felt like an unrelenting arctic wind. Helium further sapped everyone's body heat, and the divers' standard neoprene wet suits, compressed and thinned under pressure, provided little warmth. They all felt cold to the core. Barth was sure he had never been so cold.

Upon arrival he and Cannon took a few minutes to don their breathing gear, called the Mark IX, the prototype specifically designed for making deep dives from the PTC. It was a gas-recycling rig similar to the finicky Mark VI, except that its breathing gas was delivered through a hose connected to the tanks around the outside of the PTC. Umbilical-fed gear, despite the drawbacks of dragging around the equivalent of a garden hose, was essential at depths of six hundred feet, where the pressure approached twenty atmospheres. Scuba tanks, even large ones, would be depleted in a matter of minutes at such depth.

Like the Mark VI, the Mark IX had a pair of breathing bags worn like a vest. A compact fiberglass backpack contained a built-in canister, about the size of a thick phonebook, which had to be filled with fresh Baralyme, the sandy granules that scrubbed carbon dioxide from the exhaled breathing gases that the rig captured and recycled. Because it had no tanks, the Mark IX was compact and also lightweight—less than fifty pounds. By comparison, the bulky prototype rig that had been designed for use from the Sealab III habitat weighed a hundred pounds more.

Barth and Cannon dropped fins-first through the hatch in the floor of the PTC and through the short trunk. They landed on the "cage," a box made of steel mesh containing a winch motor and anchor that could be used to secure the PTC to the bottom. The cage hung a few feet underneath the PTC, like the basket under a hot air balloon, so that its top doubled as a stoop. The water was cold, in the mid-forties Fahrenheit. With their umbilicals trailing behind them, Barth and Cannon swam over to the lab. Along its sixty-foot length were various valves, clamps, and vents to check and adjust as part of the "unbuttoning" procedures for parking the lab on the seabed. One of Barth's jobs was to "blow the skirt," which meant releasing overpressure from inside the lab to create a little air pocket, like a porch, just below the hatch in the dive station, one of the two new sections added to the lab's underbelly.

Standing a couple of steps up a ladder, Barth turned a valve to blow the skirt, then yanked a few levers to unlock the spring-loaded hatch, a heavy steel square of about four feet. With a good shove the hatch should pop open, but it wouldn't budge. Barth banged haplessly on the hatch, succeeding only in bruising his hands. At that point he and Cannon had been in the water about twelve minutes, each working as quickly as they could through their separate unbuttoning checklists. It seemed to Barth that his buddy should have rejoined him at the hatch by now. Concerned, Barth left the hatch and swam around the habitat looking for him. All the while, Barth's breathing rig was giving him trouble. He found himself huffing and puffing hard and something didn't feel right. The water was dark, though much clearer than they had expected. Cannon was nowhere to be seen, and Barth was beginning to feel uneasy about being alone on the ocean floor, unable to open the hatch, unable to communicate with anyone, unable to swim for the surface, wondering where the hell his buddy went. And why such difficulty breathing? He swam for the PTC and felt as though he might black out before he reached it.

Unbeknownst to Barth, Berry Cannon had experienced a similar shortness of breath a few minutes earlier and was already back at the PTC, where he had popped up through the hatch in the floor of the PTC with a dazed look on his lean face. He was breathing like a runner who just sprinted the last mile of a marathon. Blackburn and Reaves hauled him up through the liquid looking glass and into the cramped capsule.

Blackburn was asking, in his Chipmunk falsetto, "Where's Barth?"

Minutes seemed to pass before Cannon regained his composure enough to respond. He didn't know where Barth was. He had left the habitat without informing Barth, an uncharacteristic breach of buddy diving etiquette. Fearing that Barth might be in trouble, Blackburn quickly strapped on a diving rig and dropped fins-first through the hatch onto the cage. No sooner had Blackburn jumped in than Barth appeared, breathing hard like Cannon. Blackie and Barth climbed back up into the PTC. Barth kept repeating something but Blackburn couldn't understand. Distortion from helium speech was more pronounced at six hundred feet, where they were breathing a gas mixture that was more than 90 percent helium.

The topside commanders, via a camera monitoring the Sealab III hatch area, had watched a grainy black-and-white image of Barth as he struggled with the hatch and then silently swam off screen. The divers had no voice communication in the water, so no one could know exactly where Barth went, or whether he was in trouble. But topside soon heard from the PTC that everyone was back in the capsule. A voice over the intercom told the divers to stay put. They were going to be brought back to the surface. Barth and the others closed the hatch and buttoned up the PTC for the return trip to the *Elk River*. The foursome left the bottom at 7:25, about an hour after they had arrived. Inside the command van, a converted trailer on the main deck of the support ship, assorted Chipmunk utterances, often incomprehensible, could be heard coming over the intercom. It was clear, though, that the four divers were all cold, too cold to say much of anything during their ride up. As on the way to the bottom, the four divers were again chilled by the lack of heat and the frigid breeze from the scrubber fan. The return trip to their decompression chamber on the *Elk River* took about ninety minutes. The actual passage through the water took less time than the process of raising the pod through the moon pool, moving it with the gantry crane and mating it to the chamber hatch with the crucial airtight seal, as on the *Berkone*.

All four divers in the PTC were shaking badly from the cold by the time they reached the *Elk River*. It was almost nine o'clock that night when they finally lifted the hatch in the floor and left the PTC. One by one, they climbed down through a short connecting trunk into the deck decompression chamber below, a horizontal cylinder twenty-four feet long, with a bathroom, a couple of bunks, and just enough headroom to stand upright. The contrast in temperature as they entered the chamber was so great it felt as though they were dropping into a blast furnace.

On the opposite side of the continent, in Caribbean waters as clear and balmy as the Red Sea, things were going much more smoothly on the first day of Project Tektite I, an undersea habitat run by the Office of Naval Research, Sealab's former lead agency, along with NASA and the Department of the Interior. The $2.5 million project's duration was to be sixty days, the same as Sealab III, but Tektite was not nearly as deep, or as expensive, although news reports barely differentiated between the barrier-breaking six-hundred-foot target depth of Sealab and Tektite's markedly less perilous forty-three feet, a depth similar to Cousteau's early trials. A direct swim for the surface was unlikely to kill anyone. There was no ultralight helium in the Tektite atmosphere to complicate matters and decompression would take less than a day, compared to nearly a week for Sealab III. The Tektite habitat, a pair of interconnected vertical tanks two stories tall and twelve feet in diameter, sat on a rectangular base, like upside-down bell jars on a shoebox. In a secluded cove on the south side of St. John in the Virgin Islands, four aquanauts, all Interior Department scientists, shared roomy quarters pleasantly reminiscent of Cousteau's Starfish House. They planned to make scuba dives and carry out marine studies, while NASA's interest was in observing the effects of isolation to help predict human behavior on long-duration space flights.

Back on the *Elk River,* as the four chilled Sealab aquanauts recuperated in the chamber, about a dozen topside commanders gathered at about midnight in a small conference room to decide on their next move. Jack Tomsky, George Bond, Walt Mazzone, and Scott Carpenter were among those hashing over a variety of possibilities, including halting the project to raise the habitat, fix the leaks back at the shipyard, and start over again. They decided that another dive to open the habitat could and should be made. Once inside the divers should be able to plug those leaks and the sixty-day quest could proceed on schedule.

Barth had gotten as far as blowing the skirt and twisting the levers to unlock the hatch. The main problem seemed to be that when he tried to push the hatch open, the pressure inside the lab was higher than outside. Tomsky and the topside commanders had kept their eyes on the gauges in the command van and ordered more gas be pumped to the habitat to keep the pressure up and the water out—the finger-in-the-dike scenario. Equalizing the inside pressure with the outside could be tricky from six hundred feet away and there was a tendency to err on the side of maintaining a higher pressure inside the lab to stave off the leaks. The steel hatch was heavy to begin with and proved to be a bear even dur-ing practice openings in shallow, less debilitating conditions. Blowing the skirt, as Barth did, should have helped equalize the pressure above and below the hatch, but it didn't take much additional internal pressure to turn the four-foot-square hatch into a two-ton beast that only Super-man could have lifted. As Barth was trying to get in, the lab had become overpressured by as much as eight pounds per square inch; the hatch might as well have had a Sherman tank parked on it.

Timing an intentional drop in Sealab's pressure with the exact moment a diver went to open the hatch could be difficult, especially without any voice communications once a diver was in the water. One suggestion was to have the diver give a thumbs-up, or some such sign, into the closed-circuit TV camera outside the hatch. Topside could then shut off the gas flow and allow a deliberate drop in the lab's internal pressure. The diver should be able to pop open the hatch and the topside crew could then crank up the pressure again to minimize the flooding. But should Barth and Cannon be the ones to try again? What about Blackie or Reaves? By now the aquanauts to make up the rest of Team 1, the five who were put under pressure more gradually, were ready in the second chamber nearby. Why not send some of them to open the lab?

The topside commanders turned to Captain Bond for an estimate of how long Barth and Cannon should have to wait before they could safely get back in the cold water. Research into adequate recovery times from thermal stress was lacking, Bond said, but his best guess, with a few caveats, was that the recuperating divers should be able to make another dive by one o'clock in the morning, about four hours after they had returned from the first dive. They would be cold and they would be very uncomfortable, Bond said, but they could get through it.

Tomsky said he would prefer that Barth take Blackburn to open the hatch. He reasoned that Blackburn was bigger and stronger than Cannon, but Mazzone pointed out that Barth and Cannon had trained together and Cannon, as team engineer, was most familiar with the internal electrical system and how to get at the leaks. Carpenter agreed that Barth and Cannon ought to dive again but not until the following morning. They needed time to rest. If the commanders wanted the best out of the divers, they shouldn't send them back to the bottom at two in the morning, Carpenter said. Others argued that it made more sense just to send fresh divers. The skirt had been blown and virtually all that remained to be done was to open the hatch. Tomsky ultimately sidestepped the issue and asked Mazzone, the diving operations officer, to talk over the matter with Barth, who as team leader would have to decide whether his foursome should go down again, and who should make the opening dive.

Inside the decompression chamber, Barth, Cannon, Blackburn, and Reaves had stripped off their gear and were trying to warm themselves. Reaves hadn't even gotten wet, but like Barth he had never been colder in his life, colder even than on a salvage dive he once made under several inches of ice. As the four divers regrouped, there wasn't much conversation among them. Barth and Cannon talked a little about the troubles they had had on the first dive, and the feeling that they weren't getting any good satisfaction from the breathing rig. They agreed that cold was the likely culprit—severe cold could cause disorientation, among other things—but they also reasoned that their own nerves may have been in overdrive as they went about their work, a whopping six hundred feet down. They had practiced the unbuttoning procedures before, but never in deep water or such daunting conditions. Cannon said he was sorry about having gone back to the PTC without telling Barth, but Barth said there was no need to apologize. They had both been busy. Nothing all that drastic had come of it anyway.

From inside the chamber, Barth spoke with the topside commanders, mainly Mazzone, and shared a few observations but didn't dwell on what happened to him and Cannon as any kind of insurmountable problem. Barth was typically unfazed. He had come to expect the unexpected. The difficulty with the breathing rig and the entry hatch were just the latest examples of Murphy's Law underwater.

When Blackburn asked Cannon what had been wrong, why he was so dazed upon returning to the PTC and breathing so hard, Cannon shrugged off the question. He said it was likely just the cold. Working divers were not apt to complain about discomfort or hardship, and neither Cannon nor the others would have wanted to miss out on the opportunity to take a historic place on the deep frontier. Blackburn understood; he had had to work the system to get himself a coveted spot on this third Sealab venture. The three other divers had all been on Sealab II.

To help restore the divers' core body temperature, it was recommended that they eat, drink hot fluids, take hot showers—long enough to make them sweat—and get some rest. Soup and steak were passed in through the oven-sized medical lock, but with the possibility of a second dive looming there was a lot to do—setting up the PTC, running through a long checklist of valves, and changing all the power supply batteries. No one seemed to eat much or get more than an hour of sleep, and only Blackburn got around to taking a hot shower.

After catching an hour-long catnap, Blackburn checked over the two Mark IX rigs that Barth and Cannon had used on the first dive. Blackburn's years of diving with explosive ordnance disposal teams made him more familiar than most with the mechanics of this type of gas-recycling rig, which were frequently used in the kind of underwater work he did. In the early 1960s Blackie was on the Navy team responsible for testing and refining the Mark VI, which was subsequently chosen for use on the first two Sealabs, and the Mark IX was similar to the Mark VI, with similar peculiarities that meant a diver had to be alert to subtle signs of danger, without the benefit of warning lights or alarms.

As Blackburn checked over the rigs it occurred to him that a restriction in the inhalation hose might have been to blame for Barth's and Cannon's breathing difficulties. Among the recent refinements to the Mark IX, a stiffener had been inserted to strengthen a section of the inhalation hose, but it also narrowed the passageway, possibly impeding the gas flow. Blackburn scrawled a note with this concern and put it and the rigs' two used Baralyme canisters into the oven-sized medical lock, through which they could be passed to the crew working outside. Replacement canisters were soon sent back into the chamber through the lock, and Blackie put the two rigs back together.

About half past one in the morning, several hours after the first dive and with Sealab III's gas supply dwindling, barges arrived with a backup supply of helium. In the meantime a flooding alarm had gone off, signaling that seawater was finding a way in around the edges of the entry hatch. A camera inside the habitat showed at least six inches of seawater covered the floor of the dive station. Once the new helium supply was hooked up, they cranked up the pressure inside the lab and forced the water out.

After the conference room meeting Mazzone went down to the two decompression chambers, which were shoehorned into dusky, low-ceilinged quarters below the main deck, and discussed the situation with Barth. Barth assured Mazzone and the topside commanders that his group was in good physical and mental shape. There was no reason to change horses in midstream, as Barth put it. He and Cannon were ready to make the opening dive again. They had already done most of the unbuttoning and should be able to swim quickly to the hatch, open it, get inside the habitat, and get to work fixing the leaks. Mazzone gave them one other instruction: "Don't be a hero. If you have any problems return to the capsule. Don't take any unnecessary chances to get into the habitat."

To fend off the cold on the second dive Barth and Cannon were supplied with a new type of hot-water wet suit, known as the Wiswell suit. It looked something like a standard neoprene wet suit but was looser fitting, with little plastic tubes running like blood vessels throughout the rubbery material that bathed the diver in hot water delivered through an

umbilical. It had proven to be effective and was certainly an improvement over the electrically heated suit they tried out during Sealab II.
Because the suit was designed only for use from Sealab III, and not from
the PTC, the commanders decided to jury-rig an umbilical to pump hot
water to the PTC. Then Barth and Cannon could grab the hoses and attach them to the Wiswell suits. For all Sealab III's more generous funding
and expanded army of personnel, there was still room for jury-rigging,
just as in the program's early days.

Once they had all wriggled back into their suits and gone through
the final check-offs, they climbed up the short ladder leading into the
PTC still mated to the top of their chamber. They took seats around the
periphery, squeezing in with the Mark IX rigs amid the pipes and valves
inside the pod. They secured the circular hatch in the middle of the floor
and could feel the temperature falling almost as soon as their capsule was
cut off from the chamber's heat.

The PTC was detached from the decompression chamber shortly after
three o'clock in the morning on Monday, February 17, 1969. As on the
first dive, the gantry crane on the *Elk River* lifted the white pod from its
perch, just behind the ship's gray superstructure, and rolled aft over the
moon pool. The divers had all been on the job for twenty hours straight.
With Captain Bond's approval, another Navy doctor had locked in some
amphetamine tablets—pep pills—to combat fatigue, but none of the four
divers opted to take them. The PTC blowers again chilled them with an
unrelenting arctic breeze. Barth put a foul-weather jacket over his head
and shared a blanket with Cannon. They all felt even colder than they had
on the first ride down. By the time they reached the bottom they were
shaking badly. It was four-thirty in the morning, and they had been sealed
inside the frosty PTC for more than an hour.

The water was surprisingly clear, and through the PTC's little portholes
they could catch glimpses of the bubbling lab, lit up like a monument in
a deserted town square after nightfall. Barth and Cannon, shaking like the
others, were eager to get hold of the hot-water hoses and hook them up to
their Wiswell suits. They pulled open the hatch in the floor of their pod and
reached down to grab the jury-rigged hoses outside, but out of the hoses
came only a cool trickle. Without hot water, their loose-fitting suits would
afford less protection from the cold than a standard wet suit. Still, they
shouldn't have to be in the water very long. Barth and Cannon went ahead
and donned their diving rigs. From just below the open hatch, they also
grabbed the ends of the umbilical hoses that would supply each rig with

the proper mix of helium and oxygen. Cannon wore no hood or gloves, as was his custom. Barth plunged through the open hatch. The water creeping down his neck felt warm. The sensation puzzled him, but even at temperatures in the mid-forties, the water felt warm because he was so cold. Cannon followed Barth, dropping fins-first into the trunk and onto the cage. As Cannon hit the water his head bobbed up and he made such a desperate gasping sound that Blackburn thought he might decide not to dive after all. But Cannon dropped back through the looking glass and met up with Barth on the winch cage, their stoop below the PTC. It was now shortly after five in the morning, two hours after this second bone-chilling ride to the bottom began. Barth felt a tap on his shoulder and Cannon pointed toward the floodlit habitat nearby. Despite the cold and the ticking clock, the two paused for a moment to admire the oasis of light on the seabed and the impressive structure that would soon be their home in the sea.

With their umbilicals trailing them like garden hoses, Barth and Cannon swam the rest of the way to Sealab III, another thirty feet down to the bottom. Cannon split off from Barth, apparently to check on something else before swimming to the hatch. Barth went straight to the hatch, as Mazzone had instructed, to give it another try. Watching on a TV monitor, those in the command van on the *Elk River* saw Barth swim into view, step up the short ladder, and push on the hatch. The recent flooding had broken the seal, the skirt was still blown, and with the plan to coordinate a drop in the lab's pressure, Barth should be able to get in this time. Barth gave it a try, but the damn hatch wouldn't budge. In training, Barth had found that a little extra leverage could help pop a stubborn hatch and he had had a crowbar installed nearby, just in case. A little plaque with instructions on how to use the crowbar was even posted, as if an aquanaut might need to study it before proceeding.

Tomsky, Bond, Mazzone, Carpenter—everyone watching the topside TV monitors saw Barth swim off screen. They figured he was going for the crowbar this time. Fatigue and frayed nerves contributed to a subdued tension, as did conflicting opinions as to what they should be doing topside as they watched Barth struggle with the hatch—Hold pressure steady? Shut off the gas? Allow some water to flood the dive station? Whatever suggestions were made by those present, Commander Tomsky had the final say. Gauges in the command van seemed to be showing that the pressures inside the lab and out had equalized. Given those readings, Tomsky could not understand why that hatch was still stuck. It took Barth about a minute to undo the two wing nuts holding the crowbar in place—a clumsy method, Carpenter thought, and another example of something that more time and money could have improved.

No sooner had Barth grabbed the crowbar than he heard a strange grunting sound. At that moment Blackburn and Reaves were startled by a yell, or maybe it was a scream, that came from somewhere outside the capsule. Barth turned and saw Cannon on the opposite side of the habitat. At first glance his buddy looked like a man jumping rope—legs pumping, arms jerking. Barth dropped the crowbar and the topside commanders saw him reappear on their monitors, but then swim away, stirring up silty clouds as he vanished into stygian darkness. Bond immediately sensed trouble.

Barth soon returned to view under the habitat, this time lugging Cannon, who appeared to be convulsing. Barth tried to lift Cannon into the skirt beneath the hatch to get him a breath. Impossible. So he grabbed the Mark IX's buddy breather and tried to force it into Cannon's mouth. He jabbed it once, twice, a third time. The escaping gas created a hail of bubbles but Cannon's jaw was shut tight. Barth grabbed Cannon by the collar and began to swim for the PTC, disappearing from the camera's view. As Barth pulled Cannon along the sea floor, stirring up more clouds, their umbilicals caught on something. Barth struggled to free them, got another few feet closer to the PTC, and was jerked to a halt again. Now Barth was winded, as he'd been on the first dive, getting no good satisfaction from the rig, and sensed a ringing in his ears. He thought he might pass out. He triggered the Mark IX's bypass, using a lever at waist level to inject more fresh gas into the rig. Something still didn't feel right. At that point there was little choice but to leave Cannon and swim for the PTC, which was still a few paces away and about twenty feet overhead. Barth wrangled some more with his own umbilical before swimming up, up, up. The capsule seemed as if it were a mile away.

Inside the PTC, Captain Mazzone's voice came over the intercom with an order to get another diver into the water. Blackburn and Reaves couldn't be sure what was going on, but Blackie quickly strapped on a Mark IX. He was about to jump into the water when Barth popped through the looking glass, out of breath, but managed to squeak out the words that Berry was in trouble. Blackie squeezed past Barth to get through the trunk and down onto the winch cage. It took him a minute to get clear of all the umbilicals and he hurriedly swam off the cage and around the side of the PTC to unspool his own umbilical.

In the pools of light cast from the habitat, Blackburn could see Cannon's mouthpiece floating over his head, emitting a stream of bubbles.

Cannon was on his back, his shoulders resting on the lab's hefty main umbilical as if he were napping on a chaise longue. As Blackburn swam down he kept triggering the bypass to overcome the nagging sensation that he wasn't getting enough gas from his rig. Blackburn was six-foot-two, perhaps the biggest and fittest of all the aquanauts, with a steely resolve born of the dangerous and delicate work of disarming underwater mines and bombs.

Blackie first tried to force Cannon's mouthpiece into his mouth, but like Barth had no luck. Swimming for the PTC, dragging more than two hundred pounds of Berry Cannon and gear, wasn't going to work. Blackburn began to feel short of breath, but managed to wrap one arm around Cannon and with his free arm grabbed hold of his own umbilical, as if he were a rock climber clutching a rope. As he kicked hard to propel himself and Cannon, he also pulled himself up the umbilical, which formed a steep hypotenuse to the PTC from where Cannon had been lying atop Sealab's umbilical on the sea floor. Blackburn, his mind racing, wondered if the line was strong enough to support his weight and Cannon's. If the umbilical snapped, he and his unconscious buddy would sink back to the ocean floor, doomed.

For everyone watching and waiting topside, scant information trickled in from the PTC, in Chipmunk garble. Barth had apparently returned, and they knew Blackburn was in the water. For those empty minutes, mission control could only wait and pray. The silence ended when Blackburn broke through the liquid looking glass. He was breathing hard, with a fury he had never before experienced. He held Cannon upright in the trunk and wrestled him out of his Mark IX, then hung the rig on a protruding valve, shed his own gear, and hung it up next to Cannon's. Blackie slid past Cannon and climbed up into the PTC. Together, he, Barth, and Reaves were able to hoist Cannon into the capsule.

The open hatch created a hole three feet wide in the floor, leaving only a narrow peripheral ledge to stand on—with one diver unconscious and limp as a rag doll. Blackburn held Cannon out of the way while Reaves and Barth shut and sealed the PTC hatch. Reaves had never seen a dead diver but Cannon sure looked like one—motionless, expressionless, eyes like glass. Reaves put his face next to Cannon's and sensed no trace of breath, nor was there a discernible pulse. They put an emergency mask over Cannon's nose and mouth that fed Cannon a fresh flow of their helium-rich gas.

The mask didn't seem to help, so Blackburn started in with

mouth-to-mouth resuscitation. Cannon's face was still warm. The mouth-to-mouth was getting a good exchange of gas into Cannon's lungs—a hopeful sign. In the awkward circular space, Reaves got down on his knees to give Cannon external heart massage while Blackburn kept up the mouth-to-mouth and Barth ran the PTC. Questions from topside were spilling over the intercom—*What is your internal pressure gauge reading? What is your oxygen reading?* Walt Mazzone donned a headset and was in constant communication, and at the same time was getting the decompression chamber ready to receive them. Mazzone and the others finally heard that the PTC hatch was shut and sealed—"Take us up! Take us up!"—and Tomsky ordered the capsule raised. Within a few minutes a Chipmunk voice could be heard saying, "Faster, faster," but it might be an hour or more before the PTC was safely mated to the chamber hatch. And then what? Their decompression from six hundred feet would take the better part of a week.

Everyone in the PTC, except for Cannon, was shaking from the cold. They noticed a froth coming from Cannon's nose and mouth. Amid the cold confusion, they sensed that their friend had slipped away. They were all tired and shaky and not in the best condition to make medical assessments, but about fifteen minutes after the PTC left the bottom, Captain Bond and the others heard a grim, helium-spiked pronouncement: "Berry Cannon is dead."

14

AN INVESTIGATION

An effort to revive Berry Cannon continued, even after Barth and the others reported over the intercom that they believed their friend was dead. Topside commanders ordered them to keep up their attempted resuscitation, and they did so until the end of their hour-long ride back, when their PTC was mated with the decompression chamber. At that point they finally stopped working on Cannon so they could open the hatch in the floor. They lowered Cannon's body through the hatch and passed it down into the chamber, where they laid it to rest on one of the bunks. Barth took a moment to close his friend's eyes.

During the PTC's nail-biting ascent there had been a discussion topside about putting George Bond or one of the other doctors on the *Elk River* under pressure inside the decompression chamber's outer lock so the doctor could rendezvous with the aquanauts inside the chamber's main quarters. But Cannon was presumably going to be dead on arrival. More than an hour had passed since the convulsion. Bond didn't think a doctor's presence would serve much purpose. Commander Tomsky, who had the final word, agreed that they should not complicate an already complicated situation by locking in a doctor.

Barth, Blackburn, and Reaves, and five other Team 1 aquanauts in the second chamber would have to spend the next six days undergoing decompression. But the three weary aquanauts zipped Cannon into a body bag and carried it through the bulkhead hatch and into the adjoining outer lock. They then returned to the main chamber, closed and sealed the hatch. Since Cannon was already dead, he was brought rapidly back

to sea level pressure. His body released myriad gas bubbles, causing significant puffing and bloating.

When the PTC had surfaced through the *Elk River* moon pool, three
of the Mark IX rigs were found dangling around the winch cage, the stoop
below the PTC. In the rush to lift Cannon into the capsule, the three
rigs were left outside—the two Blackburn had hung up in the trunk, just
below the PTC hatch, and the one Barth had doffed and tossed down the
hatch to get it out of the way. Only one rig was still dry and inside the
capsule—the rig that Reaves never got to use.

An autopsy report revealed that carbon dioxide poisoning triggered
Cannon's death and the rigs became key pieces of evidence in the Navy's
board of investigation, led by a panel of three officers, plus a legal counsel.
No sooner had the aquanauts emerged from their week-long decom-
pression than the board held its first hearing at the Deep Submergence
Systems Project Technical Office at Ballast Point, the Navy Submarine
Support Facility on San Diego Bay.

Pursuant to long-standing military tradition, board hearings were a
fact-finding mission. They were a formal, courtroom-meets-boardroom
hybrid, with military lawyers representing the "parties of interest," but
with the parties themselves free to take part in the discussion and ask
questions of each other and the many witnesses who were called to testify under oath. Jack Tomsky, the on-scene commander, Captain Walter
Mazzone, the diving operations officer, and Paul A. Wells, a senior chief
mineman and experienced diver who was in charge of the Mark IX diving equipment, were named as parties of interest. This entitled them to
be present throughout the proceeding, introduce evidence, examine and
object to any evidence submitted, testify, and ask questions.

On the board's second day, Captain Bond asked that he, too, be designated a party. Bond felt that as principal investigator he ought to be able
to participate in the proceedings, even if he risked being found at fault.
The three-member board agreed, but Bond was the only one of the four
parties of interest who chose not to be represented by legal counsel. They
were not sitting as defendants, per se, but they all knew that the board
could recommend disciplinary action or even trial by court-martial.

The board held sessions for fifteen days, through the middle of March,
often meeting late into the evening. More than eighty documentary exhibits were submitted, from the hefty *U.S. Navy Diving Manual* to the
scrap of paper on which Richard Blackburn had scrawled his concerns
about the Mark IX rigs. Barth, Blackburn, and Reaves were among the

nearly fifty people who were called to testify. Most were witnesses to some part of the action on board the *Elk River*; others, like Dr. Robert Workman, provided expert testimony.

The central testimony was that of the three divers who were with Cannon in the PTC. One by one, they recalled the details of their ordeal: the crippling cold, the fatigue, nearly passing out with the Mark IX, the stuck hatch. The failure of the jury-rigged hot water for the wet suits. Cannon's convulsion. The attempted resuscitation. The grim uncertainty as to whether their friend was dead or alive.

Barth voiced measured frustration about such things as the Mark IX and the continual snagging of his umbilical. Describing the second dive, and his struggle to save Berry Cannon, Barth said: "It is quite trying to know when a man is in your hands and he is not breathing, he is in difficulty, and he is dying, we will say, and you can't do much good because you are continually getting fouled on something on the bottom." But in further testimony Barth defended the operation, and bristled at the unflattering picture he believed that reporters had been painting of Sealab III since the accident. News of the tragedy had made all three television network newscasts and stories were splashed onto front pages across the land, often accompanied by pictures of Cannon's chiseled face. The board hearings were open to reporters and in the following days the papers continued to cover the mysterious death of this heretofore unknown aquanaut. Barth, who was never a great fan of the press, didn't like all the reports, as was apparent from his testimony. "I'm not happy with what I read in the newspapers because it makes us out to be a bunch of stumble-bums, actually, that didn't know what we were doing, and this is not the case.

"We had a bunch of people who worked damn hard for a number of years, to get this program on the road, and we were fighting the problems that everybody has in the building of the piece of equipment as big as Sealab and the deck decompression chambers and trying to get the thing on the road with some of the problems we had. We had the PTC flooding out; we had this happen to us and that happen to us, and it makes it look like nobody tried too hard to do a good job and I think everybody out there in the Sealab program broke their backs to get it going and tried darn hard to get it done. I am sorry that some things were said by the press, because to me they don't seem to be the truth. *Period.*"

Barth's reaction was a lot like Bond's after his Sealab II interview with *The New York Times*. Relatively little had been written about Sealab's

pioneering successes over the years, or the related breakthroughs in saturation diving, but Cannon's death was widely reported. When *Apollo 1*
caught fire on its launch pad two years before, killing three astronauts,
everyone at least recognized that, tragic fire notwithstanding, the space
program had been on a triumphant roll. Even as Barth and the others
testified about Sealab III, *Apollo 9* was launched.

Unlike Barth's, Richard Blackburn's testimony was laced with sharp
criticism, which generated headlines of the kind that Barth found irksome,
such as one in the *Los Angeles Times*: "Fatal Dive Should Never Have Been
Made, Sealab Probe Told." Blackburn's opening statement enumerated a
number of technical problems that may have contributed to the troubles
with their Mark IX rigs—arcane regulator pressure settings and also the
possible increase in breathing resistance from the stiffener added to the inhalation hose, as he had noted after the first dive. He testified that the Mark
IX had not been properly evaluated, and their training with the rig was so
limited that most divers didn't know much more than how to put it on and
where to find the bypass. The divers had all trained in less than sixty feet of
water, and most didn't go any deeper than thirty feet. It was as if they had
trained for the Tektite project instead of Sealab.

"Training looked good on paper but that is all," Blackburn said. He and
others had sensed this, and occasionally someone spoke up. But this was
the United States Navy, and "no can do" is not a popular refrain. Besides,
the Navy aquanauts and the civilians, too, Cannon included, had been
eager to get to the bottom. In the aftermath, though, many were having
second thoughts similar to those of Blackburn, who declared: "I believe
that the second dive should never have been made, but it was consistent
with the type of treatment we have been having all along. All the divers
in Sealab have been worked too hard for too long, and then not given any
consideration or compensation for their hard work. We were all pushed to
the point where mistakes were inevitable. It is a horrible thing for a man to
have to die to slow this outfit down enough for a realistic look at what the
divers have to contend with."

It was an unvarnished, gutsy statement for an enlisted man to make
before senior officers. Few others would be as direct, and many were not
called to testify at all. More than a week later, Scott Carpenter offered his
opinion that the program was underfunded. Money would have bought
not only time but men, and they needed both. He also told the board
that people were pushed harder in the Sealab program than he had been
pushed anywhere else.

Tomsky and his legal counsel were quick to respond to these and other criticisms. Tomsky maintained, for example, that the program received all the funding he had asked for, and that people were pushed only within reason. Mazzone often just listened to such testimony but occasionally piped up, as when Tomsky indicated that Barth and Blackburn were the ones originally slated to make the first dive and open the habitat as part of the unbuttoning process.

"May I clear a point?" Mazzone asked. "Barth and Cannon have always been working together as a team for the unbuttoning. This was the original team for the unbuttoning and they were so trained."

On the eleventh day, Bond volunteered to testify. After he presented an abridged history of Sealab, he found himself pelted with questions about why he had advised against locking in a doctor to meet the divers inside the pressurized deck chamber after the first or second dives. Dr. Bond defended his decision—the final call was ultimately Tomsky's to make—but his testimony was not without a touch of emotion. The tragedy had hit him hard. Bond considered Berry Cannon a great friend. But more than an hour had passed between the time Cannon collapsed and his body was lowered into the decompression chamber. He had been limp and unresponsive the entire time, but some suggested that there was a slim possibility that he was still alive when he reached the chamber. Without a doctor there to personally examine him, how were Bond or any of the doctors certain that Cannon had been dead before his body was locked out? The morbid implication was that Bond may have recommended the explosive, fatal decompression of a living human being.

The board's investigation was deliberately and narrowly focused on the circumstances of Berry Cannon's death. Any testimony that seemed to wander into broader issues was usually cut short. Lieutenant Commander David Martin Harrell, an engineer on the Deep Submergence Systems Project staff who was among the designated aquanauts, drew immediate objections when his testimony turned to the habitat leaks. Only cursory attention was paid to anything deemed the least bit peripheral to Cannon's death, most notably the problem with opening the hatch, but many crew members believed that at least some of the major snafus should be looked into as part of the investigation. It could be argued, after all, that if corners had not been cut, if money and men had been more plentiful, if the habitat had not been leaking like a sieve, if the hatch had opened promptly and properly, Berry Cannon might still be alive and the program on track.

On the fourth and fifth days, Barth was called on to narrate the silent black-and-white film that showed him trying to open the hatch on the first and second dives, and then struggling to help Berry Cannon. To relive those moments might have rattled some, but Barth spoke dispassionately, not because he didn't care about having lost a friend. It was just his pragmatic way. Risks had to be taken. Since Genesis, Barth had taken them himself. Unfortunately, there is always the prospect of death with these types of endeavors, which Berry Cannon knew as well as anyone. The footage of Barth and Cannon from the closed-circuit television camera became for Sealab III what the Zapruder film had been to President Kennedy's assassination: a grainy visual record scrutinized for answers.

In their statements on the day Cannon died, Sealab officials were insistent that whatever may have gone wrong, Cannon's equipment had functioned normally. They said the cause of death was an apparent heart attack, but that must have been a best guess, since an autopsy had yet to be performed. But within a few days the Mark IX rig had become a smoking gun. One of the three rigs found hanging outside the PTC had a canister with no Baralyme, the sandy carbon dioxide absorbent. Without it, a diver wearing the Mark IX wouldn't get far before he had trouble breathing. But the empty canister theory raised as many questions as it answered. Complicating matters was a chicken-and-egg quality to the findings of the autopsy, performed within a day of Cannon's death with Dr. Bond observing.

Dr. Francis Luibel, the pathologist with the San Diego County Medical Examiner's Office, testified that, as stated in the autopsy report, carbon dioxide poisoning triggered Cannon's demise, and death was then due to acute pulmonary edema and congestion, brought on by a combination of cardiac and respiratory failure.

Luibel also testified, however, that it was not entirely clear which came first—the heart and respiratory failure or the troubling fluid on the lungs. That left some question as to what triggered the fatal chain of physiological events. In his testimony, Luibel ruled out a few possibilities, namely drowning, but indicated that asphyxiation—caused either by a total loss of breathing gas, or a fatal gas mix—could have initiated Cannon's death. In that case, a faulty rig, not necessarily one with an empty Baralyme canister, could have been responsible. Carbon dioxide poisoning, as might occur from using a rig with an empty Baralyme canister, could also do damage to the heart and lungs. But, as Dr. Luibel testified, there were many unknowns about what happens when a person is poisoned by carbon dioxide. So it was hard to say for sure that CO_2 poisoning was solely to blame for triggering Cannon's death, although that's what was stated in the autopsy report.

Lieutenant Commander Lawrence Raymond, who was to have been an aquanaut with the second team, had witnessed portions of the fatal dive on the support ship's TV monitors and had overheard the divers' helium-spiked reports coming from the PTC. He was called to testify as an expert on thermal factors. Dr. Raymond left little doubt that cold-induced stress played a large, possibly even decisive role, in Cannon's death. He said the circumstances surrounding Cannon's death were likely further complicated by the high pressure and the heat-sapping helium, whose combined effects were not entirely known.

Dr. Workman testified that if a diver breathing through an empty canister kept up a moderate level of physical activity, he could be dead within five to ten minutes. In Cannon's stressful situation dangerous levels of carbon dioxide would have been reached even sooner. However, Workman and Raymond both noted that if Cannon was riding the bypass, as Barth and Blackburn found they had to, that might have extended his breathing time on the rig.

A further possibility was that Cannon was electrocuted. Sealab III had some grounding problems, and Cannon, the electronics engineer, knew that the worst of the leaking appeared to be around the electrical stuffing tubes, along the side of the habitat where power and other lines were connected. It was conceivable that before going to the hatch Cannon had split from Barth to take a look for himself at the bubbling exterior leaks—and a 440-volt jolt through the conductive saltwater could have sparked his demise. Some sort of vocal outburst such as Cannon made would be more likely in response to electric shock than to breathing a bad gas mixture, but no discussion about the yell that Blackburn and Reaves heard, nor the grunting sound that drew Barth's attention to Cannon, was entered into the record. Electrocution was ultimately ruled out, as were asphyxiation and thermal stress. Since none of those was considered the most likely cause of death, then there was no need to dwell on the grounding problems, the failed jury-rigging of the hot water suits, the unheated PTC, the finicky Mark IX, the leaks, or even the hatch problem. Carbon dioxide poisoning was singled out as Cannon's prime killer. With that settled, the board members still had to figure out how one Mark IX wound up with an empty canister.

The attention shifted to the diver in charge of the equipment, Paul

A. Wells, a soft-spoken senior chief petty officer known to all his buddies as "P.A." Wells, a diver for eighteen years, was universally admired for his integrity, reliability, and conscientiousness. He had been a plankton-eating member of the third Sealab II team and thoroughly impressed Bob Sheats, the exacting master diver and team leader. Wells was in his early forties and had spent most of his Navy career working with explosive ordnance and diving, like Blackburn. During World War II Wells had received a Purple Heart for service in the Marine Corps. The gold on his left sleeve signified an award for good conduct. Most important to his fellow divers was that a diving rig set up by P.A. was a rig you could trust. Everyone had complete confidence in P. A. Wells.

The board focused on the handling and preparation of the Mark IX rigs and their interchangeable canisters. On the ninth day Wells, who oversaw the rigs, was called as a witness and asked to recall his every action from the time he reported to the diving station on the *Elk River* on the morning of February 14, through the fatal dive three days later. Wells was not alone in the diving station, a workshop area on the *Elk River* that could bustle like a pit stop and was "sometimes utter chaos," as Bill Schleigh, one aquanaut-to-be, told the board. Those in the diving station didn't always recognize each other, a testament to how many more people were involved than during the first two Sealabs. Not everyone could remember exactly who gave what to whom, who did which job, or even who had assisted whom in filling canisters. Wells thought one of the aquanauts, Sam Huss, had given him a hand filling canisters, but Huss had no such memory. One person thought to have helped fill a canister could at first be identified only as "a huge man," a description that fit two likely candidates.

Wells was certain that he had personally placed loaded canisters into three rigs. About the fourth he could not be so sure. He remembered putting all four rigs set up that Friday into a diving station storage bin. On Sunday morning, about thirty-six hours later, Wells learned that they were going to start the first dive sooner than planned to try to fix the leaks and that four Mark IX rigs were needed right away. Wells lifted them out of the storage bin, where they appeared to have been undisturbed. He had checked all the rigs previously and had no reason to believe anything had changed.

It was not common practice, but ordinarily the conscientious diver would have double-checked the canisters by opening each rig's fiberglass backpack and unscrewing the lid of the phonebook-sized steel canister inside to verify that it was topped off with fresh Baralyme granules. To check each one would have taken only a minute or so, but Wells told the board that he'd gotten the distinct impression that this was "a hurry-up affair," and he skipped the additional check. He carried each rig to the medical lock of the chamber where Barth, Blackburn, Cannon, and Reaves were being pressurized to six hundred feet on an expedited schedule.

In response to a question from Bond, Wells seemed to poke a hole in the empty canister theory when he pointed out that if a Mark IX canister had been completely empty, he would have been able to tell just by lifting the rig. Some in the room were skeptical, but Wells maintained that if a canister had been empty, he would have been tipped off by the lesser weight—about 17 percent less, eight pounds of Baralyme on a forty-seven-pound rig. Captain Mazzone offered a brief supportive comment and said the weight difference could be noticeable. Others privately agreed, and no one who knew Wells doubted his word.

Under the cold and harried circumstances, neither Cannon nor apparently anyone else in the PTC checked the canisters. After the first dive, Blackburn had put refilled canisters into the two rigs that Barth and Cannon used. That left two rigs that conceivably might have had empty canisters. If the empty canister theory was right, a deadly game of rig roulette was still on.

Before the board of investigation began, all four Mark IX rigs were packed up, with Wells's assistance, and then sent to Washington for analysis at the Experimental Diving Unit. It was this packing episode that cast the darkest shadow of doubt on Wells, at least for the three board members, including Commander William Leibold, who had watched closely as Wells packed the rigs, as had Bond and at least one other Sealab officer.

Under pointed questioning from Leibold and Commander Tomsky's legal counsel, Wells acknowledged that in handling the rigs to pack them up he had not noticed any difference between the weight of the rig with the empty canister and the others. Wells did suggest that this might have been because the rig had some water in its canvas breathing bag. But Commander Leibold, a highly decorated war veteran who recently completed a tour as commanding officer of the Navy dive school and Experimental Diving Unit, disagreed: "You had water in the bags, but not much. I don't believe it's eight pounds so, then, is it possible that you wouldn't know if there was an empty or full canister?"

"Possibly, but I feel right now, this time, here today, I could pick one out of six," Wells said.

"But on the particular day that you did pick it up, you didn't notice it?"

"No, not as you say, no."

Wells offered no further explanation or excuses.

When the packed rigs arrived at the Unit, investigators went to work on several fronts, although their analyses did not include a look into whether an experienced diver could reliably distinguish, by weight, between Mark IX rigs with full and empty canisters. But the investigators did find that of the four rigs—numbered 1, 4, 5, and 8 on their fiberglass backpacks—three were in satisfactory, though not perfect condition for a dive at six hundred feet, but rig no. 5 was not, mainly because of its empty canister. But who wore rig no. 5? No record had been kept. To try to determine the number of Cannon's rig, attention turned to the footage of Barth and Cannon at the hatch. This closed-circuit video was transferred to higher-quality film, enhanced and analyzed, frame by frame, to isolate the clearest possible view of the black, four-inch-high number painted on Cannon's white backpack. A panel of eight expert interpreters was then asked to name the number they saw. Five said it appeared to be a 3, two said it was a 5, and one said it was an 8.

The only sure thing was that no one had used rig no. 8, the one left in the PTC. Presumably Reaves, had he made a dive, would have donned it. In the video of the second dive it appeared clear that Barth wore rig no. 1. That winnowed the options for Cannon, and also Blackburn, down to rigs 4 and 5. Presented with the choice of whether Cannon's rig was numbered 4 or 5, the expert interpreters agreed it could not be a 4. Blackburn, who was never seen on camera, thought it was entirely possible that he had worn no. 5. His dive to save Cannon had been relatively brief, and during those harrowing minutes his extreme shortness of breath had forced him to ride the bypass constantly and circumvent the rig's gas recycling system—including, perhaps, an empty canister. Barth, too, found himself having to ride the bypass on both his first and second dives. If not an empty canister, other factors, like the hose stiffener, could have been to blame, as Blackburn had told the board. One or more of these factors could have conspired against Berry Cannon, abetted by the severe cold and stress.

The possibility of tampering was also floated in the EDU report, which recommended the board look into what happened to the apparatus between the time of the accident and the analysis done four days later at EDU. Commander Tomsky strongly believed a saboteur may have been responsible for the empty canister before the rigs were ever locked into the chamber. While the opportunity for tampering existed, the board found the probability very small. Commander Tomsky had previously raised the specter of sabotage in the mysterious case of a valve that controlled the flow of pure oxygen to emergency breathing masks inside the deck decompression chamber. That valve, which was opened and closed by one of the many faucetlike controls around the outside of the chamber, should have remained closed while the aquanauts were inside but was twice found open. It had no direct bearing on Cannon's death, but nonetheless fueled suspicions: If a saboteur really lurked among them, then the same malicious soul who opened the deck decompression chamber valve might also have been responsible for the empty canister. As for the open valve, if the decompressing divers had used the emergency system while still exposed to more than two or three atmospheres of pressure, the inhaled oxygen would have been toxic. At the very least it could have sent them all into convulsions.

After a diving officer's routine check found the valve cranked open the second time, Tomsky tried but failed to get the Office of Naval Intelligence to launch a full investigation. Tomsky, still concerned, had Captain Mazzone set up a security watch. Mazzone obliged but he didn't subscribe to the sabotage theory. To him, the valve opening seemed like the kind of mistake he'd seen before with numerous hands on board who were new to saturation diving and its specialized equipment.

After fourteen lengthy sessions, the board and its legal counsel took ten days before convening a final session, on March 24, 1969, to take some further testimony from key witnesses. Over the next couple of weeks, the board produced forty-seven unanimous opinions based on its sixty-six findings of fact. Virtually no one involved in the operation escaped criticism. The board settled on the empty canister theory and the resultant carbon dioxide poisoning as the most likely explanations for Cannon's death, but conceded the theory was not certain. Its nineteen recommendations touched on many of the problems that had become obvious during the hearing, among them the need for better testing of the Mark IX, properly heated dive suits, strict adherence to the buddy system, a tighter organizational structure, and of course a hatch that was easier to open. The board also noted that Barth, Blackburn, Cannon, and Reaves should be officially commended for their efforts. The board's harshest recommendations led to punitive letters of admonition for Tomsky and Wells.

Less severe than a letter of reprimand and certainly preferable to being recommended for trial by court-martial, the admonition was nonetheless a bitter retirement gift for Tomsky and he appealed the decision. He wrote a four-page letter to the secretary of the navy and argued that his admonition should be withdrawn. "None of the alleged derelictions on my part have been shown to have in any way contributed to the matter under investigation, i.e., the accidental death of Aquanaut Berry L. Cannon," Tomsky wrote. "Had the Sealab III exercise been successful, I am certain that my superiors would have considered the organization and management of the operation as near perfect. The exercise was not successful due to material difficulties and failures, which had no relation whatsoever to my organization or management, and to an accident which would have been prevented had existing regulations and procedures been followed."

He added: "Significance is attached to events which, without the benefit of hindsight, would be virtually unremarkable." The "material difficulties and failures" that plagued Sealab III were clearly an issue. Tomsky was not alone in lamenting that the leaking lab, reengineered and repainted traffic-light yellow, was, in effect, a lemon. Captain Bond gave the *Los Angeles Times* an interview midway through the board of investigation in which he complained about the various equipment failures, calling them "unbelievable." Still, if more rigorous tests of this three-hundred-ton lemon and its related hardware had been allowed, the various material difficulties and failures might have been revealed and addressed before anyone got hurt.

Tomsky considered requesting a full court-martial to pursue the matter further, and perhaps get a more favorable outcome for himself, but decided against it. Such a trial, he believed, would only hurt the morale of the Man-in-the-Sea personnel and generate some more undesirable headlines. As Tomsky and everyone knew, the Navy had weathered not only the Sealab III accident and investigation but a two-month-long inquiry into North Korea's capture of the U.S. spy ship *Pueblo* in January 1968, a high-profile proceeding that took place in a nearby San Diego courtroom and came to a somber end just as the Sealab hearings did.

P. A. Wells accepted his punitive letter, but Tomsky counted himself among the many who did not believe Wells was to blame. Tomsky's boss, Captain William Nicholson, the head of the Deep Submergence Systems Project, wrote in a memo to his superiors that Wells's admonition was "extreme and unnecessary," and urged that it be reduced to a less severe "letter of caution." Nicholson indicated that Wells would take the punishment hard, and he worried about "the probable impact of a 'punitive letter' to the self-censure of such a conscientious person as Chief Wells." He was right to be concerned. Wells withdrew into a self-imposed exile, avoiding even his many longtime Navy friends. Bob Barth couldn't help feeling that the Navy had lost two excellent men, not just one, on that fateful day. Years later, Wells had settled in Panama City, Florida, not far from Alligator Bayou, where Sealab I was built. He lived in a mobile home park called Venture Out. But as far as anyone could tell, he rarely did.

It is safe to say that few were satisfied with the outcome of the investigation. Wells had been made a scapegoat and the empty canister theory left too many questions unresolved. Many felt that the real problems plaguing the project had been glossed over, like the troubled habitat and the rush to meet the February deadline. Tomsky remained convinced that sabotage should have been given greater credence—for the empty canister, the oxygen valve, perhaps even some of the program's other setbacks and delays.

In the investigation's wake, the Navy offered public assurances that Sealab III would be repaired and the undersea experiment would go on, much as the Apollo mission had gone on despite the tragic launch pad fire. Berry Cannon would have wanted it that way, his friends knew. Funeral services for Cannon had been held on February 19, in Chula Vista, his wife's nearby hometown, an exurban enclave between San Diego and the Mexican border where houses were just beginning to outnumber citrus groves. Mary Lou Cannon sometimes made the trip west from the couple's Panama City home and stayed at her parents' place with their sons—Patrick, age nine, Kevin, who had just turned five, and two-year-old Neal. She was in her parents' little kitchen getting breakfast ready for the kids when a Navy chaplain appeared at the door of the bungalow on the morning of February 17, within hours of Berry's predawn accident.

The utter shock of that moment turned Mary Lou's heart inside out. At the funeral, Mary Lou found herself thinking that the person in the casket didn't even look like Berry. *Maybe it wasn't!* She half-believed, wanted to believe, that her husband's death was a necessary hoax, part of an elaborate Navy scheme to keep his true mission a secret. For the better part of a year she privately clung to this notion and the fragile hope that he would one day return.

Several aquanauts attended the small service at Humphrey Mortuary, but Cannon's buddies from Team 1 were still locked in a chamber, undergoing decompression on the *Elk River.* A Navy chaplain read these words: "There is a mystery in death; there is a mystery in the sea. Both are in some degree incomprehensible and unfathomable. Some must probe the mystery of the sea even at the price of probing the mystery of death . . ."

The next day Cannon's body, accompanied by his family and a couple of aquanaut friends, was flown to central Florida for burial at Wacahoota Baptist Cemetery, not far from Cannon's rural hometown of Williston, where Berry had been raised by his grandmother, while his mother worked odd jobs. The family never had much money, but Berry had always been a good student. He graduated near the top of his high school class and was captain of the football team. Now, as the first American aquanaut to give his life for the quest to live in the sea, Cannon was brought home. A simple gray headstone marked his gravesite.

A few days later, after Barth and the rest of Team 1 were released from their week-long decompression, the whole gang gathered at Barth's place near Pacific Beach. Shock and disappointment still hung in the air like a gloomy coastal fog, though many of the aquanauts were still hopeful they'd get their chance to break depth barriers and play their parts in a historic outpost on the continental shelf. Their dampened spirits were lifted when Dick Cooper, a young civilian aquanaut with a doctorate in biological oceanography, showed up with dozens of lobsters, specially flown in from Maine. They were supposed to have been Cooper's subjects in the planned attempt to start a colony of the crustaceans in deep, cold Pacific waters similar to their Atlantic environs. With Sealab III on hold, Cooper figured the homeless creatures might as well be the main course at what felt like a last supper. Some of the really big ones—caught deep in the Atlantic, in submarine canyons beyond the usual reach of fishermen—weighed as much as thirty pounds. A handful of the guys took those whoppers to Barth's garage, injected them with formaldehyde, and lacquered and mounted them, as if creating consolation prizes for the big adventure that got away.

After the accident, Tomsky and the *Elk River* crew had no more than a day in which to improvise a way to raise Sealab III. Through a combination of jury-rigging and quick thinking, the kind that had often saved the earlier Sealabs, they lifted the leaking lab to the surface before the ocean swallowed it up. Sealab III was then brought back to Long Beach and towed to San Francisco for repairs. While Navy planners mulled Sealab's future, two Americans walked on the moon and were returned safely to the earth, just as President Kennedy had pledged. Mankind seemed better equipped to reach the distant moon than the outskirts of the continental shelf.

The Navy continued to express a desire to revive Sealab III but many of the aquanauts and other personnel were already moving on to other assignments. Soon after the board of investigation concluded, a discouraged George Bond flew off to the Virgin Islands to have a look at Project Tektite as it reached its successful conclusion. In the fall the Navy said funding cuts would force the postponement of Sealab III until the following year. By the end of 1970, the Navy quietly announced that Sealab III was in mothballs and the Man-in-the-Sea program had been canceled.

Bob Barth and most of the others had long since figured out that the Sealab program died with Berry Cannon, but Barth was dismayed by the program's unceremonious demise. They had all worked hard to prepare for Sealab III. The living depth of six hundred feet would have been a giant leap down for mankind. Seasoned divers like Blackburn and scientists like Dick Cooper were on the brink of a historic mission only to be turned away with scant explanation. It didn't seem right. Death was always a possibility in their line of work, Barth believed, and should not have been allowed to defeat them. But what most Sealab personnel didn't know was that Sealab III was not an end but a beginning, and Mary Lou Cannon's grief-infused thoughts about hoaxes and secret missions were not as far-fetched as they might have seemed.

15

THE OIL PATCH

By the mid-1960s, saturation diving, and even the futuristic notion of living in the sea, was no longer the sole province of scientific dreamers, inventors, or undersea explorers. The business of drilling oil and gas from the world's continental shelves was growing into an industry, and saturation know-how arrived just in time for the offshore industry to take its hunt for hydrocarbons into deeper waters. It was as if George Bond's old exploration and exploitation mantra had been adopted by an industry with the money and the motivation to push saturation diving to whatever ocean depths the market, and the human body, would bear.

By the summer of 1973, a diver like Alan "Doc" Helvey could be found working more than four hundred feet down, almost fifteen atmospheres away, at the bottom of the frigid and notoriously tempestuous North Sea. Helvey was an easygoing ex-Navy diver and medic—the nickname "Doc" had stuck with him after he left the Navy to join the growing legion of commercial divers. Helvey was attracted to commercial diving by the kind of strike-it-rich talk he had heard about working in the oil patch. There was indeed a gold-rush quality to what was happening offshore, and not just for the titans of industry. A commercial diver, especially a saturation diver, could earn in a day what a similarly skilled plumber or mechanic or factory worker would earn in a week, maybe even more, depending on the depth and duration of the dive.

Working at a depth of 440 feet would have been unthinkable only a few years before. Less than a decade had passed since Ed Link had sent Jon Lindbergh and Robert Sténuit to a similar depth for two tenuous days, and in much less hostile Bahamian waters. Helvey was not only surviving but wrangling with pipes and related fixtures on the kind of outsized underwater plumbing job familiar to commercial divers like him. Helvey and his diving partner were in an oil field called the Forties, a sprawling patch of seabed covering an area larger than the island of Manhattan that lies a hundred miles northeast of Aberdeen, Scotland. Oil was discovered in the Forties in 1970, a first for the United Kingdom sector of the North Sea. The very first North Sea oil strike had come a year before, in a field to the southeast called Ekofisk, in the Norwegian sector. These and other North Sea discoveries yielded big-money contracts for a variety of offshore businesses, including those that supplied working divers, like Helvey's employer, Taylor Diving & Salvage. Taylor Diving had been started in the late 1950s in New Orleans by a couple of retired Navy divers. From its humble origins on the Gulf Coast, Taylor had grown into one of the largest and most influential diving enterprises in the world. The Gulf of Mexico remained a hotbed for offshore drilling, but after the North Sea discoveries, the deep frontier between Britain and Norway attracted businesses and divers, including Americans like Helvey.

Taylor Diving had become a major player in the construction field, which required some of the longest, deepest dives to do jobs like laying pipeline at sea. Maintenance and repair of the pipelines were also essential to offshore operations. Pipelines connected multiple wellheads on the sea floor with their mother platforms—industrial islets that often looked like car engines perched above the waterline on stilts. Pipelines ran between these islets and to production facilities on distant shores. Parts of the Gulf of Mexico had become so crisscrossed with pipelines that sandbags were dropped on the bottom and divers wrestled them into place as protective cushions between older pipelines and the new ones crossing over them. By the early 1970s, an estimated fifteen thousand miles of marine pipeline had been laid worldwide, enough to wrap more than halfway around the globe. Many more miles were on the way.

Much of what commercial divers were called on to do in the offshore industry might best be described as otherworldly construction work. The surveying, digging, hauling, dredging, cutting, welding, plumbing, inspecting, repairing—all of the kinds of jobs found at a typical construction site—had to be done with the challenges and constraints of working in dark, deep, and shifting water. Divers coaxed heavy parts into place with chain lassos. They handled bolts the size of baseball bats and tightened them with bazooka-sized hydraulic wrenches. They monitored dredging machines that were as noisy as locomotives and slid like giant sleds along the ocean floor, spewing water and stirring up silt, and they inspected pipes and other structures for cracks or leaks. Saturation diving proved especially useful for jobs that had to be done deeper than about 150 feet and were known to take hours rather than minutes to complete.

The specialized pipe assembly job that Doc Helvey was doing in the North Sea was his first experience as a saturation diver. He liked the unique demands of the work, and he liked the high pay. As the project neared its end, with several days of decompression approaching, he had good reason to hope that this job would not be his last. But Helvey's friend and diving partner, Paul Havlena, who was also in his late twenties, was looking forward to getting out of the diving business as soon as this job was done. Commercial diving tended to be a younger man's profession—virtually all of the roughly three hundred divers working for Taylor at the time were under the age of thirty-five. Havlena might have been expected to stick around but he had other plans. He had been accepted to college, had paid his debts, and by signing on for this one last saturation dive, he would have enough in the bank to buy a new Corvette. Pretty soon all he'd have to do was study and chase girls. That's what he had gleefully been telling Doc Helvey, anyway.

Helvey and Havlena were getting ready to start on their next round of sea floor tasks when a ruthless North Sea storm ripped through the area. The pipe-laying barge from which they were making their dives didn't have time to drop its pipe and make a run for the shore and safety, so when this unexpected storm hit, a line of heavy pipe was still attached at the barge's stern. The pipe stretched some fifty yards behind the barge, cradled in a U-shaped pontoon like a hot dog in a bun. Once the pipe reached the end of the pontoon, it gradually angled down to the sea floor, like gargantuan fishing line. But as the barge got badly roiled, sections of the freshly laid pipeline buckled. The resulting damage required a repair job at a depth of 320 feet, and it showcased the oil patch diver's brand of useful work.

While some tasks required Swiss-watch precision, often in a hostile environment, others relied more on nerve and brute strength. Sometimes, as much as anything, a job required a human hand, along with a pair of knowledgeable eyes and the good sense needed to assess a situation and confer with supervisors at the surface. Initially Helvey and Havlena had to walk the bottom to locate the buckles in the pipeline. Then, with his feet braced firmly in the muck, one of them aimed a water jet at the spot to be repaired, using the force of the jet to dig a ditch under the pipeline so that a ring of the pipe's outer surface had clearance all the way around. This particular pipeline, with a fairly typical width of three feet, and most other such pipe was coated with a protective shell of tar and concrete several inches thick with steel mesh, like thick chicken wire, embedded within it. With sledgehammers and axes, Helvey and Havlena took turns hacking away at the shell until they had cleared a ring of bare, uncoated steel around the pipe. Taking repeated whacks at the pipe through the water, while wearing restrictive hot-water suits and helmets, was exhausting work and could go on for hours.

Once the coating was removed, they could start cutting out the damaged section. An electric torch was lowered from the surface and switched on and off by surface tenders—"make it hot," a diver like Helvey would say, in his Chipmunk voice, when he was ready to start "burning" the pipe. When he pulled the torch's trigger, a stream of pure oxygen shot through a steel rod in a blast of heat of up to twelve thousand degrees Fahrenheit, turning metal to mush, even underwater. A diver who was a skilled burner might cut his way around a sizable pipe in under a half-hour, although the job could take longer. All the while, the diver-as-burner was juggling a combination of heat, oxygen, and hydrogen that could explode if he wasn't careful because of the ambient high pressure of his underwater surroundings—yet another reason this deep work wasn't for everyone.

When the cutting was finished, Helvey and Havlena removed the buckled sections by hooking them up to lifting lines lowered from the barge, coordinating the procedures with topside. A two-diver saturation team typically split an eight-hour shift. While one worked in the water, the other kept an eye on things from inside a transfer capsule, a pressurized pod much like the PTC that had transported the aquanauts to and from Sealab III. This familiar kind of capsule was called a bell, short for diving bell, as various incarnations of diver elevators had been called for years. Somewhere mid-shift the divers usually traded places, and the partner who had waited in the bell picked up where the man in the water had left off.

Paul Havlena was wrapping up a shift outside the bell that day, his lean frame encased in the latest hot-water-infused wet suit. (Pumping hot water into a suit had proved more effective than heating it with electricity.) Havlena wore a dive helmet, though not the iconic, bulbous hardhat, but one that looked like the space helmets seen in movies—bucket-shaped fiberglass with a square glass pane set into the front. It was a new model, with two-way communications built in, a feature that had only recently been perfected. Trailing Havlena was a duct-taped braid of umbilical hoses; one fed hot water to his suit and two others pumped breathing gas to and from his helmet as part of a gas-recycling system.

A saturation diver was usually pretty tired out toward the end of a shift like this, ready to climb back into the bell and ride it like an elevator to a decompression chamber at the surface. Everything seemed to be fine as Doc Helvey waited inside the bell for Havlena—until a helium-spiked cry for help over the intercom shattered the calm. Then came silence. Alarmed, Helvey started pulling in his buddy's umbilical, hand over hand, through the sloshing open hatch in the floor of the bell. The bell was hanging thirty feet or so above the seabed, and Helvey strained against the combined weight of his buddy, his buddy's gear, and all that hot water in his suit.

Another diver essentially parachuted in to help by making an unusually rapid descent from the surface, the kind that can pop eardrums or worse. He followed a connecting line down to the bell, like a firefighter sliding down a pole thirty stories tall. He found Havlena stuck around the clump, a weight that hung below the bell, and freed his tangled umbilical. Once that was done he and Helvey were able to pull Havlena into the bell. As their pod was being winched up to the surface, they gave him mouth-to-mouth resuscitation and heart massage, a distressingly similar scenario to Berry Cannon's last dive, four years earlier. Cannon's buddies had held out hope, however fleeting, that their unconscious friend might pull through, but Doc Helvey knew right away that his buddy was already gone.

Taylor Diving researchers later determined that Havlena's breathing gear, a new gas-conserving design that pushed and pulled gas through dual umbilicals, had malfunctioned. The pressure imbalance in the dive helmet had caused Havlena's lungs to collapse, and that's what killed him. But even after rigorous lab tests, the source of the problem remained a mystery, much like the mystery of Cannon's empty canister.

In the hours after the accident, all Doc Helvey knew for sure was that his friend was dead. As was done with Berry Cannon, Havlena's body was locked out of the main decompression chamber, which sat on board the pipe-laying barge at the surface. Helvey remained inside the chamber, where he would spend several days decompressing, wondering whether he should ever dive again. Paul Havlena's death, while tragic, would be one of just two reported in the North Sea that year. Some later years would see more fatalities, but statistics suggested that for all the elements working against them, commercial divers were no more likely to die on the job than coal miners or construction workers.

The offshore industry joined in the efforts to crack existing depth barriers in anticipation of sending divers down to even deeper subsea fields, a thousand feet below the surface and perhaps deeper still. If anyone had asked Doc Helvey that day whether he thought he might eventually dive to work at a thousand feet, he would have shaken his head in disbelief, and not just because of the inexplicable accident that had just killed his friend. A thousand feet was almost a fifth of a mile beneath the surface, and some thirty atmospheres away. Hannes Keller's ill-fated dive years before had underscored the awesome nature of the thousand-foot barrier. It was perhaps possible to break it—in the open sea and not just in a laboratory—but a working diver had to be able to remain there safely for hours, not just a few troubled minutes.

The undersea workforce had already come a long way since old-style hardhat divers took on the first jobs in the nascent offshore industry, which was born in shallow waters not so long after the first oil well was tapped on land in 1859, in Pennsylvania. As the nineteenth century came to a close, piers were being built to reach underwater oil fields just beyond the shorelines of Southern California. A similar evolution soon began in the Gulf of Mexico. Drilling initially took place in the swamps, bayous, and shallow coastal bays along the Louisiana coast. Divers were not in great demand until after World War II, when more sophisticated, stand-alone platforms began to appear, both off the California coast and in the Gulf of Mexico. Depths were modest in those early years, and most dives could be handled the old-fashioned way, with hardhat gear and nothing more exotic to breathe than compressed air. Out of necessity, some divers might push the conventional depth limits to three hundred feet or more, mainly by summoning well-honed powers of concentration to fend off the inevitable nitrogen-induced narcotic haze just long enough to finish a specific task.

The use of helium, which the U.S. Navy demonstrated during the *Squalus* rescue in 1939, was introduced commercially in the early 1960s to meet an increasing need for deeper, clearheaded dives, especially along the coast of Southern California, where the sea floor typically slopes downward much more precipitously than along the Gulf Coast. The early diving companies sometimes took approaches similar to that of Hannes Keller, devising their own methods—novel breathing gas mixtures, revamped decompression tables, customized gear—in order to maximize the brief bottom times possible when making deeper conventional bounce dives from the surface. A few companies continued to push the bounce diving limits down to depths of four or even six hundred feet or more, for jobs that were not terribly time-consuming. Yet for deep jobs that would take hours or days to complete instead of minutes, there was a case to be made for this new method called saturation diving. A marine version of the space race, fueled by business interests, got off to a fast start.

As Sealab II was getting under way in 1965, the first commercial saturation dives were about to be carried out by the Underseas Division of Westinghouse. Westinghouse engineers took a tip from the work being done for Sealab and built a saturation system called Cachalot—French for a deep-diving whale. The main difference from Sealab was that the Cachalot system consisted of everything but a sea floor habitat. For the jobs Westinghouse had in mind, there seemed to be no reason to go to the trouble and expense of housing saturation divers underwater. The divers could live, austerely and under pressure, in the kind of decompression chambers used on the Sealab support ships. Then, rather than living in a sea floor habitat, they could just commute to and from their underwater job site in a pressurized diving bell.

The Cachalot system, with its special bell that could be mated to a chamber, was first used in 1965 to make repairs to the Smith Mountain Dam in central Virginia. A Connecticut outfit called Marine Contractors Inc. supplied a team of eight divers who rode inside the bell in pairs to the foot of the dam at locations as deep as two hundred feet. There they worked on damaged trash racks, the large steel grates that filter water as it passes through the dam. Within three months the Smith Mountain saturation divers, with only a few close calls, had successfully demonstrated the usefulness and efficiency of saturation diving in the commercial sector. By more conventional means, this job might have taken two years and cost a lot more.

Westinghouse's initial enthusiasm for saturation diving eventually faded as its corporate focus changed, but in the meantime the company got another early jump on the competition. The next year its Cachalot system was used to make the first commercial saturation dives at sea. Working at depths down to 240 feet in the Gulf of Mexico, three pairs of divers from Marine Contractors were saturated for six days each to dismantle the remains of two Gulf Oil well structures that Hurricane Betsy had torn apart as if they were Tinkertoys. The hazardous and otherworldly demolition job required the divers to close eight wells with cement, attach hooks and explosive charges, and do ample cutting to clear out the tangle of debris.

A year later, in 1967, Westinghouse staged a demonstration forty-five

miles off Grand Isle, south of New Orleans, in which two divers were
saturated at 350 feet. As part of the demonstration, they were lowered
in their bell for an excursion dive and the divers cracked another barrier
by working for an hour at a daunting six hundred feet before returning to
350 feet. Other companies soon followed these examples, and a few years
later when Doc Helvey made his inaugural saturation dive in the North
Sea the bell-and-chamber approach to commercial saturation diving,
without any habitat, was the standard.

The friendly rivals Bond and Cousteau had set the stage for saturation
diving, yet neither of them jumped headlong into the offshore industry.
They preferred undersea projects that focused on science and exploration
rather than commercial applications. For Ed Link, however, industrial interests were never far from his entrepreneurial mind and by 1964 he helped
set up a new company, Ocean Systems Inc., to develop a whole range of
underwater equipment for commercial use, and to provide services like
saturation diving.

Ocean Systems was an early corporate entrant into the modern offshore industry and a direct outgrowth of Link's Man-in-Sea program.
Rather than disband the loose-knit team of divers, scientists, and engineers
he had assembled for his experiments, Link transformed his team into a
company and formed partnerships with Union Carbide, which was chiefly
concerned with pressurized gases and their uses, and General Precision
Inc., the parent company of Link Aviation Inc., Link's original enterprise.
Among the Ocean Systems personnel that inaugural year were a number
of former Submersible Portable Inflatable Dwelling participants, including
Jon Lindbergh and Robert Sténuit. The Linde Division of Union Carbide,
at Tonawanda, New York, would be home to the new company's diving
research facility. It lacked a test chamber, however, so in a somewhat symbolic passing of the torch, Link had the chamber aboard *Sea Diver* moved
to a vacant Linde building. There, the staff would soon begin to take some
shots at the depth barrier, such as with an experimental day-long dive to
450 feet, with Robert Sténuit serving as one of the subjects.

To broaden Ocean Systems' business ties Link had recruited as a top
manager his old friend E. C. Stephan, the admiral who headed the Deep
Submergence Systems Review Group and was soon to retire from the
Navy. Link would be a board director, a marine consultant, and a minority
stockholder, but he refused to take a regular desk job with the company.
He had turned sixty-one that summer and wanted to be able to spend
time living aboard *Sea Diver*, developing and testing his own inventions.

George Bond, who once accepted foodstuffs as payment from his impoverished backwoods patients, didn't have Ed Link's entrepreneurial drive. After the Sealab program ended, Bond would mainly stay involved in deep sea diving through a smattering of activities more in the mold of Sealab that were focused on scientific ventures and living in the sea. Project Tektite in the Virgin Islands, where Bond paid a visit after the demise of Sealab III, was one such example. But Bond's influence, and that of Sealab and saturation diving, loomed large, not just in terms of methods or technology but in terms of people, especially divers, who migrated from the U.S. Navy to the offshore industry. Like a next generation of pioneers, some brought specific knowledge of saturation diving with them and became leaders in the industry effort to crack barriers and enable divers to work deeper. Among the most notable migrants were Dr. Robert Workman and Ken Wallace. Workman, Bond's learned, low-key decompression maestro, and Wallace, a Navy master diver, had both been assigned to the Experimental Diving Unit when Bond and his New London acolytes showed up in April 1963 to run their week-long, one-hundred-foot chamber test—the second of the three human Genesis experiments.

As a Navy master diver, Wallace understood and appreciated the work of a commercial diver. Diving was practically a Wallace family business. As a teenager he had quit school to work in Boston-area shipyards with his father and four older brothers. After his brothers joined the Navy during World War II, Wallace did, too. He was sworn in on his seventeenth birthday in 1943, and took part in the Normandy invasion the following year. He got out of the Navy after the war but rejoined in 1950, and this time was trained as a diver. By the early 1960s he made master diver, an achievement he considered the most significant of his life.

Both Wallace and Workman were involved with the experimental dives and research being done as part of the Sealab program, bolstering their own knowledge along the way. Others, too, with experience gleaned from the Unit and from Sealab would migrate offshore, including Bob Barth. The Unit itself remained an important research hub, and because it was a public entity it passed along findings about methods and equipment to interested commercial operators, including engineers from Link's Ocean Systems.

By the time of Sealab II, in 1965, Dr. Workman and Wallace were working on the side as consultants to Taylor Diving & Salvage Co. Their friend Mark Banjavich was a retired Navy master diver who had cofounded Taylor in the late 1950s with another former Navy diver, Edward Lee

"Hempy" Taylor, and Jean Valz, a colorful French émigré. Valz was not much of a diver but his personable manner and skills as a classically trained pianist opened some important doors when the three set up an office in New Orleans.

After his partners moved on, Banjavich became Taylor Diving's president and mainstay. He had no personal experience with saturation diving, having left the Navy just as George Bond arrived. But he was a smart and savvy entrepreneur who could plainly see its potential in the oil patch, and that was one reason he brought in Dr. Workman and Ken Wallace. Taylor's rapid growth, and its ability to spread the gospel of saturation diving, came in large measure from its early partnership with the venerable Houston construction company Brown & Root, whose marine group was building offshore oil-drilling platforms and pipelines. Brown & Root turned to Taylor for divers, thanks to the efforts of the genial Jean Valz. While Valz was eking out a living by playing piano at Lafitte's Blacksmith Shop, a famed French Quarter watering hole, he got to know some company managers who frequented the bar and talked up his Navy pals and their excellent new business, Taylor Diving & Salvage.

Similar diving outfits were popping up along the Gulf Coast, all eager to supply underwater manpower to the offshore industry. Some early diving enterprises came and went; others were "mudhole" companies that worked the mucky fringes of Gulf marshland. They often had no more than a dozen employees and limited themselves to shallower jobs that could be handled with basic gear and conventional dives. By the early 1970s there were about a hundred of these smaller companies, but the bigger jobs and deeper dives required a level of expertise, equipment, and resources that only the larger, well-financed dive companies could afford.

With its lucrative ties to Brown & Root, Taylor grew swiftly and the partners moved their headquarters from the *Jesting*—an old schooner that Banjavich and Taylor had fixed up and sailed from New London—into a building near the boat's dock on Lake Pontchartrain. By the early 1960s they had installed a pressure chamber, similar to those at the Unit, for training, testing equipment, and cracking depth barriers. At the time of the Smith Mountain dives in 1965, Taylor was almost finished building a bell-and-chamber system similar to Cachalot—so similar that Taylor, then a mere David to the Westinghouse Goliath, sued Westinghouse for allegedly copying its design. The companies eventually settled out of court. This was business, after all, not just a friendly rivalry or some experimental fantasy, and words like "proprietary" were seeping into the lexicon. Taylor Diving tried out its own saturation system in the Gulf of Mexico in the summer of 1967, just as Westinghouse had launched its demonstration off Grand Isle. Ken Wallace and Bob McArdle, another Navy master diver who was among the early Taylor recruits, trained company divers at Taylor's test chamber and then supervised the company's first working saturation dives, which reached depths of 220 feet.

Two years after Taylor's Gulf debut, the company deepened its stake in saturation diving with the opening of its Hydrospace Research Center, a multimillion-dollar pressure complex at its new headquarters in Belle Chasse, on a canal just south of New Orleans. The new complex had chambers that could simulate depths to 2,200 feet, deeper than any private or military manned test facility in the world. Whether divers would ever be able to work at such depth was an open question, but with test chambers like this one, the answers might be found, and more barriers broken. The research center was also a place for training divers, testing equipment, and doing demonstrations to sell clients on Taylor techniques.

In 1970, a year after the Hydrospace Center opened, Dr. Workman retired from the Navy and took over as full-time head of Taylor's burgeoning research division. Ken Wallace, who had already retired from the Navy and joined Taylor full-time, had risen to vice president of the company and within a few years would take the reins from Banjavich as company president. Taylor, with its many Navy ties, would add many former Navy divers to its payroll. Yet at Taylor and elsewhere there were also plenty of divers trained in civilian diving schools, and others who, as Ken Wallace once did, learned on the job in commercial settings. The bigger companies usually required a period of apprenticeship, even for a first-class Navy diver like Doc Helvey. Clannish tensions occasionally bubbled up between ex-Navy divers and those who came up through the civilian ranks, but everyone knew that in the water cooperation was essential.

Whatever the background of their personnel, American-based diving companies like Taylor, Santa Fe International, Sub Sea International, Ocean Systems, Oceaneering International, J. Ray McDermott & Co., and Martech International became some of the biggest providers of underwater manpower. But there were others emerging around the world, including a formidable company founded by a onetime protégé of Jacques Cousteau named Henri Germain Delauze. As an adventuresome young diver in the early 1950s, Delauze had seized an opportunity to offer his services to the celebrated explorer. He spent several years with the

Cousteau team, but personality conflicts developed as the tenacious and ebullient Delauze angled for a higher-profile role. Delauze did not fit the mold of Cousteau's inner circle of skilled sailors and divers, many of whom stuck by their famed boss for years. They were grateful to be part of the burgeoning Cousteau legend and the adventures that came with it. They also tended to be a modest, easygoing bunch, or were at least temperamentally unfazed by living in Cousteau's long shadow. Henri Delauze was not hardwired for life in a shadow. When the *Calypso* sailed on the journey that would become the basis of *The Silent World*, Delauze was left behind at the office in Marseille.

Delauze gave up on the Cousteau team and at age twenty-seven took a post with a big French civil engineering firm, making a name for himself while running the underwater operations for the construction of a highway tunnel under the port of Havana, Cuba. Delauze received a prestigious Fulbright scholarship and spent a year studying at the University of California at Berkeley. By 1961 he was ambitiously preparing to parlay his diving and engineering experience into an offshore enterprise, perhaps even a legend, of his own.

Delauze met again with Cousteau, this time in Monaco, where Cousteau had by then been director of the Oceanographic Museum for several years. Delauze told JYC about his plans for a company specializing in diving and other services for the offshore industry. Cousteau countered with a tempting offer to put Delauze in charge of U.S. Divers, the American manufacturer of the Aqualung and other diving gear. It was a legitimate opportunity and few would have said no to Cousteau, who had just made the cover of *Time* magazine. But Delauze had his own plans, which began with renting a couple of small offices and a warehouse near the Old Port in Marseille, not far from the Cousteau team's main research hub. The Old Port offices would be the first headquarters for Delauze's company, the Compagnie Maritime d'Expertises, or COMEX. Now Delauze would take diving to depths that Cousteau could only dream about.

Where the cause of pursuing deeper dives was concerned, Delauze was not strictly business; he indulged his pioneering streak. The ability to send divers to work at deeper depths might prove to be good for business, but for Delauze depth barriers were there to be broken, and he relished the prospect of having his company break them first. He lured some key Cousteau team members to Comex, most notably Dr. Xavier Fructus, a top French physiologist who did work for Conshelf and brought to Comex's research program the kind of intellectual heft that Dr. Workman brought to Taylor's.

Like Taylor Diving and Link's Ocean Systems, Comex had its own test chamber to run experimental dives. Researchers at Comex and elsewhere were taking more direct aim at the thousand-foot barrier. How to get divers safely to such depths and beyond was still unclear. Human subjects had broached the thousand-foot barrier only a couple of times since the tragic Hannes Keller dive, and only briefly. The U.S. Navy made its first thousand-foot saturation dive in early 1968, at the Experimental Diving Unit, as part of the preparations for Sealab III. Soon thereafter Walt Mazzone supervised a thirteen-minute excursion dive from 825 to 1,025 feet, also at the Unit. By the end of the year, the first substantial duration at a thousand feet was reached during a three-day test carried out by the Navy with scientists at Duke University, one of the world's top hubs of hyperbaric research. As experience began to accumulate, researchers could see that when approaching simulated depths of one thousand feet, both human and animal subjects exhibited troubling symptoms, like some new form of nitrogen narcosis—except that it seemed to have something to do with breathing all that helium.

British researchers at the Royal Navy Physiological Laboratory working on deep submarine escape methods in the mid-1960s first observed some unexpected reactions in their test subjects as they reached simulated depths of between six and eight hundred feet. Most noticeably, the subjects began to shake. Their arms, hands, and sometimes their whole bodies twitched and jittered uncontrollably, as if they'd been left out in the cold. This phenomenon was initially called "helium tremors," and researchers puzzled over what was happening, among them Henri Delauze, who personally volunteered to be locked in a chamber for one of the early experimental dives. This would be the first of a series, called Physalie, run at the Comex hyperbaric center beginning in mid-1968. It marked a starting point for studies that zeroed in on the physiological effects of crossing the thousand-foot barrier. Joining Delauze in the Comex chamber was the American hyperbaric researcher Ralph Brauer, who signed on as a collaborator with Comex investigators, one of whom was Dr. Fructus, Cousteau's former medical adviser. As Delauze and Brauer hunkered down for their journey across the thousand-foot barrier, the pressure continued to rise, but before the duo got to the equivalent of about eleven hundred feet, Brauer began to experience the telltale signs of muscle tremors and visible shaking. They reversed course after a matter of minutes at their maximum depth.

As the Comex series and other experiments continued, still more troubling signs emerged when human subjects were exposed to the pressures of four hundred pounds per square inch or more at depths of a thousand-plus feet: lethargy, fatigue, sleepiness, dizziness, nausea, vomiting, even delusion and hallucination. Researchers were finding that the high pressure was somehow causing a hyperexcitability in the central nervous system. Neural transmitters seemed to be rapid-firing and synapses overloading, which could lead to convulsions—a nasty lingering threat that had been seen in animals exposed to very high pressure. As with nitrogen narcosis, symptoms varied from diver to diver—Delauze did not exhibit the same pronounced tremors as his American chamber mate.

Scientists debated what to call this new physiological phenomenon—"helium tremors" wasn't quite right—and a name finally stuck: High-Pressure Nervous Syndrome, or HPNS. HPNS was physiologically the opposite of nitrogen narcosis. It was like a shift into neural overdrive that produces wide-ranging effects, often inexplicable and counterintuitive. Narcosis, on the other hand, slowed all systems down, ultimately causing a person to pass out. While problematic in its own right, narcosis was at least fairly predictable—a martini-party drunkenness that could be alleviated either by substituting helium for nitrogen, or just by returning to a lesser depth. HPNS was largely unpredictable, like drinking some kind of LSD-laced triple espresso. And whether in the sea or in a test chamber, a badly stricken diver saturated at thirty atmospheres is a long way from a hospital, literally days of decompression away.

The Comex deep-diving series Physalie, begun with Delauze and Brauer, ushered in an active period of latter-day Genesis experiments, at Comex and elsewhere, in search of a solution to HPNS. Volunteer test subjects and scientists at the top research centers in the United States, Britain, and France tried out various forms of alchemy, mainly different gas mixtures and pressurization rates, to get divers beyond thousand-foot depths without causing them to shake, vomit, convulse, or anything else. In hindsight, it might appear that HPNS was to blame for Berry Cannon's demise, but at six hundred feet, Cannon would not yet have been exposed to sufficiently high pressure to bring on HPNS-induced convulsions.

Slowing the rate of pressurization en route to great depths was one approach that seemed to help divers dodge HPNS, but that could also make deep dives less practical. Not only would a diver face a long decompression after a dive—about ten days from a thousand feet—but it might also take a few days before the dive to ease past the thousand-foot barrier. Even the successful tests, like the one to eleven hundred feet that Dr. Workman ran in 1970 at Taylor Diving's new Hydrospace Research Center, showed that with HPNS there were no guarantees. But there was still hope. A couple of years later, in mid-1972, Comex researchers reached an astonishing new mark by pressurizing two men to 610 meters, a simulated depth of slightly more than two thousand feet. The divers were held at that depth, equal to more than a third of a mile, for an hour and twenty minutes—not terribly long, but a major breakthrough. This exposure of human beings to life at sixty atmospheres, under some nine hundred pounds per square inch of pressure, constituted the deepest experimental dive to date—and got the Comex divers featured in Rolex advertisements for wearing the ultra-pressure-proof watch called the Sea-Dweller.

Within a year, the U.S. Navy borrowed Taylor Diving's research chamber for a similarly deep test of its own—an indication that, despite Sealab's demise, the Navy had not lost interest in deep saturation dives. The Navy's need to borrow Taylor's state-of-the-art chamber also underscored the leading role the private sector had taken in furthering deep-diving and saturation research. At Taylor, Navy researchers pressurized six divers to a depth of sixteen hundred feet. To simulate a real dive to such depth, the test subjects went in and out of the wet pot for seven days. An eighteen-day decompression followed—nearly three weeks confined to a chamber the size of a walk-in closet, breathing an ever-changing blend of gases. The divers came out all right, but they had suffered many of the troubling, unpredictable effects of HPNS, including a newly identified one: dyspnea, characterized by labored breathing and the frightening sensation of not getting enough oxygen with each breath, as happens in thin air at high altitudes. The perplexing fact was that blood samples from the test subjects showed normal arterial gases. There was no actual shortage of oxygen, just the sensation—and another mysterious neural effect of HPNS.

A number of latter-day Genesis experiments like this one continued to focus on the persistent problem of HPNS and on cracking the thousand-foot barrier, but industry titans also wanted readily applicable methods and equipment. One major innovation took the concept of a sea floor habitat and cleverly adapted it to specific commercial needs. It was a hybrid habitat that had the portability of Link's inflatable dwelling and the sturdy, roomier structure of Sealab. This new industrial habitat provided a dry, pressurized shelter, but it was intended as a workshop only. Saturation divers would still spend their off-hours in chambers at the surface, under pressure. They could then commute by bell to the sea floor, set up a habitat, and spend entire shifts working inside until the job was done. Engineers

at Ocean Systems produced and tested the first one, in 1967, but Taylor
Diving was right behind them and very soon came out with its own, as did
some of the other larger companies, like Comex.

There were some differences in design, but these hyperbaric work-
shops, called underwater welding habitats, were all built with a specific
purpose in mind: the essential lining up, cutting, and especially the
welding together of pipeline sections. None of these utilitarian welding
habitats was as eye-catching as Jacques Cousteau's Starfish House, and
each was just one-quarter the size of Sealab. But they got the job done,
and brought about a useful variation on the theme of Dominion over the
Seas. The typical habitat looked like a steel shed, about twelve feet square,
often with a slightly domed roof and arched slots like the doors of a dog-
house on two facing sides. The slots fit over a pipeline lying on the sea
floor, similar to the way the fenders of a car fit over the wheels. Divers on
the bottom coordinated with the surface ship as the habitat was lowered
so it would not damage the pipe—a potentially million-dollar mistake.

Most habitats were designed with no permanent floor, so each one
was like a box-shaped bucket turned upside down. The two divers who
guided the hyperbaric hut into place had a few more preparatory steps to
take before a breathable artificial atmosphere was shot in. The incoming
gas forced out the seawater that had filled the habitat when it was put on
the bottom. Bubbles burst forth through a grated floor, which the divers
put into place, until all the water was expelled, leaving the habitat dry. An
exposed section of pipe, or often the ends of two pipelines to be welded
together, ran through the middle of its austere interior.

Inside a habitat, divers who had trained as welders—or sometimes
welders who had trained as divers—joined the sections of pipe. Underwater welding techniques had markedly improved since World War II,
but conventional underwater welds, while fine for some jobs, could not
meet the more stringent standards set for the far-flung arteries that carried the lifeblood of the offshore oil industry. That's why pipelines were
typically pieced together and welded, assembly-line style, on pipe-laying
barges, like the one from which Doc Helvey made his first saturation
dives. Several sections of pipe, each about forty feet long, could be lined
up, welded together, and X-rayed to verify the quality of the welds. Once
that was done, the newly fused pipeline was pushed off the barge's stern
into the hot dog bun of a pontoon. From there, divers helped ease the
pipeline onto the seabed.

But the logistics of offshore pipe laying still left numerous pipe ends that could be joined only after they were laid on the seabed, and that's where the habitats came in—once Ocean Systems engineers figured out that welds of equally high quality as those made on a lay barge could also be made under pressure, as long as the welding was done somewhere dry. One common and crucial underwater pipe connection, for example, which became a big moneymaker for Taylor, was between a pipeline laid on the seabed and a vertical conduit that ran like a giant downspout from the top of an offshore platform. At the foot of the platform, this vertical pipe section turned at a right angle, its open end facing out to sea and in the direction of the end of the pipeline lying on the seabed. There were also times, usually to speed up the pipe-laying process, when more than one barge would lay long segments of a pipeline, and those segments would then have to be lined up and connected out at sea. Just as Doc Helvey acted as ringmaster to coordinate the hooking up and removal of damaged pipe sections, a diver on the bottom had to coordinate the lining up of pipes, giving instructions to crane operators at the surface to yank a tethered pipe end, weighing tons, one way or the other.

This process of lining up two ends of pipe had to be executed with nearly improbable precision, given the heft of the pipe and often shifting seas that further complicated the task. There could be no slamming of pipe into pipe, and the job wasn't done until the two circular ends faced each other almost perfectly. On a smaller scale it would be like trying to line up two ends of garden hose by attaching a string to the end of one hose, climbing to the roof of a tall building, then taking instructions from someone down below as to which way to pull the string. Ideally there would be no stiff breeze—the equivalent of a strong sea current—to make the alignment even harder to get right.

Once two pipe ends were properly aligned, the diver wielded a bazooka-like hydraulic wrench and started bolting together the lined-up flanges, the collars around the end of a pipe with holes for bolts. The drawback to flanges was that they weren't as long-lasting as welds, and had a greater risk of leaking. This put potentially weak links in an otherwise scrupulously welded pipeline. But then the underwater welding habitat was brought to market, and soon followed the discovery of oil in the North Sea. Taylor Diving found a major market for its habitats and became the dominant provider of hyperbaric welding, and Comex evolved into one of its top competitors.

Because the precise lining up of pipe was critical to the subsequent welding process, Taylor and others came up with variations on another key piece of heavy equipment. This setup featured two pairs of mechanical vise grips sized to grab pipe up to about four feet in diameter. The grips hung like lobster claws from the rafters of a steel structure resembling the framework for a big three-car garage. Taylor called its seventy-foot-long version a Submersible Pipe Alignment Rig, or SPAR. With divers on the bottom, the SPAR could be lowered from a barge over two ends of pipe. A welding habitat fit into the center of the SPAR framework, and a diver operated the big clawlike grips on either side of the habitat the way a construction worker would operate heavy machinery. A second diver gave directions to the one at the controls and the two of them worked together to move the pipes around until they lined up inside the habitat. Then the divers could both get to work in the pressurized dry confines of their sea floor workshop.

The first thousand-foot contract came in the early 1970s from Conoco for exploratory work in the North Sea. No one had ever worked at that depth, but Delauze and Comex had spent more time and money than most preparing to crack the thousand-foot barrier. Comex was clearly a leading candidate, and Delauze was dismayed when the contract went to Oceaneering International, an agile competitor formed a few years earlier by a union of upstart diver-entrepreneurs out of California and Canada.

Fast-growing Oceaneering was a study in the offshore industry's frontier spirit. Its California-based founders, a posse of former abalone divers, broke with Ocean Systems after deciding they didn't care for its corporate attitude or the retired admiral that Ed Link put in charge, his friend E. C. Stephan, the former chief of the Deep Submergence Systems Review Group. When Oceaneering won the contract for the first thousand-foot dive, it had never before put divers to work at such depth. The company principals had to play catch-up just to get a saturation diving system built that would handle a thousand-foot job, but as it turned out the dives for Conoco never exceeded about seven hundred feet.

Within a couple of years, in 1975, Henri Delauze and Comex were the first to work at a thousand feet after all. They were hired by BP Canada to recover a hefty stack of valves about thirty feet tall that looked something like a small rocket pieced together from fat fire hydrants—a blowout preventer. The assemblage was worth $10 million and had to be abandoned off the coast of Labrador before it could be properly placed on a sea floor wellhead to serve its fail-safe function of controlling pressure. Attempts to raise it had failed. Comex divers worked from out of their bell, which was lowered from its mother ship to the job site at a record working depth of

1,070 feet. The six divers, working in two teams, made several dives and spent four hours in water that was a few degrees below freezing. Warmed by their hot-water suits, the divers were able to work through the check-list of mechanical preparations they had practiced in the Comex test pool back in Marseille. They then went about hooking up the heavy fixture for its successful lift back to the surface. Comex even shot and produced an in-house film about the project. The footage would never have anything like the marquee value of a Cousteau show, but it was another testament to the achievements of Henri Delauze, France's less celebrated diver.

Within a couple of years divers were gearing up for another series of thousand-foot-plus dives that were also notable for durations far longer than those at Labrador. Shell Oil had an offshore structure of unprec-edented size in the works that would require days of undersea labor in the Gulf of Mexico. The deepest offshore platform ever built, called Cognac, was about to be put together on the sea floor. The platform was named for a prospect Shell had acquired and it was to be built at a site about fifteen miles south of the Mississippi River Delta.

Cognac's main supporting structure, known as the "jacket," was going to be set on the bottom of the submerged Mississippi Canyon, some thirty atmospheres away, at a depth of 1,025 feet. The original plans for Cognac called for much of the deep work to be done remotely, by mechanical devices, hydraulics, and electronics. Divers were to be more of a backup brigade for system failures. Instead, with the growing confidence born out of diving research and oil field experience, it was decided that divers should be put on the front lines. Taylor Diving & Salvage won the coveted contract for the Cognac diving operation, though its parent company, Brown & Root, lost the bid to build the jacket structure, a prestigious deal worth $275 million. That honor went to J. Ray McDermott & Co., a major competitor in marine construction, which had a diving subsidiary of its own, thereby creating an unusual alliance of not entirely friendly corporate rivals.

The Cognac jacket, a densely latticed steel tower, was built on land, just west of New Orleans near Morgan City. The jacket would be too tall to handle and transport in one piece, so it was designed in three sections—base, middle, and top—to be stacked one at a time. Beginning with the base, each section of the prefabricated jacket was to be floated into the Gulf on football-field-sized barges from shore and then lowered to the designated site on the canyon floor. When fully assembled, the Cognac jacket would look like an industrial version of the Eiffel Tower, its matrix of steel beams

perhaps less ornate but every bit as intricate. With the platform and der-
ricks that would top the jacket above the waterline Cognac would reach a
height of more than eleven hundred feet, taller than the Parisian landmark
it evoked. At nearly sixty thousand tons, Cognac would weigh five times as
much as the Eiffel Tower, a heft designed to support two drilling rigs and
withstand both the everyday surges of a thousand feet of seawater and the
occasional wrath of a Gulf hurricane.

Two saturation diving systems, each with its bell and chamber, were installed aboard a McDermott barge, along with a supply of several hundred
thousand cubic feet of breathing gases. It was the largest saturation spread
ever assembled. Through a rigorous screening process worthy of NASA,
Taylor selected twelve divers, half of whom might as well have been called
the Cognac Six. They never thought of themselves in such elite terms, but
like the Mercury Seven astronauts who rode the first rockets into space,
the Cognac Six divers would be the first of the twelve to ride a bell down
to a thousand feet for a series of day-long shifts scheduled to go on for a
month. Fittingly, this 1977 dive coincided with the twenty-year anniversary
of George Bond's rejected "Proposal for Underwater Research," although
it's unlikely anyone knew to mark the date. The Cognac Six, along with
the six others who would follow them, plus one tender for each group and
two dozen members of a topside support crew, were chosen from a field
of more than four hundred experienced divers and supervisory personnel.
One of the Six was Doc Helvey.

After the death of his buddy in the North Sea, Helvey had seriously
considered quitting diving for good. With encouragement from fellow
divers, including Ken Wallace, then Taylor's vice president, he decided
to stick with it. He enjoyed the otherworldly work, and over the next
few years spent many more hours in saturation. Still, when the need for
divers on the Cognac project was announced, Helvey did not rush to
the front of the line. It occurred to him that this job might just be too
risky, too deep, too dangerous, not worth it—although the money would
be incredibly good. As a diver for Cognac, Helvey would gross almost a
thousand dollars a day.

Fat paychecks aside, diving to work at a thousand feet, for days at a
time, was not merely a job. It was more like a *mission*, a deep-sea version
of . . . the moon landing. Helvey didn't want to miss out. Risky, yes, but
risks of all kinds came with any underwater job. Doc Helvey knew that
as well as anyone, and he was also the designated medic for his six-man
Cognac team. If anyone started to exhibit any signs of HPNS, it was up to

the topside controllers to try to alleviate the symptoms by adjusting the pressure and gas mixture, but there were other reasons to have medics on a diving team. Helvey once saved another diver's badly crushed finger by deftly stitching it back together while saturated in a chamber at seven hundred feet. From that depth and pressure it would take the better part of a week to decompress and get to a hospital. How wonderfully fleeting a typical spaceflight was by comparison. As Cognac got under way, Doc Helvey couldn't help but wonder who would tend to *him* if he had a medical emergency. But that unlikely scenario wasn't going to change his mind about taking on the job and the *mission*.

So on a fall day in 1977, the Cognac Six and their tender were sealed into a spartan chamber barely the size of a school bus. Laboratory tests had shown that a slow compression seemed to be the best way to avoid HPNS, so over the course of a full day, the pressure edged up toward 450 pounds per square inch. Because gases become more concentrated under pressure, the oxygen content at a thousand feet had to be held close to a paltry one percent to approximate the normal percentage at sea level. Determining and maintaining safe oxygen levels had been an early challenge for Bond, Mazzone, and Workman during Genesis. Too much oxygen was of course toxic, and at a thousand feet, with so little oxygen in the mix, there was even less wiggle room than at lesser depths. Most of the rest of the artificial atmosphere pumped in consisted of helium.

When Helvey's turn to dive came he made the hour-long descent to the bottom in a diving bell with his partner, Ronald Schwary, an old friend. Schwary was one of the Gulf Coast frontier divers who had found ways to get his offshore training on the job. The two divers were already wearing their hot-water suits. Pretty much all that remained to be done before leaving the bell was to pull on their helmets, a popular design with a snug-fitting neoprene hood affixed to a face mask that looked borrowed from a motorcycle helmet. Helvey dropped through the open hatch in the floor first, falling through liquid space as if he'd taken a slow-motion leap from a rooftop.

He landed on the bottom and to his surprise sank to his knees in fine silt. He had never experienced such a thick sedimentary blanket. Then there was the darkness—not just darkness, but a total absence of light, except for the gauzy glow from the bell overhead and his own handheld light. The sound of his own breathing was more resonant because of the snug fit of his helmet. With each exhalation Helvey produced a burst of bubbles—this rig did no gas recycling—but breathing gas was pumped directly to his headgear through an umbilical. Scuba tanks were not an option at such depth and pressure, where each breath sucks far more gas out of the tank than at lesser depths. At thirty atmospheres, even a large tank would be rapidly depleted. This is why Helvey hoped that he would not be needing his "bailout bottle," a small tank that could be switched on in an emergency. At this depth it wouldn't last more than a few breaths—barely enough to swim for the bell. Direct surfacing was obviously out of the question.

Each Cognac diver, upon leaving the bell, would have his Neil Armstrong moment, even if their first steps on the bottom weren't televised. Now it was Helvey's turn, and he kicked up a particulate fog as he shuffled along the seabed. The current was just strong enough to blow the silt clouds away, but fortunately not strong enough to thwart his progress. When the fog lifted, the visibility was spectacular, when there was enough light to see. When Helvey shone his light into the void, the water seemed to twinkle like a starry night sky. That twinkling turned out to be reflections from schools of silvery ribbon fish. They were twice as big as any Helvey had ever seen—up to ten feet long. Spooky-looking angler fish, that toothy, jut-jawed species resembling a swimming skull, also appeared from out of the dark. Most astonishing was the sheer size of the grouper, a great frowning football of a creature that looks pretty much the way a child would draw a fish. The grouper were huge, like Volkswagen Beetles with fins, and could swallow fish that were about as big as Helvey.

When Helvey took a first pneumofathometer reading, the depth gauge said he was standing 1,030 feet below the surface—five feet deeper than anticipated. The slightly greater depth wasn't terribly significant, although Helvey was tickled to find out that he was even deeper than he thought he would be. For the moment, Doc Helvey was the deepest man in the world. More than thirty atmospheres! It was a breathtaking depth, with seawater applying some 440 pounds per square inch of pressure. Not only that, but Helvey and the rest of the Cognac Six—Schwary, Alan Andersen, Mike Cooke, Fred Miller, and John Propeck—were going to be spending more time working at greater depth and pressure than Homo sapiens had ever spent before. And they would have much more to do than plant flags or collect rocks. Survival, too, could be counted as a major achievement—putting six men down and getting six back, to paraphrase Captain Bond's sober remark about Sealab II's greatest success. But for these divers survival alone would not suffice. For one thing, if their work didn't prove to be useful enough, market forces would drive them out of business.

Helvey and the others spent their first shifts going through what amounted to an industrial dress rehearsal on the seabed. A barge had been sunk that was filled with mock-ups of the various tasks to be performed on the jacket. All the tools and procedures had been previously tested, in chambers and on shore, but this final run-through gave the divers an opportunity to make certain that everything worked the way it was supposed to. Burning at a thousand feet had never been done before, for example, but they confirmed that the technical hurdles had been overcome. Helvey was not the only diver who initially experienced some nerve-racking bouts of dyspnea, still a mysterious symptom of HPNS, but the divers learned to pace themselves in a way that eliminated their shortness of breath and the disconcerting sensation of lacking oxygen even when lab tests showed that they had all that they needed.

Following the seabed rehearsals came the lowering of Cognac's massive square base, a three-acre forest of crisscrossing steel beams, the thickest ones vertical, like the trunks of old-growth redwood trees nearly twenty stories tall. Once the base was lowered to the bottom, the divers filled critical roles as ringmasters, and one of their first jobs was to guide two dozen steel piles, one by one, into individual "sleeves," a vertical length of pipe affixed at intervals around the base. The piles themselves were six hundred feet long, seven feet in diameter, and weighed thousands of pounds. Each one had to be carefully dangled from the surface and slid into its designated sleeve. A pile driver, which the divers helped swing into position, hammered the pile into place, nailing the base to the sea floor. The divers had also learned how and where to hook up the steel lines through which grout would be pumped into the sleeves to cement the piles into place and solidify the base. Throughout the mission the divers were going to be shadowed by a new remote-controlled device that everyone called a "flying eyeball." These unmanned orbs, operated from the surface, were about the size of a big beach ball, functioned as mobile klieg lights, and were equipped with cameras so that topside supervisors could watch subsea operations for themselves rather than rely exclusively on garbled Chipmunk narrations.

In preparation for the insertion of a pile into its sleeve, one of the early jobs, Doc Helvey first had to unbolt a lid from the top of the sleeve, a heavy disk like a manhole cover that was seven feet in diameter. The lids were in place to act as parachutes, slowing the base's descent by preventing water from passing freely through the sleeves. Once the base was safely on the bottom—Helvey took a walk around the steel forest to

verify its positioning—the lids had to come off. The nuts and bolts holding them in place were about a half-foot wide, and heavy. The divers used
hydraulic wrenches to wrestle each one loose, and when the bolts were
out, the divers orchestrated the lid's removal by crane. Because the divers were then working at the top of a sleeve, they were at a depth of only
nine hundred feet or so—*only!* Upon arrival at a lid, Helvey often found
the VW-sized grouper lingering at his workstation. When he shooed them
away, they would seem to vanish into the darkness. But when he shone his
light around, Helvey found them parked nearby, sometimes right overhead. When Helvey peered out of the diving bell one day, he could see
his bell partner, Ronald Schwary, tiptoeing along a horizontal beam. The
grouper tagging along looked as if they had found their pied piper.

During their off-hours, within the austere confines of their chamber,
the Cognac divers played cards, read, ate, slept. Sleep came especially
easily after an eight-hour shift on the bottom. The distinction between
days and nights faded, as was often the case for saturation divers who
made commutes around the clock, from their chambers to the sunless
depths and back. The Cognac Six would live this way for thirty-one days.
The second chamber, right next to theirs on the sprawling barge, housed
the second six, who would be ready to take the next month-long shift.
The two teams would then trade off once more, racking up a total of
more than 120 days in saturation.

If it had suited the needs of the multimillion-dollar project, there's
little doubt that saturation divers could have spent their days and nights
on the sea floor in some kind of industrial version of Sealab. As it was,
each team had to wait out a ten-day decompression to return to the surface, sealed inside their spartan chamber. In a way that was the hardest
part, or at least a reminder that saturation diving was a poor career choice
for claustrophobes. Doc Helvey happily spent much of his decompression playing pinochle with his chamber mates. By the time the chamber
door could finally swing open, he'd won $147—at a penny a point. No
ticker-tape parade was in store, although considering the magnitude of the
dive, of the thousand-foot Cognac mission, a parade might have been in
order, or at least some triumphant headlines. But for the Cognac Six, the
refreshing embrace of a Gulf breeze would have to suffice.

After the jacket's base was set, the middle and top sections were lowered into place beginning in the spring of 1978. The base had required
the deepest work, but the process of attaching the three-hundred-foot-
tall middle section to the base called for dives to about eight hundred

feet. To put the sections together the divers coordinated the lining up and fastening of jumbo-sized pegs protruding from the bottom of the midsection as they were lowered into matching holes in the top of the base. A Shell Oil spokesman likened the joining of the Cognac sections to docking two spacecraft—an image fresh in the public's mind from the historic Apollo-Soyuz test in 1975.

Once the midsection was lowered onto the base, and after the top section was lowered into place on the midsection, the divers did more plumbing and grouting to secure the structure, in addition to hooking up electrical and hydraulic connections. Working for hours at a thousand feet, or even eight hundred or five hundred feet as sections were added—well, barriers had surely been broken in the two decades since Hannes Keller struggled to do nothing more than plant flags a thousand feet down.

Of course Cognac was never just about achieving milestones in deep-sea diving or in marine construction, although when finished in 1978, ahead of schedule, it was the world's tallest offshore platform. But like many other offshore platforms that boasted impressive feats of engineering and diving, Cognac was strictly business. It was there to produce the lifeblood of the industrial world, and to turn a profit. Drilling soon began on the first of more than sixty planned wells that would feed into the structure, adding to the more than 2,700 offshore wells drilled that year. Another twenty-seven hundred wells were drilled in 1979. Before the 1970s were over, almost a quarter of the world's oil and gas would come from offshore oil fields, and within a few years the offshore contribution to fossil fuel supplies would approach one-third.

As long as saturation divers continued to serve the offshore industry's needs, divers like Doc Helvey might expect to be summoned to ever greater depths. Yet the unmanned, robotic flying eyeballs used in setting Cognac were a sign of an impending change in divers' roles. Such devices evolved into more sophisticated remotely operated vehicles, called ROVs, that were designed to work with divers, and also to replace them, much as robots had begun to replace workers on auto assembly lines. The savings on underwater jobs would be even greater, since ROVs required neither meals nor costly breathing gases. Nor did ROVs have to spend idle days decompressing in artificial atmospheres. Furthermore, an ROV had the same kind of practical advantage as a probe launched into outer space: It could go to depths far beyond the reach of a diver, to oil fields the industry was hoping to explore more than four thousand feet below the surface.

Tapping deeper fields heightened industry interest in ROVs—Taylor added an ROV test tank to its facility in the mid-1980s—but the rise of ROVs also contributed to a decline in demand for saturation diving, as did the global slump in oil prices that hit in the early 1980s, not long after Cognac's completion. At that point the entire offshore industry was experiencing the business equivalent of the bends, and the resulting pain was acute among providers of diving services. The market picked up a bit toward the end of the decade, but by then many divers, including saturation divers, had gone looking for other frontiers. Companies large and small were either going out of business or merging. Taylor Diving & Salvage, which had been such a prominent force throughout the boom times of the 1960s and 1970s, was sold off and disbanded as the decade of the 1980s unfolded. Ocean Systems, which Ed Link helped bring into being, was ultimately swallowed by Oceaneering International, one of the big companies to survive the downturn and thrive, in no small part because of its ROV business.

Despite the painful reversals of fortune, the offshore industry's embrace of saturation diving had done a lot to break depth barriers. Cognac had put a very tall exclamation point on the thousand-foot mark, and companies like Comex, another that weathered the downturn, had not given up on solving the mysteries of High-Pressure Nervous Syndrome or on figuring out how to send divers to depths well beyond a thousand feet. The U.S. Navy, too, carried on the deep-diving revolution, even if the demise of the Sealab program made it appear otherwise. George Bond, Jacques Cousteau, and Ed Link had clearly started something, but their own quests were coming to an end.

16

THE RIVALS PRESS ON

As the private sector began to put saturation diving to use in the mid-1960s, the public sector seemed to be on the verge of doing the same, but with an eye toward science and exploration rather than industry. In 1967, during the period between Sealabs II and III, President Johnson appointed a blue-ribbon commission to consider a range of oceanographic issues and chart a more deliberate course for the nation's undersea activities. Two years later, the commission's fifteen members produced a three-hundred-page report called "Our Nation and the Sea; a Plan for National Action," but often referred to as the Stratton report, after the commission's chairman, Dr. Julius Stratton, a physicist, electrical engineer, and former president of the Massachusetts Institute of Technology. Among the report's key recommendations was the establishment of an independent civilian agency, like NASA, that would also report directly to the president. That brought hope to those who believed that if the United States was going to put the kind of concerted effort, and money, into manned undersea research and exploration that NASA was pouring into human space flight, it needed such an agency—a "wet NASA," some liked to call it.

In addition, the Stratton report made a case for some major undersea initiatives, including one that sounded remarkably similar to a certain "Proposal for Underwater Research." It recommended that over the next decade half a billion dollars be used to establish laboratories on the continental shelf. "This project is based on the premise that, if man is to conquer the sea, he must go into the sea," the report said. "The Navy Sealab and French Conshelf projects have been impressive demonstrations of the ability of man to go into the sea for short periods of time. The next step is to make it possible for man to stay with a degree of permanence and to provide him with facilities for research and development at continental shelf depths."

What a difference a decade had made for the undersea gospel according to George Bond. The report envisioned both fixed and portable labs, and set an ambitious depth goal of two thousand feet for free-swimming divers. If divers could reach that depth, they would have access to both the continental shelf and the next topographic level down, the continental slope, which together comprise a vast frontier known for a richness of resources. To achieve that goal, the commission members recommended that additional millions be spent on research, bolstering private sector efforts to break depth barriers and speed up the downward progress.

In late 1970, the year after the sudden demise of Sealab III and the first oil strike in the North Sea, the National Oceanic and Atmospheric Administration was formed from a smorgasbord of government agencies and departments with ocean-related interests. The acronym NOAA, pronounced *Noah*, was fitting since the new agency's responsibilities would be about as diverse as the animals on the Ark, covering issues from the deep sea to the far reaches of the atmosphere. After Richard Nixon's victory in the presidential election of 1968, NOAA wound up under the umbrella of the Department of Commerce in the new administration, to the chagrin of those who favored the fresh outlook and clout that being an independent agency like NASA could have brought. Still, it was at least a start.

In the meantime, the friendly rivals pressed on in their individual ways. While none of them was immune to the need for funding, they were essentially unfettered by specific industrial or military demands. After Conshelf Three, Jacques Cousteau was dreaming up plans for what seemed to be the next logical step in the quest to live in the sea: a mobile habitat. Conshelf Three had achieved a greater measure of autonomy from the surface, but in terms of mobility, its occupants could only lower their spheroid habitat to the bottom and raise it to the surface. The habitat still had to be transported to a desired location at sea, and once on the bottom, it couldn't go anywhere for the duration of its stay. For in situ scientific research, or perhaps for a salvage job or archaeological excavations, stationary habitats like Conshelf or Sealab could suffice—or perhaps even the copycat habitats that were popping up around the world, many of them projects of poorly funded diving enthusiasts. A couple of scrappy, tanklike habitats initiated in the mid-1960s by diving clubs in Czechoslo-vakia, of all places, were fairly typical. The Soviet Union produced quite a few modest designs, but these and some better-equipped prototypes, including a few American-made habitats like Tektite, were rarely capable of functioning at depths of more than a few atmospheres. Many of them made Sealab I, even with its train axle ballast, look like a wonder of space age engineering.

Cousteau wanted to try to improve on the concept of stationary habitats by building an undersea base that was more like a mobile home, one that divers could swim out of and reenter, an idea that had been around at least since Jules Verne's fantastical *Nautilus*. Cousteau, an inveterate fund-raiser who was never averse to being first, succeeded in selling the French government's oceanographic agency and the French Petroleum Institute on his plan to build a mobile habitat he would call Argyronète, named for a type of diving spider. Argyronète might someday be the underwater equivalent of Cousteau's *Calypso*, and a real-life *Nautilus* submarine equipped with a built-in mini-habitat. Saturation divers would be able to live in that sealed-off, pressurized compartment, much as commercial divers spent their off-hours in chambers. The divers could then swim out of Argyronète down to at least a thousand feet. But Argyronète was running way over budget and its prime sponsors backed out. Cousteau had been able to move quickly to start building his version of Captain Nemo's vessel, but by the early 1970s its hull sat unfinished in a storage hangar on the waterfront near Marseille.

Ed Link, too, had been thinking along the lines of habitat mobility—his inflatable dwelling had been portable, after all, if not entirely mobile—and the energetic inventor promptly produced a first prototype in the mid-1960s, called *Deep Diver*, which may even have been an inspiration for Argyronète. Link's approach was decidedly more modest than what Cousteau had in mind, more like a diver taxi than a fully equipped mobile home, and Link could be reasonably sure to finish what he started. *Deep Diver* was a mini-submarine, like the stout little research sub *Alvin*, launched in 1964 and improved in ensuing years, or *Star I*, the pint-sized rescue sub that visited Sealab I. Mini-subs, also known as submersibles, typically had to be launched from a mother ship. They could move about freely, under their own power, for a number of hours before they had to return to their waiting mothers. But submersibles might achieve even greater feats if saturation divers could come and go from them. This they could do from a small lockout compartment built in to a section of the twenty-two-foot-long *Deep Diver* hull. Link designed the compartment to handle depths down to 1,250 feet—forward-looking indeed, since dives to only half that depth pushed the safe and practical limits of the day.

In 1967, the same year President Johnson set up the Stratton commission and a couple of years after the *Deep Diver* debut, two divers locked out of Link's mini-sub at seven hundred feet in the same Bahamian waters where the portable inflatable dwelling had its trial. The dive lasted just fifteen minutes, long enough for the divers to collect some flora and fauna before sitting out thirty hours of decompression. Even a more conventional dive to seven hundred feet would have been remarkable for the time, and here was a very deep dive, albeit a brief one, that was made from a submersible.

Deep Diver was used for some contract work by Ocean Systems, the commercial diving company Link helped form, but the craft wasn't catching on with industry the way underwater welding habitats had. Structural problems with the *Deep Diver* hull ultimately put the submersible out of commission, but by then Ed Link had plans for an improved vessel he would call the *Johnson-Sea-Link*, in honor of his partnership with his friend and fellow marine enthusiast J. Seward Johnson, the septuagenarian scion of the Band-Aid maker Johnson & Johnson. Johnson had founded the Harbor Branch Foundation in 1970 and Link suggested that it set up its headquarters for oceanographic and engineering research at Fort Pierce, Florida, where Link had bought land to establish his own coastal base, called Link Port.

The *Johnson-Sea-Link*, often called *Sea-Link* for short, was very different looking from *Deep Diver* but similar in size, and once again, equipped with a lockout compartment for two divers. At a length of just twenty-three feet, one-quarter the size of Cousteau's still unfinished Argyronète hull, the *Sea-Link* looked less like a baby submarine than a helicopter without rotors. Rather than encase his vessel in a sleek hull, Link decided to leave its parts accessible for repair and so outdated components could be readily replaced and the entire craft improved over time.

The *Sea-Link*'s forward section was like a two-seater cockpit, with room enough for a pilot and observer. They sat side by side in a transparent acrylic sphere nearly six feet around. The metallic tail section behind the sphere housed the battery-powered motors and the two-diver lockout chamber. *Sea-Link* could reach depths of three thousand feet and was designed to release divers down to fifteen hundred. Link was, once again, aiming well into the future and he had created this craft for under half a million dollars—a bargain compared to the several million dollars Cousteau poured into Argyronète. *Sea-Link*, launched in 1971 and commissioned to the Smithsonian, was made available to marine researchers from universities and scientific institutions. Over the next two years the unusual-looking submersible was used for dozens of successful undersea projects. Then came the dive that would forever disturb Ed Link's peace of mind. It began on Father's Day, 1973.

Link and his wife, Marion, were on board their research boat, the ninety-one-foot *Sea Diver*, which was equipped as the *Sea-Link*'s mother ship. Link added a crane to lift the submersible on and off of the deck for transport. Shortly before ten that morning, fifteen miles out at sea from Key West, they learned that the *Sea-Link* was stuck 360 feet below. Operating a kind of mechanical skewer from the cockpit, the *Sea-Link* pilot was trying to recover a small fish trap from a man-made reef formed by a pair of scuttled Navy ships. In its maneuvers Link's submersible got hooked on a cable.

At the controls was thirty-year-old Archibald "Jock" Menzies, an experienced diver who had piloted the *Sea-Link* on about a hundred previous outings. Seated next to him in the cockpit was Robert Meek, a pressure physiologist who was twenty-seven and on board as a scientific observer. Two others were sealed in the diver lockout chamber at the rear of the craft. One was Albert "Smoky" Stover, the fifty-one-year-old veteran submersible pilot who had visited the site of Sealab I to test the *Star I* mini-sub. The other was the Link's younger son, Clayton Link, who was thirty-one and had joined his parents on many an expedition, including the first saturation dives at Villefranche. He served in the Navy as a diving officer aboard a submarine rescue ship, did a commercial stint with Ocean Systems, and had lately become the Smithsonian's director of diving.

When Ed Link heard his craft was stuck he quickly got on the radio to summon a Navy submarine rescue ship, the USS *Tringa*, which was docked at the Key West base. Another Navy rescue team and a couple of commercial vessels got the Mayday call about *Sea-Link* and they, too, were on the way. With walkie-talkie in hand, his craggy face glistening with sweat under the visor of his cap, Ed Link orchestrated the rescue from the deck of *Sea Diver*. The sea was calm. Ed Link was calm. Navy rescuers were on the way. Link and his *Sea Diver* crew calculated that the atmosphere in the diver lockout compartment ought to last two and a half days, about sixty-one hours. The cockpit's air supply wouldn't last as long, but should be good for forty-two hours. Link was confident that his son and the three others would soon be returned safely to the surface.

As the minutes dragged into hours, it would become agonizingly clear that the breathing-gas supply would last only about half as long as they initially thought. Murphy's Law seemed to be governing virtually every aspect of the rescue. The *Tringa* was delayed getting to the site and it was almost eleven o'clock that night before two Navy hardhat divers, wearing bulbous Mark V helmets and cumbersome old-style dive suits of canvas and rubber, were ready to descend to the *Sea-Link*. Floodlights illuminated patches of dark seawater. The divers made it to three hundred feet, still a couple of atmospheres shy of the ensnared submersible, but they couldn't get through the wreckage of the scuttled ships. They had to turn back.

Pilot Menzies and physiologist Meek had no choice but to wait for help. There was no way to open the hatch on their acrylic sphere underwater. Stover and Clayton Link twice considered locking out of their compartment, first before the *Tringa* arrived and again several hours later. They hadn't planned to dive outside the submersible during this brief, routine trip and wore only shorts and T-shirts. Still, they could either try to free their sub while making some breath-holding swims, or they could escape to the surface by doing a blow and go, like Bond and Cyril Tuckfield years before, when they made their record submarine escape in these same waters, from almost the same depth. But Stover and Link were already feeling chilled in their aluminum compartment and didn't want to risk further complications by swimming into very cold water, a strong current, and the tangle of debris. Help was on the way, after all. Ed Link, his topside crew, and the trapped divers all agreed that it would be best to hold out for rescuers.

As the divers waited, the temperature in their compartment continued to drop into the mid-forties Fahrenheit, the temperature of the surrounding water. Pilot Menzies and Meek were better insulated from the chill by the four-inch-thick sphere, but both compartments were running short of breathable air. Over the radio, Menzies kept telling rescuers he didn't know how much longer he could hold out. The two men in the *Sea-Link* cockpit and the two in the diving compartment were also battling faulty carbon dioxide scrubbers and a hazardous rise in CO_2. Each duo had to come up with some creative jury-rigging to stave off suffocation, like the imperiled crew of *Apollo 13* three years before. Pilot Menzies took off his shirt, filled it like a sack with Baralyme, the sandy carbon dioxide absorbent, and held it in front of a fan to bring down the CO_2 level. The falling temperatures further compromised the scrubbers so Stover and Link increased the pressure in their lockout compartment. That warmed their compartment and themselves, and also improved their scrubber's efficiency.

About one-thirty in the morning, a Monday, another dive from the *Tringa* had to be aborted. Five hours later, a brisk current thwarted a dive by a rescue team flown in from San Diego. Several hours after that, another dive had to be called off. Around noon, another submersible, the little two-man Perry Cubmarine, arrived on a Navy ship and was sent down but couldn't do anything to free the *Sea-Link*. A commercial salvage ship, the *A.B. Wood II*, came with a remotely operated vehicle similar to Taylor Diving's flying eyeball. With the aide of the ROV's camera, a grappling hook was hitched on to one of the *Sea-Link*'s propeller shrouds. Just before five in the afternoon, more than thirty hours after the ordeal began, the ship's crane lifted the entire nine-ton submersible to the surface.

The hatch in the top of the sphere could be opened right away. Menzies and Meek were pulled out, like figurines from a snow globe. They were cold and weary but they were going to be all right. The *Sea-Link* was hoisted on board its mother ship, with Stover and Link still sealed in the diving compartment. Through the portholes, the two could be seen, seated and slumped over. Were they dead or merely unconscious? The pressure inside had reached the ambient pressure of their depth at 360 feet as the two had tried to warm the compartment by increasing the pressure. Link and his crew flushed the sealed compartment with fresh helium and oxygen. Without any time allowed for decompression, the two unconscious men would be dead for sure. Ed Link had scarcely slept since he first called in the Navy rescuers. One can only imagine how wrenching it must have been for Ed Link to peer through a porthole and see his son and his friend Stover, but not be able to open the hatch immediately and rush to their aid.

By ten the next morning, forty-eight hours after the *Sea-Link*'s initial call for help, doctors on board *Sea Diver* concluded that Al Stover and Clayton Link were dead and the effort to warm and ventilate the chamber was called off. Later in the day Ed and Marion Link were in their stateroom, holding tightly to one another, tears flowing. Clayton Link left behind a young son. Al Stover had seven sons; his eighth had been killed during paratrooper training.

Ed Link, who had helped lay the groundwork for improved rescue and other undersea technologies, was haunted by thoughts of what he might have done differently. In 1968, four years after the Deep Submergence Systems Review Group sessions, the nuclear sub USS *Scorpion* disappeared with its ninety-nine-man crew somewhere out in the Atlantic and the Navy had needed to borrow Link's first submersible, *Deep Diver*, to search for survivors. Now, five years later, Navy crews couldn't even reach Link's stranded craft in waters less than four hundred feet deep. The Coast Guard investigators blamed pilot error and the *Sea-Link*'s modular construction, which gave it irregular shapes and projections that were more prone to ensnarement. They also noted that Stover and Clayton Link "displayed an incredible casualness in their preparations" for the dive. But a former deputy director of the Deep Submergence Systems Project, F. R. Haselton, in a pointed op-ed piece in *The Washington Post*, chastised the Navy for its "feeble attempts to rescue these aquanauts stranded a mere three hundred feet beneath the sea," and called the situation "nationally embarrassing."

The best medicine for Ed Link after his son's death seemed to be a return to his drawing board. He spent the next two years working to develop rescue equipment and came up with a tethered, unmanned submersible he called CORD, for Cabled Observation and Rescue Device. It had lights, TV cameras, hydraulic claws, and cutters: just the sort of device that could have set the *Sea-Link* free. Link continued to lobby for improved rescue operations and followed through with his plans to produce a second *Johnson-Sea-Link* submersible that looked a lot like the first.

As Captain Bond approached retirement in the early 1970s, he often played the role of elder saturation statesman, preaching his gospel about Dominion over the Seas and undersea living. He was point man for a graduate program called Scientist in the Sea, jointly sponsored by the Naval Coastal Systems Laboratory, the State University System of Florida, and NOAA, then in its infancy, which taught young researchers to use diving as a tool for their undersea projects. In summer 1973, the summer of Ed Link's tragic loss, Papa Topside watched over a dozen graduate students who were living in a habitat called Hydrolab. Hydrolab was one of the more successful examples of the modest Sealab- and Conshelf-inspired habitats that popped up around the world in the 1960s and early 1970s, most of them limited to relatively shallow waters. Hydrolab, one of a handful made in the United States, was a sixteen-foot-long horizontal cylinder—a third the size of Sealab II—with room for three or four aquanauts to live for about a week at depths of between forty and sixty feet—not exactly barrier-breaking, but the quest to live in the sea was at least being kept alive. Born out of an academic interest in sea floor living, Hydrolab became the most used habitat in the world. It was even featured on *Primus*, a short-lived underwater TV series that never caught on the way *Sea Hunt* had in the 1950s and 1960s with Lloyd Bridges as the scuba-diving investigator Mike Nelson.

For the past couple of years Hydrolab had been stationed about a mile off Freeport, Grand Bahama Island, at a depth of about fifty feet, similar to the depth and hospitable conditions for Tektite I. The Tektite habitat was later refurbished, renamed Tektite II, and went on to house a number of scientific diving teams in the early 1970s at the same shallow, secluded cove at St. John in the Virgin Islands.

Hydrolab and Tektite had both been made possible by assorted public and private partnerships that were the principal means of pursuing undersea living outside of the U.S. Navy in the pre-NOAA years. Tektite had been hatched by the Navy's Office of Naval Research, NASA, and the Department of the Interior with General Electric as the habitat builder. Hydrolab came about through a joint effort of Perry Submarine Builders and engineers from Florida Atlantic University, with consultation from none other than Ed Link. NOAA lent its support to the habitat's operation in the early 1970s after the agency was formed.

But because NOAA was not the wet NASA that some had hoped for, and despite the lofty goals set forth in the Stratton report, it still took a bureaucratic village, sustained by disparate pools of money, to build and run a habitat, or to undertake any undersea venture whose purpose was more focused on science and exploration than on meeting specific industrial or military needs. That's why one American entrepreneur even took a shot at turning a habitat into a viable business. Taylor A. "Tap" Pryor, founder of the Makapuu Oceanic Center in Hawaii and a member of the Stratton commission, built and tested a habitat for hire in 1970 that he hoped would cater to wide-ranging interests—industrial, military, and scientific—and pay for itself. A couple of former Sealab aquanauts, along with Bond's old friend Charlie Aquadro, were part of the group working with Pryor but by early 1973, as NASA was getting ready to launch its $3 billion Skylab, market forces turned Tap Pryor's fledgling undersea business belly-up.

By this time George Bond had relocated to Panama City, Florida, where the Navy was moving the historic Experimental Diving Unit from its brick home on the Anacostia River to the grounds of the former Mine Defense Lab. Bond was serving as a technical adviser for the Ocean Simulation Facility, a major new hyperbaric complex. He and his wife had rented a house on the water where they got frequent visits from a "pet" alligator they called George Foote III—Bond's older son was George Foote Bond Jr., then in his late twenties. To the astonishment of Bond's grandchildren, the gator came when called, presumably a result of having hot dogs tossed its way from the grill.

When the multimillion-dollar Ocean Simulation Facility opened in 1975, it gave the Navy a state-of-the-art chamber of its own—in recent years it had had to borrow others, like Taylor's Hydrospace Research Center, for some deep experimental dives. The facility's centerpiece was an egg-shaped chamber three stories tall that could hold up to 55,000 gallons of water, like a massive wet pot, and could be pressurized to simulate ocean depths down to 2,250 feet. That was fifty feet deeper than the Taylor chambers and a depth that saturation divers were only just beginning to approach. The OSF was installed in a cavernous building a stone's throw from Alligator Bayou, where Sealab I was first lowered into the water.

The investment in the OSF showed the Navy could be as serious as the offshore industry about staying in the business of cracking depth barriers. The Navy had also developed setups that were similar to those used by the offshore industry, such as portable saturation systems that could be placed on any ship. Two new submarine rescue ships, the *Pigeon* and the *Ortolan*, were equipped with a bell and chamber for saturation divers. But Bond still believed that sea floor habitats should be a Navy priority, and that the Navy was best equipped to handle their continued development, even if civilian agencies and partnerships might eventually take up the call, or entrepreneurs like Tap Pryor. The Navy had at least turned the *Elk River*, the Sealab III support ship, into a saturation diving school, and Bond still had hopes of reviving the Sealab III habitat itself, even as he worried that it was in danger of being scrapped.

In the meantime, with his retirement date approaching, Papa Topside got one more chance to watch over a habitat. Toward the end of 1975, the year the Ocean Simulation Facility opened, Bond was loaned to NOAA as senior medical officer for a multinational project bulkily called FISSHH, for First International Saturation Study of Herring and Hydroacoustics. Here was an arcane application of saturation diving and undersea living that was largely scientific, not commercial or military, the kind of civilian project showcased during Sealab, like the weather, current, and plankton studies. During FISSHH divers living in a habitat planned to study the mating habits of herring along with issues related to sound transmission in water. They would have ample time to observe schools of herring, much as bird watchers might study birds.

The FISSHH project was another product of partnerships, beginning with its German-built habitat called Helgoland. A Sealab-style cylinder built in the 1960s, Helgoland had been refined several times since its first test at seventy-five feet in the North Sea, about a mile from its namesake island in July 1969—the month of the *Apollo 11* moon landing. A few months later oil was discovered in the North Sea. The German habitat was transported to American waters on a Polish factory ship and the sea floor study brought together teams of West German and American aquanauts, along with several dozen topside participants from both countries, including Captain Bond and a few former Sealab aquanauts. There were also to be observers from Poland and the Soviet Union, a collaboration like the rendezvous that summer of American and Russian spacecraft that showed how science could occasionally transcend Cold War politics.

The study, Helgoland's deepest trial, got under way in the fall of 1975, about eight miles east of Rockport, Massachusetts. At a depth of just over one hundred feet it was hardly barrier-breaking, and no costly helium had to be added to the breathing mix. Five teams of four aquanauts were lined up to live in the lab for two weeks each, but almost from the start Helgoland was plagued by rough weather and equipment failures—a hot-water heater went on the fritz, one section of the habitat sprang a minor leak, and the entire cylinder swayed back and forth with each passing swell, a Rockin' Ritz to rival the old Tiltin' Hilton. There were struggles with oxygen levels creeping too high, posing a toxic threat. On top of that, it turned out that the herring were spawning too far from Helgoland to be readily observed—an inadvertent argument for mobile habitats like Argyronète. In the end only two teams were able to complete their missions.

Helgoland was designed so that aquanauts could decompress on the bottom, inside a sealed-off section of the habitat, and then make their way to the surface through the water, following a line that ran up to a support buoy from which they could be picked up by boat. Two aquanauts had just safely surfaced on a stormy September day, but a third, Joachim Wendler, was apparently hampered by gear he carried with him as he made his ascent through the roiling Atlantic. The young but experienced German aquanaut must have taken a deep breath at a fateful moment. Bond later theorized that a passing swell could then have lifted him ten feet or more, and with that sudden change in pressure deadly bubbles forced their way from his lungs into his circulatory system. He would suffer a case of gas embolism, like someone making a submarine escape who failed to properly blow and go as they rose through the water column. In circumstances similar to those that almost killed Charlie Aquadro early in Bond's Navy career, Wendler died.

Despite the death of an aquanaut, the FISSHH program was allowed to continue, however fitfully. When it was over, Captain Bond noted that the project had as many close calls as he could recall in a similar span of time, and the experience gave him no shortage of material for his habitual writings. Bond still had at least one other diving project in store. It involved the recovery of millions of dollars' worth of precious metals and artifacts from the *Awa Maru*, a Japanese ship sunk in the East China Sea during World War II. Bond's friend Bill Bunton, a former Sealab II aquanaut, had been engaged for some time in delicate negotiations with the Chinese to enter into a joint salvage venture. Bunton had recruited Bond, Scott Carpenter, and Jon Lindbergh to take part, but the ambitious plans fell through.

Over the years George Bond had often returned to Bat Cave to relax at his old clapboard cabin retreat. In retirement he would have a lot more time to spend there, pondering life and literature, drinking in the solitude and a favorite bourbon. He liked to sleep on an old Army cot on the little back porch, just a few paces from a roaring, isolated rocky bend in the Broad River. Sleeping under the stars on that old cot had been his custom since the days when he brought his kids to the cabin, bouncing over a couple of rutty miles of rocks and dirt in his jeep. Now, at a table near the stone fireplace, Bond worked on several books, including one about saturation diving, to be called *Tomorrow the Seas*. There was one important chapter, however, that George Bond could not write. No one really could, because by the time of Sealab III the Navy had quietly altered the undersea program's primary purpose. Instead of a single-minded quest to live in the sea, Sealab became the ideal cover story for the Navy's secret preparations to send saturation divers to the front lines of the Cold War. As it turned out, Berry Cannon's accidental death added to the cover. Sealab III could be halted, publicly perceived as a failure, and allowed to fade away. Even if Papa Topside had known the Navy's secret, he would not have been allowed to reveal it.

17

THE PROJECTS

Only as Sealab III reached its premature end were certain aquanauts and support personnel informed that they would remain with the Deep Submergence Systems Project Technical Office, the DSSPTO, to work on a highly classified mission. From its inception, unbeknownst to all but a few, Sealab III was intended as a credible front for mission preparations, much as the famous, and later infamous, *Hughes Glomar Explorer* was masquerading as a massive civilian mining ship but was secretly being readied for a risky Central Intelligence Agency–led mission to raise a sunken Russian submarine from the Pacific floor, more than a thousand miles from Hawaii. Sealab III appeared to be an unabashedly public quest to live in the sea, with an international cast that included British, Canadian, and Australian aquanauts and even Philippe Cousteau, son of the celebrity diver. Planned experiments like the one to transplant dozens of Atlantic lobsters into deep Pacific waters could have produced real results, but they also provided a further smoke screen for a top secret program. A specially outfitted Navy submarine was going to be sent on a clandestine mission to the Siberian coast, where saturation divers could slip out of the sub to retrieve the remains of Russian test missiles from the sea floor.

Commander Jack Tomsky was among the few higher-ups who knew that "the projects," as the secret missions were discreetly called, were to be the true beneficiary of the methods and equipment developed for Sealab III and for saturation dives to six hundred feet or more. Tomsky oversaw the projects from the time he was brought in to head Sealab III until his planned retirement in mid-1969, after the end of Sealab and his thirty-year naval career. He was soon called back to duty, a rarity for retired officers, especially for one of Commander Tomsky's modest rank. But the Navy wanted him to continue working on the projects, and for Tomsky there was a redemptive quality in this final role. It gave him a more satisfactory coda to his long career than the harsh words he received from the board of investigation for the failure of Sealab III. Captain Bond, the father of saturation diving, would not be told about his aquanauts' new assignment with Submarine Development Group One, based at the same Ballast Point location in San Diego as DSSPTO, which Captain Mazzone had run until a replacement was brought in after Sealab III.

About a dozen divers and support personnel were just getting up to speed on the projects in the wake of Sealab III. The pieces began to fall into place when they paid their first visits to Mare Island Naval Shipyard, across the bay from San Francisco. Inside one of the buildings they were shown what, at first glance, appeared to be a cigar-shaped midget submarine, about fifty feet long, just a bit longer than Sealab I. It looked like a Deep Submergence Rescue Vehicle, the kind developed in response to the 1963 loss of the *Thresher.* A DSRV was essentially a submersible, lightweight and compact enough to be taken at a moment's notice by air, land, or sea to the site of a disabled submarine. The mini-sub was designed to piggyback on a much larger mother submarine. In the vicinity of a downed submarine, a DSRV's four-man crew could lift off and position their vessel so that the hatch in its underbelly could mate with a downed submarine's escape hatch. The DSRV would then be able to pick up two dozen stranded sailors at a time and deliver them to the waiting mother sub. It needed no diver assistance and could reach depths down to five thousand feet—major advances over the tethered McCann Rescue Chamber of the 1930s, which was still in use.

Two DSRVs were indeed in the works but this vessel at Mare Island was not one of them. It looked like one from the outside, but on the inside it was a self-contained saturation diving system—like a mini-Sealab habitat, decompression chamber, and topside control station all rolled into one. The former Sealab crew members were able to look it over and offer suggestions before everything was finished and readied for training missions. Part of the need to disguise this mini-habitat was that, like a real DSRV, it would be mounted to ride piggyback at the stern of a submarine—in plain view of the Russians and anyone else who happened to see it. Tourists on the Golden Gate Bridge, for example, might be curious about the sight of a submarine cruising along the surface, headed out to sea with an unusual mini-sub on its back. The words "U.S. Navy DSRV Simulator" were painted in block letters on the side to bolster the *Glomar*-style cover story, which went something like this: The simulator was a stand-in for a real DSRV so that its mother submarine, the specially modified USS *Halibut*, could be sea-tested while carrying the added bulk of one of the forthcoming mini-subs.

Unlike a mobile DSRV, the simulator was permanently affixed to the top of the *Halibut*. The faux rear propeller was just part of the mock-DSRV's disguise. Inside the hull were four main compartments, each roughly the same size, walled off from one another by bulkheads with hatches that could be opened and sealed shut. The compartment at the tail end was a diving station about the size of a Personnel Transfer Capsule, with hundreds of feet of umbilical hoses bundled up inside. And like a PTC the compartment had a hatch in the floor that opened to the sea. The next compartment forward contained tight quarters for four saturation divers. Besides the two bunks against each wall and the narrow passage between them, there wasn't room for much else—a toilet, a sink, some pipes and valves, all the comforts of a pressurized prison cell and no room for interpersonal conflict. The divers' meals would be prepared on the submarine and passed into the living quarters through a small medical lock in the next bulkhead forward, which separated the living quarters from the equally cramped control room. In the control room a couple of support crew members could work the valves and monitor the gauges of the saturation system, much as Bond, Mazzone, and others did during Genesis and Sealab.

The control room was reached by a hole in its floor, where a ladder came up through a short trunk, at about the midpoint of the simulator's underbelly. The trunk was affixed to the top of the mother submarine. The supposed simulator's most forward compartment was packed with gas storage tanks and the mechanisms for creating a high-pressure artificial atmosphere in the living quarters. When it came time for a diving mission, the proper mix of breathing gases was shot into the dive station, at the simulator's tail end, to raise the pressure so that the saturated divers could open the hatch door separating their living quarters from the dive station. Once inside the dry, pressurized dive station, they could don hot-water suits and the latest incarnation of the deep-water breathing rigs developed during Sealab III, which were not so different from those in commercial use. The dark water outside stopped at the hatch in the floor, forming a liquid looking glass.

Divers selected for the projects would learn to use this novel system during test runs, including some just outside San Francisco Bay, where there were depths similar to those they would encounter on missions planned for the other side of the world. Their mother submarine, the *Halibut*, also known as the "flat fish," would stealthily take them to the desolate Sea of Okhotsk in Russia's Far East. During the winter months of 1971 and 1972—as Bond was finding new projects after Sealab III, Cousteau struggled to finance his mobile *Argyronète* habitat, and not long before *Sea-Link* entombed Ed Link's son—the USS *Halibut* arrived at Okhotsk. All four divers then opened a new, secret chapter in military diving.

Once the submarine hovered over the seabed, carefully anchored into place, the divers' work began. Upon dropping through the hatch of their mini-habitat, they landed on top of the *Halibut*, shuffled a few paces, and then jumped over the side of the hull for a slow-motion fall from the equivalent height of a three-story building, with an umbilical trailing precariously behind. The precise depth, like everything else about this first mission, was classified, although it was likely in the range of four hundred feet, similar to the depths at which some commercial saturation divers were beginning to work, especially in the North Sea.

The collection process began with the lowering of a "gondola"—a huge tray built in to the belly of their mother submarine. In darkness, with their mother ship hovering nearby, the divers swam around looking for their quarry like determined revelers on a midnight treasure hunt. They might find fragments they could carry themselves, but when they needed to reel in a big carcass of a weapon, the spy divers had at their disposal winches and other gear that was packed into the gondola. They used methods not so different from those of their commercial diving counterparts on jobs that could take hours and had physical and technical difficulties akin to wrangling with pipelines in the oil fields. The main difference was that out in the Sea of Okhotsk any mistake, accident, or equipment malfunction could result in more than just injury, or even the death of a diver. If the Soviet government discovered *Halibut* divers poking around in sovereign territory, the United States could find itself saddled with an undersea corollary to the capture of Francis Gary Powers, the U-2 spy pilot shot down over Russia in 1960.

Despite the extraordinary circumstances and high stakes, there were some lighthearted moments, like the time the divers spotted an enormous crab on the bottom. It was probably a king crab, about the size of a trash can lid, with menacing claws to match. A somewhat capricious, helium-spiked challenge was issued to grab and capture the impressive crustacean, and one of the divers somehow managed to shepherd it into the submarine, where it was boiled in the galley and turned into a memorable meal. So for a moment there, at the bottom of the Sea of Okhotsk, it felt like the more free-spirited early days of Sealab, with time for the occasional prank or some impromptu local cuisine, like Bob Sheats's plankton soup.

In a matter of months, another diving team returned to Okhotsk, this time with a goal even more ambitious than collecting missile parts—so ambitious, in fact, it seemed like a plan conceived in George Bond's journals. This one was dreamed up, almost literally, in the Pentagon by Captain James Bradley, the head of undersea warfare at the Office of Naval Intelligence. Bradley had a hunch that the United States could strike the intelligence equivalent of North Sea oil by having the *Halibut* locate submerged Soviet communication cables linking the submarine base at Petropavlovsk on the Kamchatka Peninsula with the mainland several hundred miles to the west. These lines, Bradley believed, were likely to carry a steady flow of strategic conversations. Once the undersea cables were found, saturation divers could do the necessary work of tapping them with electronic surveillance devices. On board the submarine, a half-dozen personnel from the National Security Agency, headquarters of America's electronic eavesdroppers, could listen in, and all the intercepted telephone conversations could be recorded so NSA analysts could parse them back in the United States.

The first cable-tapping device worked by induction, so there was no splicing required that might cause disturbances in the line and arouse suspicions. The tap was about the size of a suitcase, and could record a few days' worth of communications at a time. The *Halibut* waited nearby, its divers living in their saturation cell, awaiting their next foray into the sea. The NSA eavesdroppers on board soon recognized that Bradley's hunch was right. Information from the taps did indeed flow like oil from a well, offering valuable insights into the minds of Soviet military planners. As a bonus, many conversations turned out not to be encrypted, probably because the Soviets assumed that hardwired undersea cables—unlike fickle airwaves—were not vulnerable to taps, especially cables close to their own, isolated shores.

In order to record for months or even a year at a time, Bradley and his intelligence staff turned to Bell Laboratories, whose engineers came up with a higher-capacity recording pod that also worked by induction and was far larger than the first pods. It resembled a twenty-foot section of pipe, two or three feet in diameter, and was crammed full of electronics that could pick up frequencies from dozens of lines within a fist-sized cable. Weighing about six tons, it made the divers' job of getting it from submarine to muck-shrouded cable as tricky as lining up and welding two ends of an oil field pipeline.

A handful of newspaper articles and books over the years shed specks of light on the projects, most notably the watershed book about American submarine espionage, *Blind Man's Bluff*, published more than a decade ago. Soon thereafter came a memoir by John Craven, a chief scientist in the Navy's Special Projects Office. Craven, as head of the Deep Submergence Systems Project, had hired Jack Tomsky to lead Sealab III and simultaneously prepare for the intelligence projects. Within a couple of years Craven relinquished his DSSP post to focus on *Halibut* and its outfitting for spy missions. His insider's account acknowledged that secret saturation dives took place, as did *United States Navy Diver: Performance Under Pressure* by Mark V. Lonsdale, a former commercial saturation diver and an expert on military diving and special operations. His history of Navy diving was published in 2005 with official blessings. There was even a History Channel documentary based on *Blind Man's Bluff.*

But almost four decades after the first *Halibut* foray into the Sea of Okhotsk, and twenty years after the dissolution of the Soviet Union, virtually everything about those early missions, including the names of the divers and any details about the dicey work they did for their country, remains classified. The Navy rejected two recent requests and two subsequent appeals for information about the projects made under the Freedom of Information Act. A third FOIA request, made in September 2010, went unanswered. The saturation dives and the American cable-tapping coups might seem historical, especially in the age of WikiLeaks, but despite the passage of time and what is known from a number of previously published accounts, the Navy steadfastly claims that the release of any relevant information, even from the early 1970s, is a matter of national security. So it is that saturation divers who carried out these secret missions feel bound by their long-standing oaths of silence and, to this day, will say little if anything about the history they made, and how they made it.

The trailblazing *Halibut* missions, initially code-named "Ivy Bells," continued through the mid-1970s. When the *Halibut* was removed from service, the cable-tapping missions were taken over by other divers and other submarines—first the *Seawolf,* and in the late 1970s the *Parche.* In the early 1980s the Soviets learned of the Okhotsk taps but the United States did not know how until a Soviet defector put U.S. officials onto the trail of a low-level National Security Agency employee, Ronald W. Pelton, whose tips to the Soviets included information about the prized Okhotsk taps. Pelton was arrested in late 1985, convicted of selling intelligence secrets to the Soviet Union, and given three consecutive life sentences, plus ten years. In the meantime, the Russians were able to pluck two recording pods from the sea floor. They put photographs of one, along with some of the data-recording innards, on display in a Moscow museum, but there was never an international uproar over the missions.

Missions to the Sea of Okhotsk did taper off, but in 1979 *Parche* placed a first tap on cables in the more heavily trafficked, and thus more treacherous Barents Sea, and others followed. Saturation divers had to work at greater depths there than in Okhotsk—five hundred and even six hundred feet, the onetime goal of Sealab III. The missions paid off by providing a significant reading of Soviet military minds. One of the major revelations gleaned from the taps, in the mid-1980s, was that Soviet forces were not designed for a first nuclear strike from the sea and were not preparing to launch one—a significant relief to Pentagon war planners, and information that may have made the world a safer place.

Intelligence of this caliber was deemed worth the substantial risk that came with each cable-tapping mission—the risk to the nuclear submarine itself, to its crew of about 150, to the divers, and to international relations. A plan to run a cable to shore and open a conduit for real-time taps, possibly between the Barents and Greenland, was ultimately deemed too expensive. But periodic taps continued in the late 1980s and early 1990s by the *Richard B. Russell* until it was retired. By then the *Parche* returned to service after a long overhaul that included a hundred-foot addition to the hull, for "research and development" equipment. This R&D function included intelligence gathering and underwater salvage, according to Lonsdale's *United States Navy Diver*, while *Blind Man's Bluff* says the sub was equipped to tap cables and to pick up pieces of hardware. In the 1990s, after the Soviet Union ceased to exist, the focus of the projects that began with Ivy Bells shifted to other parts of the world, according to the authors of *Blind Man's Bluff*. If so, this may explain the Navy's insistence on maintaining a veil of secrecy that stretches more than forty years into the past.

The *Parche* was finally decommissioned in 2004, but the USS *Jimmy Carter* came specially equipped to pick up where the *Parche* left off. The new sub, at 453 feet, could have fit bow-to-stern on a football field if not for the hundred-foot hull extension that made it that much longer than the other two subs in its class. That additional space, the same length that had been added to the *Parche* during its overhaul, is officially, if obliquely, known as the "multi-mission platform." The Defense Department has said that the platform's uses include "tactical surveillance." A 2006 press release went so far as to characterize one of the new sub's missions as "intelligence collection." A Navy spokesman, responding to a request to the Defense Department in mid-2009 for further information about the sub, made it clear that there would be no official statement as to whether the *Carter*, or any other sub, is equipped for saturation diving. Yet there seems to be very little doubt among experts that the extended hull includes the necessary chambers and systems to support saturation divers.

With a new generation of saturation divers on board, the *Jimmy Carter*—whose namesake approved some of the early cable taps—may very well continue the cable-tapping tradition. It's also possible that the choice has been made to stop placing taps, for any number of reasons, not the least of which may be that the risks have become too great, the returns too small, and the technology required to tap newer, high-capacity fiber-optic lines too problematic because they carry so much more information than old-style copper cables, all in a dizzying rush of staccato light pulses. Literally thousands of telephone conversations, e-mails, and other data transmissions flow through each hair-width fiber-optic line, and a single cable of the kind laid on the ocean floor typically houses several lines.

Still, these individual lines can be tapped, with relatively inexpensive devices and fairly straightforward procedures. The hair-width lines themselves, made of a pure glass, are encased in materials that give them the thickness of pencil lead. A small amount of that casing must be carefully stripped away, almost like stripping insulation from an electrical wire, so that another hair-width line can be placed in close proximity. Once that is done, the enormous data flow—usually moving at a rate of ten gigabits per second—can be siphoned and stored on high-capacity hard drives. A week's worth of data could fit on a cluster of drives that takes up no more space than a refrigerator.

All of this might seem extremely difficult, if not impossible, to do underwater, but technology advances briskly these days, transforming man-on-the-moon fantasies into standard features of the workaday world. Fiber-optic lines were themselves mere fantasy not long ago, but are now strung all over the globe and across oceans, making them natural targets for military eavesdroppers. To get to them, one can imagine the usefulness of a mobile workshop like the underwater welding habitats developed for the offshore industry—or, for that matter, a twenty-first-century version of Ed Link's portable inflatable dwelling. If welder divers can precisely cut and fuse unwieldy pipe ends inside a habitat, a pair of spy divers with the necessary technical expertise might just as readily use a dry sea floor workspace to cut into heavy-duty cables and then perform the delicate task of tapping the fiber-optic communication lines housed within. The divers could live down there for a while, or return to their mother submarine, perhaps making the trip on board a mini-sub something like the DSRV. In the meantime, computer hard drives, placed either on the ocean floor or inside a nearby submarine, would fill with gigabits of siphoned data. With the right computer programs, the jumble of raw data collected can be sorted and specific contents revealed, right down to an individual phone call or e-mail message.

Prior to the breakthroughs in diving made during Sealab,
the deepest dives had to be made wearing heavy gear
like this "helium hat," an adaptation of the standard
dive suit design from the early nineteenth century.

Dr. George Foote Bond, who began his medical career in the late 1940s with a practice in the backwoods of the Blue Ridge Mountains, might have seemed an unlikely father of a project like Sealab.

Commander George Bond, at right, with Cyril Tuckfield immediately after their record-setting swim to the surface from out of a submarine that was more than three hundred feet below. Bond arranged the 1959 demonstration to show that emergency escape from such depth was indeed possible with little more than a lungful of air and the proper technique.

Right: Walter Mazzone, a veteran of numerous harrowing World War II submarine patrols, in 1963 at the Medical Research Laboratory of the U.S. Navy's submarine base at New London, Connecticut, where he became an indispensable aide to Bond.

Above: Bond and Mazzone open the medical airlock that allows small items to be passed in and out of a pressurized test chamber during the final phase of Genesis, the experiment they conducted prior to Sealab to show that divers could safely live in a sea floor shelter.

Right: Bob Barth, a Navy diver and volunteer test subject throughout the entire Sealab project, prepares for an extended stay inside a pressure chamber as part of the final phase of the Genesis experiments.

The Sealab I habitat, in transit from Panama City, Florida, where it was pieced together on a shoestring budget, to Bermuda for its first major trial run in the summer of 1964.

Standing with Sealab I are (from left) Bob Barth, Captain George Bond, Lester "Andy" Anderson, Dr. Robert Thompson, and Sanders "Tiger" Manning.

Scott Carpenter, the Mercury program astronaut, at left, with Lieutenant Commander Roy Lanphear, the officer in charge of Sealab I operations. Carpenter sought to be among the first to live and work on the ocean floor.

Jon Lindbergh, left, son of famed aviator Charles Lindbergh, and Robert Sténuit. In June 1964, three weeks before Sealab I, Lindbergh and Sténuit were preparing to attempt a deep, multiday dive near the Bahamas with inventor Edwin Link's latest creation, a kind of undersea pup tent called the Submersible Portable Inflatable Dwelling, or SPID.

Andy Anderson, facing camera, shares a meal
inside Sealab I with Tiger Manning, to his right,
and Dr. Robert Thompson.

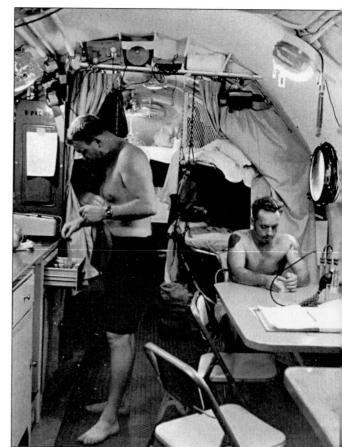

Dr. Thompson, left, and
Manning, inside Sealab I.
Behind them are the
aquanauts' four bunks.

The SS *Berkone*, with Sealab II in tow, at right, was made up of converted ships to serve as mission control during the lab's forty-five days as a sea floor base in 1965 off the coast of Southern California.

The first of three aquanaut teams, including Berry Cannon and Scott Carpenter, at left in the front row, pose with Sealab II.

Captains Bond (above) and Mazzone, taking a rare break from their Sealab II monitoring station on the support ship *Berkone*.

The Team 1 aquanauts inside Sealab II take a moment to listen over the intercom as Captain Bond gives one of his typical Sunday sermons. Standing are, from left, Billie Coffman, Fred Johler, Scott Carpenter, Robert Sonnenburg, and Berry Cannon.

The USS *Halibut*, cruising out of San Francisco Bay, carries on its afterdeck what was said to be a mobile mini-sub, but was really an affixed pressure chamber from which divers could come and go to carry out top secret missions.

The massive base section of the offshore oil platform Cognac before it was lowered for installation in the Gulf of Mexico at a record-setting depth of 1,025 feet.

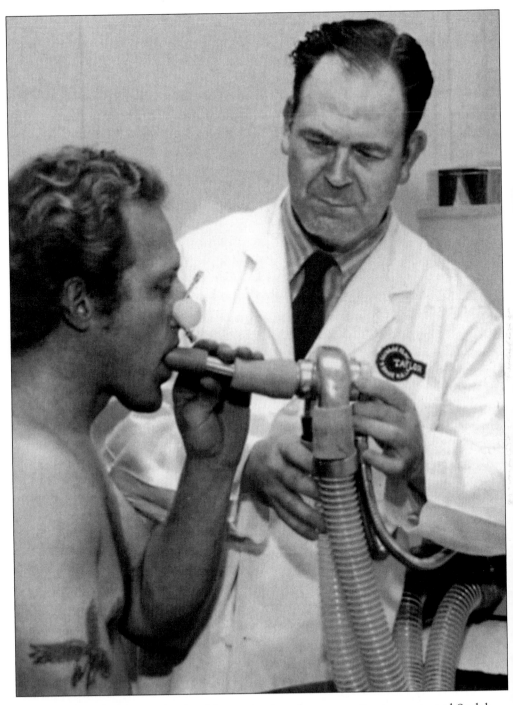

Dr. Robert Workman, a leading Navy scientist whose expertise jump-started Sealab, went on to head the research department at Taylor Diving & Salvage Co., a New Orleans–area company founded by former Navy divers to tap into the growing market for offshore oil.

Ed Link looks over a scale model of the deep-diving vessel to be called the *Johnson-Sea-Link*, in honor of Link's partnership with his friend and fellow marine enthusiast J. Seward Johnson. The vessel could deliver divers to extraordinary depths and had many successful runs before a catastrophic accident.

Henri Delauze, founder of the French commercial diving giant Comex, pictured with the SAGA, the submarine and mobile habitat project he took over after Jacques Cousteau had to abandon it.

An artist's rendition of the support ship *Elk River*, the Personnel Transfer Capsule, and Sealab III habitat, as they would appear during the actual operation.

The Sealab III support ship USS *Elk River* under way. The Personnel Transfer Capsule, a pod used to deliver divers to their habitat on the sea floor, sits between the ship's superstructure and its open moon pool. The gantry crane for moving the PTC is at the stern; the twin command vans, where top managers like George Bond and Scott Carpenter could often be found, are amidships.

The gantry crane positions the Personnel Transfer Capsule over the moon pool on the *Elk River*. This PTC system, while more advanced than the relatively primitive prototypes for the previous Sealabs, developed problems early.

Captain George Bond with Jacques-Yves Cousteau at a conference, circa 1968. The two friendly rivals were early proponents of creating sea floor stations and planned to collaborate on a film about Sealab III.

The new Sealab III habitat arrives in Long Beach by barge from San Francisco, where it was remodeled and expanded from the hull of Sealab II. Changes in the design ultimately contributed to the project's premature end.

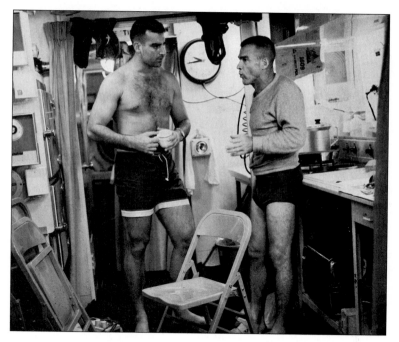

Robert Sonnenburg, left, the only aquanaut on both
the first and third teams, inside Sealab II with Robert
Sheats, the experienced master diver and Team 3 leader
who suffered an unwelcome case of the bends as
the undersea project reached its end.

Team 3 aquanaut Paul A. Wells,
at work on a sunken aircraft fuselage,
wears the gas-recycling Mark VI
diving gear favored during both
Sealab I and II, despite some
dangerous quirks. The Mark VI looks
similar to standard scuba, except for
the breathing bags worn like a vest.

18

ANSWERS AND QUESTIONS

With its Sealab program, state-of-the-art Ocean Simulation Facility, and the still shrouded projects that enabled divers to live in quasi-habitats on nuclear submarines, the Navy seemed to be the closest thing there was to a wet NASA. Indeed it was at the Ocean Simulation Facility, and at a select group of American and foreign research labs, that scientists continued to run latter-day Genesis experiments to unravel the mysteries of High-Pressure Nervous Syndrome and find answers to those old questions: *How long can a man stay down? How deep can a man go?*

In 1979, after Taylor Diving & Salvage cracked the thousand-foot barrier at sea and the Cognac platform had begun to drill its first wells, the OSF was the scene of another significant push downward. During a thirty-seven-day experimental saturation dive, six Navy divers stayed for eight days at 650 feet before easing past the thousand-foot mark en route to a depth of eighteen hundred feet. They spent nearly five days at that simulated depth before making their gradual return to the surface. Everyone survived, but not without experiencing the whole unpredictable range of HPNS effects: physical shakes, fatigue, dyspnea, dizziness, nausea, stomach cramps, vomiting, an aversion to food, and, perhaps not surprisingly, poor sleep. Some symptoms only seemed to get worse over time. A similar British experiment that year produced similar results. When the HPNS-riddled eighteen-hundred-foot test was finally over, the U.S. Navy established a diving limit of one thousand feet.

The end of the Navy test coincided with the beginning of an experimental series of four deep dives at Duke University called Atlantis and led by Peter Bennett, an Englishman-turned-American whose involvement in diving physiology dated back to the 1950s and his early observance of tremors during deep submarine escape trials in the mid-1960s at the Royal Navy Physiological Laboratory. Atlantis had the backing of the U.S. Navy and the National Institutes of Health. Henri Delauze, the Comex founder and a friend of Peter Bennett, also chipped in to cover the substantial expenses. The investment paid off. In 1981, Bennett and his Duke team broke a barrier that made the Stratton report's two-thousand-foot diving goal look like an imminent possibility. Three volunteer divers lived in Duke's new experimental chambers for more than four days under pressure equivalent to a depth of 2,132 feet, nearly a half-mile down, and a very distant sixty-five atmospheres away. As in the bygone days of Genesis, the Duke divers did a variety of tasks and exercises to simulate underwater work and were subjected to a full range of modern medical pokes and prods. During the final twenty-four hours prior to the start of a long decompression, the simulated depth was increased by 118 feet—about the height of a twelve-story building—to 2,250 feet. That was a true depth record in saturation diving, albeit in a dry chamber. Nonetheless, the successful test pointed the way down to a vastly larger undersea frontier.

The Atlantis experiment had demonstrated an improved antidote to HPNS, a new recipe for breathing gases known as "trimix." Trimix was a traditional blend of helium and oxygen but with just enough nitrogen added to quell the effects of HPNS. (One researcher, as if summoning the frontier spirit, gave his test subject a shot of Scotch to stave off HPNS. It helped.) With nitrogen the trick was to keep its concentration high enough to curtail the dreaded effects of HPNS, but not so high as to induce a wildly drunken state of narcosis. A concentration of 5 percent in trimix seemed to alleviate HPNS, especially when combined with adjustments in the dive schedule, like using slower compression rates when approaching depths of a thousand feet or more.

When the Atlantis divers reached their deepest simulated depths, they were able to do hard physical tasks and displayed an impressive ability to adapt to the extraordinary pressure. Yet living all those atmospheres away still wasn't easy. The divers found that the increased density of the pressurized gases became strangely palpable, almost as if they were breathing a liquid. The artificial atmosphere was, after all, about sixteen times as dense as ordinary air. They could also see it in the way objects like sheets of paper, or a rubber band shot through the chamber, would fall noticeably slowly, almost as if in water. In these conditions the subjects couldn't get enough of a breath by inhaling through the nose. As if they were badly congested, they would breathe through their mouths. That made eating uncomfortable, but they could still manage it. Even when physiological problems were under control, there were logistical problems, like the extra time required when using slower compression rates. To bring the Atlantis divers safely to their 2,100-foot depth took seven days, longer than a flight to the moon. After five days at depth, the divers began a decompression schedule that would last thirty-one days.

Cost-conscious industrialists had embraced saturation diving in large part for its efficiency, but these substantially longer compression and decompression rates indicated that there was a point of diminishing returns. Even if divers could do a fairly long-duration job at depths beyond two thousand feet, would it be worth the price—and the danger? The question was almost rhetorical, at least in the business sense. The divers would have to be paid, fed, and supplied with an expensive artificial atmosphere while decompressing for an entire month—doing little with their time but reading, sleeping, and playing pinochle. Remotely operated vehicles, advanced forms of the flying eyeballs used during Cognac's construction, looked like the better investment: Send a robot or a machine rather than a man.

But even in industry there were some who chose not to let the bottom line get in the way of pursuing deeper dives. After the final Atlantis dive in the early 1980s, a number of other deep forays were made, including several notable dives by Comex, whose tenacious founder, Henri Delauze, was fond of seeing diving records and barriers broken, especially by his company. In the late 1980s, Comex, in cooperation with the French navy, staged a stunningly deep demonstration in the Mediterranean, not far from company headquarters in Marseille.

In rough winter conditions, with winds gusting to fifty knots, the Comex saturation divers took turns riding in their bell to a working depth of 1,750 feet. This was nearly as deep as the U.S. Navy's troubled 1,800-foot test a decade earlier, and these Comex dives were at sea, the true proving ground. Six divers put in a total of twenty-five hours on the bottom, bolting pipe flanges and generally replicating some common oil field jobs to show that saturation divers could still be productive at this incredible depth. To trump HPNS, and also to thin the palpable density of helium and oxygen under high pressure, Comex had developed a method for adding hydrogen to the breathing mix. Although highly flammable, hydrogen is the least dense of all gases, and that means it becomes proportionately less dense and difficult to breathe under high pressure. This record-breaking open-sea dive in 1988 made good advertising for Comex's hydrogen alchemy.

Government funding of the kind that made the Atlantis dives possible had been drying up and experimental efforts to continue the push downward largely ended, at least in the United States. But the Duke University team responsible for Atlantis was able to prove the practical worth of its trimix recipe across the Atlantic, at a state-of-the-art chamber complex near Hamburg. The West German government had poured money into the new facility to gear up for future deep construction jobs in the North Sea. With the Duke team on hand as consultants, American, British, French, and German divers made more than two dozen simulated working dives over the course of seven years at the elaborate complex, a real modern-day marvel. Although these dives were not at sea, the complex provided considerable vérité. It was set up so that divers had to take all the same steps they would take at sea to get into an underwater welding habitat—moving from their pressurized living quarters into a diving bell, from the bell into a tank filled with seawater, and from the seawater into the welding habitat. Soon after the last Atlantis dive in the early 1980s, and continuing through 1990, thirteen divers in all took turns putting in many full days welding. There was lots of useful work, long-duration dives, and depths down to almost two thousand feet—the stuff of pure fantasy not many years before. HPNS was barely a problem. But market forces intervened. The offshore industry continued to favor machines over divers, at least for jobs deeper than a thousand feet, and with declining oil prices, the industry contracted. Then, as the 1980s ended, the generous West German research funding tumbled with the fall of the Berlin Wall.

The collapse of markets and governments did not stop Delauze and Comex from taking yet another stab at the depth barrier. At the end of 1992, to further demonstrate hydrogen's potential, Comex arranged to send a diver to a simulated depth of 701 meters, about 2,300 feet. This would be the latest in the experimental series called Hydra, which Comex had started some years before to develop its breathing mixtures with hydrogen. Henri Delauze was well aware that the planned experiment would be fifty feet deeper than the deepest Atlantis dive; it would be the deepest saturation dive ever attempted. In early November, three volunteer Comex divers climbed into one of the four interconnected chambers at the company's hyperbaric research center. If all went well, they would be healthy and home in time for Christmas.

In order to alleviate if not entirely avoid the various effects of HPNS, the divers were compressed slowly, with hydrogen gradually added to their artificial atmosphere. It took two weeks for them to reach a simulated depth of about 2,220 feet, thirty feet less than the Atlantis record and eighty feet shy of the target depth. Just one of the three Comex test subjects—the one known to be the least sensitive to the unpredictable effects of HPNS—crawled into an adjoining chamber and closed and sealed a hatch. He alone would experience the pressure of another two and a half atmospheres to reach a depth of 2,300 feet. Once at depth, he was put through some familiar pokes, prods, and tests. He exhibited only minor signs of HPNS, a slight tremor. He remained at depth, seventy atmospheres away, for three hours. At that point the Comex researchers eased off the pressure so he could open the hatch and rejoin his fellow divers in the adjoining chamber, where they were held at 2,200 feet or so for three days.

As pressure in the chamber was reduced, the testing continued and the three made a number of progressively shallower dives in a wet pot, to better simulate actual diving conditions. Their decompression went on this way for twenty-four days. Although symptoms of HPNS had been reduced, they were not eliminated. When the whole experiment was over, the trio had been locked in one test chamber or another for forty-two consecutive days, breathing the hydrogen-rich atmosphere much of the time. Henri Delauze proclaimed Theo Mavrostomos, who had reached 2,300 feet, to be the deepest diver in the world. Despite that distinction, he would remain lesser known than, say, the celebrated dogs and monkeys first launched into space. But his Greek origins fittingly evoked the ancient tradition of sponge diving and man's earliest attempts to dive deeper and stay longer. His Comex record still stands, although the Atlantis researchers would diplomatically contend that their experimental dive of a decade earlier, down to 2,250 feet, was at least as meaningful a record, since its duration was a full twenty-four hours, enough time to allow for saturation—and whatever effects might come from it. In any case, all of the deep forays were part of the process of finding answers to those old, even ancient, questions.

In the mid-1960s, around the time of Sealab II, George Bond had optimistically predicted that divers would reach depths of three thousand feet within a few years, at least on an experimental basis, with hydrogen likely in the mix. Jacques Cousteau had already made headlines with his pronouncement about the coming of Homo aquaticus. But as the twentieth century came to a close, researchers were seeing evidence of an ultimate pressure barrier, a depth beyond which man as a free-swimming saturation diver could not safely go.

Researchers were finding that even with just the right mix of breathing gases, pressure created the ultimate barrier, mainly because high pressure triggered HPNS. At the very least, as suggested by animal studies, a human diver was likely to go into convulsions. Dyspnea—that eerie sensation of not getting enough oxygen even when you are—along with hallucinations and other unpredictable effects might not be fatal themselves, but they were difficult to anticipate and could put a diver in danger, especially if working deep in the sea rather than being monitored in a test chamber. Still, some people, like Theo Mavrostomos, were less susceptible to high pressure than others, so some might be able to reach substantially deeper depths than others. But how much deeper?

Studies on animals indicated that Homo sapiens would not fare well, and perhaps not even survive beyond depths of somewhere between about 2,500 and three thousand feet. But no one knew for sure. With research funding reduced to a trickle, questions would linger. HPNS had always been a wild card, yet it had successfully been held in check throughout the long-duration German dives, at least down to two thousand feet. The Atlantis and Comex dives offered further glimpses into a deeper future. In some cases, too, researchers had found that HPNS symptoms would eventually pass, as if divers somehow became acclimatized to the pressure. That phenomenon, like a number of others related to HPNS, was not well understood, but it did seem to leave open the possibility that saturation divers really could go deeper still, if anyone was willing to try.

It was not out of the question that divers might somehow cross a pressure threshold and break free of HPNS, like astronauts who must endure punishing g-forces to escape the earth's gravitational field, but once in space find themselves in a comfortable state of weightlessness. Divers, however, would still have to contend with the uncomfortable and potentially deadly effect that high pressure had on their breathing gases by making the gases more dense, and thus more like breathing a liquid. In fact, the idea of breathing a liquid instead of gases, like a fetus in the womb, had been tried in the 1960s on animals with impressive results. Liquid breathing, often referred to as fluid breathing, was the concept behind Cousteau's prediction about Homo aquaticus. George Bond, too, once saw great potential in fluid breathing. It could eliminate the need for decompression, but in the end a fluid breather still couldn't escape the effects of pressure, and therefore the unpredictable effects of HPNS.

To one of the old questions—*How deep can a man go?*—there appeared to be an answer. It seemed clear from the German experiments and other research that divers could safely reach depths down to two thousand feet, and that they could live those many atmospheres away for an indefinite duration. But the other old question—*How long can a man stay down?*—begged a further question: Does that man have a place to live?

Setting up habitats across the continental shelves and slopes of the world was still possible, and at much greater depths than the few atmospheres at which Hydrolab or Tektite were placed. President Johnson's blue-ribbon panel, impressed by Sealab and Conshelf, had recommended spending half a billion dollars to set up sea-based laboratories on the continental shelf, both fixed and portable. The approach gaining momentum was to develop a mobile habitat that would be more autonomous than Ed Link's *Sea-Link*, but not as big and mission-specific as a Navy spy submarine with a built-in mini-habitat. The vessel would be more like the kind of mobile laboratory that Argyronète was supposed to become before Cousteau ran out of money.

After more than a decade in storage, the unfinished Argyronète was acquired in the early 1980s by a partnership of Comex and IFREMER, the French government's oceanic research agency. Henri Delauze, who once "carried Cousteau's bags," as he would sardonically say, was now going to take over the project his more celebrated countryman had been unable to complete. Within a few years IFREMER and Comex spent millions more francs to transform Argyronète into a vessel unlike any other in the world. When Cousteau asked for royalties for the use of its name, the partners promptly rechristened it with a more pragmatic-sounding moniker: Sous-marin d'Assistance à Grande Autonomie, or SAGA I, whose name trumpeted its highly autonomous design. The Roman numeral was an indication that Delauze hoped to build a larger version, SAGA II, if all went well.

With its built-in mini-habitat for up to six saturation divers, SAGA I provided both a hyperbaric shelter and mobility. It was equipped to remain submerged for up to four weeks and had the capacity to cover a radius of 150 miles from a home port. The divers could leave their pressurized quarters at depths down to nearly fifteen hundred feet. The sub was also equipped to release remotely operated vehicles at almost two thousand feet.

At the launch of the first SAGA in 1987, the French prime minister, Jacques Chirac, came to Marseille to see it for himself—although it wasn't so unusual for the French to give their undersea ventures the kind of attention reserved for the space program in the United States. The SAGA was a chubby-looking sub, painted a jubilant yellow that brought to mind the beloved yellow submarine in the animated Beatles movie. It was another couple of years before the SAGA went out on its first commercial assignment. The undersea worksite happened to be off the dazzling coast of Monaco, where years before Delauze had met with Cousteau and declined a tempting job offer. From its hangar in Marseille, the SAGA sailed a hundred miles to the east along the scenic Côte d'Azur to the dive site. At the end of August 1989, the sub positioned itself on the bottom for some otherworldly work at a sewer outfall, at a depth of about 330 feet. The divers then swam out of their yellow submarine to spend more than five hours on the job, followed the next day by a second shift.

Nine months later, in May 1990, the SAGA ventured eastward along the coast to Cap Benat, near Toulon, where Comex divers locked out at a trial depth of 1,040 feet, far surpassing Ed Link's old record with his first lockout submersible, *Deep Diver.* Six months after that demonstration, the SAGA sailed out to the Frioul Islands, just a few miles offshore from Marseille, in the same area where Cousteau and his *Calypso* crew recovered relics from a Roman shipwreck in the early 1950s. Delauze, too, was forever fascinated with finding and retrieving artifacts from the sea floor. In the 1960s, he had lured Robert Sténuit away from Ocean Systems, not long after Ed Link formed the company, and put the Belgian diver in charge of archaeological missions that Delauze sponsored for years through Comex. Now, out around the pale rocky islands where Cousteau shot the film footage featured years before in his American television debut, Delauze would get some airtime for himself. The SAGA slipped beneath the surface, down to a depth of two atmospheres or so, and Delauze locked out of the submarine with Nicolas Hulot, a French ecologist and host of a popular TV program, who filmed the dive for a documentary.

The yellow submarine clearly had potential, but a new director of IFREMER cut off his agency's funding. Delauze had intended to find a commercial niche for his craft, but market forces effectively drove the SAGA back into its storage hangar, where it has remained. In 2001 it was given to the city of Marseille. As one Comex employee put it, the SAGA was a solution looking for a problem. One problem was clear: not enough money. Even Henri Delauze, who was more eager than most to invest in new undersea technologies, could not foot the bill alone.

Back in the United States, some had lobbied for a wet NASA, an agency whose civilian purposes, and therefore funding, could transcend practical industrial needs and operational military requirements. Such an agency would also work with industry and the military, especially the Navy, but would be in a powerful position to pursue undersea exploration for exploration's sake—and an updated Man-in-the-Sea program that took fresh aim at depth barriers and deep frontiers previously untouched. Missions could be undertaken for purposes of exploration, or science, or simply because the inquisitive human species is a pioneering species. Why, after all, learn to fly? Why go to the moon? Why go West? Why climb Everest?

A wet NASA was not to be, but in the mid-1970s, as Argyronète languished, NOAA got a million and a half dollars to draw up preliminary plans for a mobile undersea laboratory. The American plans heeded the call of Johnson's blue-ribbon commission for Sealab-style outposts on the continental shelf and down to the continental slope. The vessel envisioned was called Oceanlab. The basic blueprint called for the mobility and autonomy of a submarine, equipped so that researchers could lock out and get directly into their undersea environment, to observe, explore, or do whatever useful work they needed to do, as free-swimming saturation divers. With a depth capability of a thousand feet or so, this mobile lab would give researchers and explorers of all stripes a way to visit thousands of acres of undersea frontier. It might also be designed to launch ROVs, remotely operated vehicles, or even small submersibles if necessary. In other words, Oceanlab might look a lot like the SAGA.

A preliminary review found no shortage of scientific interest in Oceanlab, much as there had been ample civilian interest in Sealab. A list of four hundred research proposals was compiled, research that might best be carried out directly by man in the sea, rather than by other means, like from inside a submersible. The list effectively echoed George Bond's scientific sermons about making great mineral discoveries, meteorological discoveries, medical discoveries, and discoveries about how to harvest edible protein to feed a growing population. As Bond was preaching early on, *the very survival of the human species depended on pursuing a whole range of undersea activities.*

A few years later, about the time of Ronald Reagan's landslide victory in the 1980 presidential race, the Department of Commerce, the Office of Management and Budget, Congress, and others got a look at the plans for Oceanlab and its potential price tag: upwards of $25 million. That was considerably less than the French partners spent on the SAGA I and a drop in the bucket compared to space program budgets. Some might even call it a bargain. Nonetheless, the Oceanlab plans were shelved. The nation that poured billions into the moon shot—and had lately entered the space shuttle business—was not in the mood to attempt any sort of like-minded quest in the sea, even at a fraction of the cost, even if the rewards might rival those reaped in space. Reports that the Russians were planning a mobile undersea laboratory of their own, presumably for nonmilitary purposes, did not spark a Sputnik-like reaction in the United States.

The press scarcely took note of the Oceanlab plans. And very soon, two important voices for man in the sea would no longer be heard. Ed Link died on Labor Day in 1981, at his modest home in Binghamton, New York. He was seventy-seven and had been in deteriorating health, but that had not stopped him from expressing dismay at the poor design of his wheelchair. Just a couple of weeks before his death, he had pledged to build a better one. George Bond died less than two years later, in 1983, at the age of sixty-seven. He was buried in Bat Cave, in the little hillside cemetery at the Church of the Transfiguration, his regular sanctuary during those long days as a country doctor. The Ocean Simulation Facility was dedicated to Captain Bond in 1991, and the Navy that had once shunned his ideas officially recognized him as "the father of saturation diving." Captain Walter Mazzone, who might justly be called "the other father of saturation diving," kept busy with a variety of projects after his Navy retirement, as was his energetic way. Soon after celebrating his ninety-first birthday in January 2009, he attended a final reunion of Sealab personnel. His handcrafted stained glass had long been prized at reunion raffles.

Jacques-Yves Cousteau had once talked about building at least two more Conshelf habitats with the ultimate goal of housing divers at more than six hundred feet. Instead Cousteau embarked on Argyronète, but when the money ran out, he would abandon saturation diving and undersea living to focus on popularizing ocean research and conservation through his books, movies, and television programs. Long before Cousteau died in 1997 at the age of eighty-seven, at his home in Paris, his name had become synonymous with exploring the silent undersea world, if not with the quest to live in it.

The U.S. Navy, a benevolent big brother to the friendly rivals, eased out of saturation diving, in part for reasons similar to those of the offshore industry—the high cost, advances in ROVs and other undersea technologies, including one-atmosphere diving suits, sometimes called hard suits.

The concept of hard suits had been around for years, since eighteenth-century designs that weren't much more than barrels with arm holes. Modern one-atmosphere suits look like a cross between an armored space suit and a robot costume. In essence, the suit works like a submarine that a diver can wear: It keeps the external pressure out and a comfortable single atmosphere of pressure inside, with ordinary air to breathe. That means no worries about compression, decompression, gas mixtures, narcosis, or HPNS.

At the dawn of the twenty-first century, innovative designs, improved materials, and technology combined to make possible some one-atmosphere suits that are sturdy enough to keep thousands of pounds of water pressure at bay, but also agile enough to give a diver the mobility and dexterity that had mostly been lacking. In 2006, off the coast of La Jolla, once home to the Tiltin' Hilton, a Navy diver donned one of the latest one-atmosphere suits and was lowered, like a robot on a fishing line, to a record depth of two thousand feet, for a brief stay. Like ROVs, the new hard suits are another valuable undersea tool, and have gained popularity in the Navy and in the offshore industry. Still, neither is quite the same as dispatching a free-swimming diver.

The Navy's two submarine rescue ships, the *Pigeon* and the *Ortolan*, equipped with saturation systems when they were built in the early 1970s, were taken out of service by the mid-1990s. At about that time, ironically, the Navy got involved in a multiyear undersea project for which saturation diving was the key. This was the salvage of the USS *Monitor*, the famous ironclad gunship that sank during the Civil War in a storm off North Carolina's Outer Banks. The wreck was at a depth of 240 feet. Saturation diving was clearly the best way for Navy divers to reach the sunken ship and work on it, so the Navy rented a saturation system from a commercial diving company.

The joint operation with NOAA became one of the largest archaeological recovery projects ever conducted. By the project's end, in 2002, divers working with surface crews had recovered *Monitor*'s steam engine, a significant portion of its hull, and its gun turret, along with smaller artifacts. In the meantime, the *Elk River*, the support ship for Sealab III that had been turned into a saturation diving school, had gone into mothballs and was ultimately sunk in gunnery practice in 2001. Sealab III, which might have made a worthy Smithsonian exhibit, if nothing else, had long ago been cut up for scrap, despite Captain Bond's hope that the habitat could somehow be resurrected.

After Sealab, Bob Barth was one of the many divers who found his way into the oil patch. Barth initially went to work in the Middle East for Henri Delauze and Comex, as did Bill Meeks, Barth's buddy from Sealab II. Within a couple of years they left Comex to start their own diving company in Dubai. They opened several offices around the world and ran their company through the mid-1970s before selling it.

Barth came back to the United States and eventually joined the many ex-Navy divers at Taylor Diving. When Taylor's business slumped with the rest of the offshore industry in the 1980s, Barth returned to the Navy as a civilian, working at the new Experimental Diving Unit as a diving accident investigator. For years a door to his office was marked "Dinosaur Locker," an endearing tribute, one would assume, to his years of experience. The work at EDU brought Barth back to Panama City and the grounds of the former Mine Defense Lab. Years later, in the same salvage yard where George Bond first showed Barth the old mine floats that would be transformed into Sealab I, Barth spotted the carcass of a diving bell that looked awfully familiar. He learned that the neglected bell was one of the Personnel Transfer Capsules from the *Elk River*, in fact the very PTC in which Barth rode for the ill-fated dives to open Sealab III. A relic of Navy history and of human history, it seemed to deserve a more dignified resting place.

Barth arranged to have the PTC moved down to the Florida Keys, where the experience of underwater living goes on, in relatively shallow waters, at the three remaining American habitats in regular operation. Marinelab is in a Key Largo lagoon. It started out as an engineering project by U.S. Naval Academy students and is now a tank-shaped shelter mainly used for training, teaching, and research. Nearby, at the same no-decompression depth of just under thirty feet, is the research habitat formerly known as La Chalupa.

In the 1970s La Chalupa housed a series of aquanaut teams at depths down to a hundred feet off the Puerto Rican coast, but when funding for its research programs evaporated, it was towed back to Florida, where it had been built. The habitat was bought in the 1980s by partners who remodeled it into the world's first underwater hotel, Jules' Undersea Lodge—named for Jules Verne, of course. Among the names in the guest book: former Canadian prime minister Pierre Trudeau and Steven Tyler, the frontman of the rock band Aerosmith. Both Marinelab and Jules' Undersea Lodge are run by a nonprofit foundation whose entrepreneurial chief, Ian Koblick, has long been active in the marine world and was himself an aquanaut on Tektite and on La Chalupa, for which he was also a lead designer.

The old *Elk River* PTC sat outside Koblick's offices near the lagoon for a while but finally found a new home near a third habitat, called Aquarius, the world's only undersea research station, as far as its operators know. Aquarius was built in the 1980s as a replacement for the modest but much used Hydrolab and was initially dubbed the George F. Bond. The name of the father of saturation diving was inexplicably dropped in favor of Aquarius, which is owned by NOAA and run on a minimal budget by the University of North Carolina, Wilmington. The habitat is perched on a reef that's nine miles directly south of Key Largo, in the Florida Keys National Marine Sanctuary. It's at a modest depth of about sixty-five feet, which puts the entry hatch at about fifty feet. Aquarius is a horizontal tank of sorts, forty-three feet long, about the same size as Sealab I, but with more elaborate contours and additional features like the wet porch, which gives divers a staging area a few feet below the large open hatch in the habitat's underbelly.

NASA astronauts have been regular occupants of the habitat, since living in the sea, even at a modest depth, is one of the more realistic ways to prepare for life in space. Marine researchers have logged hundreds of hours in Aquarius since its debut at this location in the early 1990s, and Navy divers have used the NOAA habitat for various training missions. When the shell of the old PTC was finally affixed to the seabed in 2007, the plan was to gradually equip the stripped-down capsule so that it worked like a diving bell similar to the one that was being designed for a new, portable chamber-and-bell setup the Navy was building. This surface-based unit, known as the Saturation Fly-Away Diving System, looks something like those in commercial use and is intended to replace the saturation systems the Navy had mothballed a decade earlier. The new saturation system is designed to make possible six-man, month-long diving missions anywhere in the world, down to a thousand feet.

The old PTC was placed to the southeast of Aquarius, about two hundred yards away, like a hidden monument to Sealab. The considerable distance put the pod at a depth of 112 feet, an inadvertent bit of symbolism in that this was about the same depth as the old submarine escape training tank at New London, where George Bond's quest began. The capsule was also far enough from the habitat to give divers a genuine sense of what it's like to work on their own from out of a bell in a remote sea, something that not even the Ocean Simulation Facility could truly simulate.

Over the years the Aquarius staff had developed way stations that looked like little gazebos, open at the bottom with pockets of breathable air in their steel canopies. Divers could pop up inside for a rest, a chat, or in an emergency. In more recent years, to streamline the process of accomplishing underwater work, divers could also replenish their scuba tanks at the elevator-sized way stations, which were set up whenever divers were working far from the habitat, sometimes as far as a thousand feet—no trivial distance in a liquid haze. The *Elk River* PTC was itself some six hundred feet distant from Aquarius, across a patchwork of coral thickets and sandy groves. A way station was being set up in early May 2009 to create a haven midway between the habitat and the PTC.

Dewey Smith, a veteran diver and former Navy medic who had been with Aquarius for a couple of years, was in the water to assist two Navy divers who were using an underwater jackhammer to anchor the way station into place, at a depth of about eighty-five feet. As the two Navy divers, Bill Dodd and Corey Seymour, worked the jackhammer, Smith was a few paces away, tending to other equipment. By this time the three had been working around the habitat for almost an hour. It was a routine day on the bottom, as much as any day spent underwater is ever routine. Smith came over at some point to let Dodd and Seymour know that he was going back to the habitat but would return to the worksite shortly. During a brief exchange, in which the divers used hand signals and shouts through their helmets to communicate, Smith made it clear that he was okay. He didn't say why he was going back, but there was nothing unusual about making a pit stop at the habitat—whether to pick up a tool, or to use the outhouse, or any number of reasons—so this was no cause for alarm.

Unlike Smith, the Navy divers were getting their air through umbilicals and also had voice communication with Aquarius. So they let those in the habitat know that Smith had just begun the hundred-yard trek back to the base. On a working dive like this one, the divers all wore boots, not fins, and got around by hiking across the bottom more than by swimming.

The two Navy divers went back to work, wrestling with the noisy jackhammer, their backs turned toward the habitat. Minutes had passed—maybe five, maybe ten; it was hard to say—when diver Seymour caught a glimpse of an odd sight in the distance, a darkish object of some kind about fifteen yards away. In another second he realized it was Dewey Smith, lying motionless on his side in the calm, clear, merciless water. Then came a surreal replay of the tragic events of forty years before, when Bob Barth had spotted Berry Cannon. Seymour found that Smith's mouthpiece was out, but replacing it was no use. He hurriedly picked up Smith, cradling the unconscious diver in his arms, and began a hundred-yard dash to the habitat—although a dash takes place in slower motion when carrying a buddy across a scrubby reef through a medium much thicker than air.

After a couple of minutes they had almost reached the habitat. They had maybe thirty yards to go when Seymour's umbilical got fouled. He was jerked to a halt and could go no further, but Bill Dodd was able to carry Smith the rest of the way while Seymour freed up his own umbilical and made sure Dodd's didn't get stuck. With the help of the others inside the habitat, they got Smith onto the wet porch, up through the hatch, and inside Aquarius. Attempts to resuscitate Smith were ultimately unsuccessful. About three-thirty that afternoon, a Navy medical officer swam down and pronounced Smith dead.

Smith, who was thirty-six, had been given a clean bill of health just prior to beginning this latest Aquarius mission. A triathlete, he was physically fit and known to be a meticulous diver, always attentive to his gear and proper procedures. Even after subsequent tests and investigations, no one knew for sure what had gone wrong. Looming large in the accident's aftermath were questions about the future of Aquarius. The first death of an aquanaut at the world's only sea floor base didn't register much of a blip in the frenetic twenty-first-century news cycle, but it did bring about a heightened sense of caution and a slowdown in Aquarius's underwater activities. Amid escalating talk in Washington about cutting the federal budget, some worry that the NOAA habitat program could lose its shoe-string funding, a lifeline that's already less than $2 million a year. The Lord giveth and the Lord taketh away, as Captain Bond might have observed, with a puff of his pipe. Even with Aquarius keeping the faith, the gospel of living in the sea could still use some preachers.

APPENDIX

SEALAB AQUANAUT ROSTERS

SEALAB I AQUANAUTS

Lester E. Anderson

Robert A. Barth

Sanders W. Manning

Robert E. Thompson

SEALAB II AQUANAUTS

Team 1

M. Scott Carpenter
(team leader)

Berry L. Cannon

Thomas A. Clarke

Billie L. Coffman

Wilbur H. Eaton

Frederick J. Johler

Earl "A" Murray

Jay D. Skidmore

Robert E. Sonnenburg
(Teams 1 and 3)

Cyril J. Tuckfield

Team 2

M. Scott Carpenter
(team leader)

Robert A. Barth

Howard L. Buckner

Kenneth J. Conda

George B. Dowling

Arthur O. Flechsig

Glen L. Iley

Wallace T. Jenkins

John F. Reaves

William H. Tolbert

Team 3

Robert C. Sheats
(team leader)

William J. Bunton

Charles M. Coggeshall

Richard Grigg

John J. Lyons

William D. Meeks

Lavern R. Meisky

Robert E. Sonnenburg
(Teams 1 and 3)

John Morgan Wells

Paul A. Wells

SEALAB III AQUANAUTS

Frederick W. Armstrong
Robert A. Barth
Richard C. Bird
Richard M. Blackburn
Robert A. Bornholdt
Mark E. Bradley
William J. Bunton
Frank Buski Jr.
Laurence T. Bussey
Berry L. Cannon
Derek J. Clark
Kenneth J. Conda
Richard A. Cooper
George B. Dowling
Wilbur H. Eaton
Matthew C. Eggar
Richard A. Garrahan
Leo C. Gies
Lawrence W. Hallanger
David Martin Harrell
Samuel E. Huss
Wallace T. Jenkins
Duane N. Jensen
John C. Kleckner
Cyril F. Lafferty
Lawrence M. Lafontaine

Paul G. Linaweaver Jr.
Fernando Lugo
William P. Lukeman
James E. McDole
James M. Melder
Keith H. Moore
Jack W. Morey
Jay W. Myers
James H. Osborn
Andres Pruna
William C. Ramsey
Lawrence W. Raymond
Frank L. Reando
John F. Reaves
Terrel W. Reedy
Don C. Risk
N. Terrel Robinson
Irwin C. Rudin
William J. Schleigh
Donald J. Schmitt
Richard R. Sutton
Cyril J. Tuckfield
James Vorosmarti Jr.
Paul A. Wells
William W. Winters

ACKNOWLEDGMENTS

A work of nonfiction like this one would have been impossible to write without the cooperation of many people, especially those whose personal knowledge and firsthand experiences form this book's essential blueprint. I learned early on that if I had any hope of telling the forgotten story of Sealab and of saturation diving, I would need to seek out a former Navy diver named Bob Barth.

I first reached Bob by phone at his office at the Navy Experimental Diving Unit, and soon thereafter, in December 2001, with the nation still reeling from the attacks of September 11, I flew to Panama City, Florida, to meet him and to start working through a long list of questions. Bob is not always a man of many words, but those he chooses he does not mince. He makes no secret of his dim view of the media, and by extension those who call themselves journalists, like me. But something in our encounter apparently convinced him that he ought to give me a chance, and that I might actually write a book that did justice to the U.S. Navy program in which he was involved from beginning to end.

Bob and I would meet again several times in person, and by phone and e-mail he would continue to field my seemingly interminable questions with admirable candor, remarkable patience, and, not surprisingly, a peppering of his sardonic wit—samples of which can be found in the pages of his memoir, *Sea Dwellers*, which lends his singular voice to the Sealab experience. Bob also opened his extensive Rolodex to me—a major help in reaching many others, perhaps most critically Captain Walter Mazzone, who was by then in his early eighties and as sharp as the day George Bond first met him. Walt Mazzone is a story in himself, as indicated in the book, and the true embodiment of the Greatest Generation. It's been my privilege to get to know him. He, too, would patiently field many questions, in person, by telephone and e-mail, including questions about the late Captain George Bond, with whom Mazzone worked more closely than anyone.

As central as Barth, Bond, and Mazzone were to Sealab and the development of saturation diving, many others had a hand in shaping the far-flung pieces of a story that spanned the decade of the 1960s and beyond, both above and below the waterline. To help gather those pieces, I had the good fortune of attending four annual reunions of Sealab personnel, and others close to the program, between 2002 and 2005, as explained in the endnotes. These were extraordinary opportunities to meet people I needed to know, to put names to faces, and faces to names, and generally fill in the gaps in my reporting by doing interviews on the spot, or by following up later—in person, by phone or e-mail or some combination. I benefited, too, from just listening in as old friends and shipmates talked and reminisced.

Not everyone showed up at reunions and there were many people I had to catch up with at other times and places. The names of those I interviewed can often be found in the text or cited as sources in the notes, or both, but those mentions rarely describe the full extent of everyone's contributions to my research and reporting. Many sources are not specifically cited at all, but I am equally grateful for their time and assistance. Their collective input and insights made a big difference in my ability to tell the Sealab story as fully and fairly as possible. So, for the record, in addition to Bob Barth and Walt Mazzone, I would like to thank, in alphabetical order, the following people, most of whom were personally involved in one or more of the Sealab experiments or related activities: Charles Aquadro, Frank "Tex" Atkinson, Richard Bird, Richard Blackburn, Jim Bladh, Robert Bornholdt, Robert Bornmann, Glenn "Tex" Brewer, Bill Bunton, Scott Carpenter, Derek "Nobby" Clark, Tom Clarke, Billie Coffman, Charles Coggeshall, Richard Cooper, John Craven, Bill Culpepper, George Dowling, Matthew Eggar, Richard Grigg, Martin Harrell, John Harter, Wally Jenkins, Charles E. Johnson, Cyril Lafferty, William R. Leibold, Paul Linaweaver, Fernando Lugo, Mal MacKinnon, Jim McCarthy, Bill Meeks, Lewis Melson, James W. Miller, Keith Moore, Jim Osborn, Andres Pruna, Andreas B. Rechnitzer, Jack Reedy, Don Risk, Jack Schmitt, Robert Thompson, Jack Tomsky, Cyril Tuckfield, James Vorosmarti, Ken Wallace, and William Winters. My sources would often remind me, with good-natured stoicism, that no one in this group was getting any younger, and that I had better get all my interviews done sooner rather than later.

I'm sorry to say that more than a few people on the list have not lived to see the publication of this book, but I'm grateful to have found them in time to learn from them and to hear their stories. To anyone I may have overlooked—my apologies.

Bornmann, Linaweaver, and Vorosmarti, who were all captains in the medical corps upon leaving the Navy and continued to work as doctors, provided additional assistance in bringing me up to speed on the various physiological effects of diving. Dr. Bornmann also shared his recollections about working on the first saturation dive at sea with Ed Link, who died years before work on this book began. Robert Sténuit, the Belgian diver who worked with Link, answered questions that helped me better understand the published accounts I had read about him and his work. Helen Siiteri gladly shared some observations and forwarded some useful papers that were still in her possession from the early 1990s when she edited *Papa Topside: The Sealab Chronicles of Capt. George F. Bond, USN.*

Many who were associated with Sealab went on to work in the off-shore oil industry and described those subsequent experiences to me. I also turned to a number of others to learn about oil field diving, including the crowd at the sixteenth annual divers' reunion held in 2007 at Dick Ransome's roadside bar in Bush, Louisiana. Among those I met was Harry Connelly, who took time during the day-long cookout to answer many questions. He later loaned me his rare collection of *Undercurrents*, a short-lived trade magazine that provided a valuable window on commercial diving, circa 1970. Jack Horning, another attendee, shared memories and also photographs and mementos. Lad Handelman, Drew Michel, John Roat, R. J. Steckel, Bob Tallant, Geoff Thielst, and Dennis Webb were not at Dick's bar that day, but gave me a lot of useful information in separate interviews. Special thanks to Alan "Doc" Helvey, who was exceptionally cordial and patient throughout a series of interviews about his commercial diving experiences, notably those described in Chapter 15.

A number of key Sealab participants died years before I began my research, but family members kindly responded to my inquiries and contributed important details, and often documentation, to bolster the recollections of friends and shipmates. In that regard I am especially grateful to Mary Lou Cannon for speaking openly with me about a heartbreaking time. Lois Birkner Workman, married fifty-five years to Dr. Robert Workman when he died in 1998, filled me in on her husband's life and times, as only a wife could, and offered insight into his mien. Sherry Anderson, the eldest of Lester "Andy" Anderson's five daughters, and Rose Anderson, Lester's wife of nearly fifty years when he died in 1999, were equally helpful, as were the two sons of Robert Sheats, Phil and Carl Sheats, and also Phil's wife, Bonnie.

Captain George Bond died nearly twenty years before work on this book began. But in his stead I found gracious hosts in George Bond Jr. and his wife, Barbara, who had just built a new home in western North Carolina, not far from Bat Cave, and George Jr. introduced me to the hamlet that was so much a part of his father's life. There, in October 2003, I was able to meet and interview some old-time friends and family members, including Ellen Moorehead, née Barrino, the late Dr. Bond's sister-in-law, and Ellen's daughter, Lynn, who lived for a time during her formative years with the Bond family. William A. Burch, a former Bat Cave postmaster, took the time to drive the bumpy back road to Bond's old cabin retreat and share stories about his good friend Dr. Bond.

The Bonds insisted on putting me up in their new guest room while I spent several days sifting through a fantastic array of old files, papers, clippings, and photographs stored in their basement. Theirs was one of a number of private collections in which I found documentation I probably never would have found anywhere else, and which proved critical to pinning down numerous facts about Dr. Bond and about Sealab.

By the time I visited the Bonds I had learned that official records pertaining to Sealab were spotty at best. This I determined with the assistance of David Osborne, an intrepid Library of Congress researcher who took it upon himself to scour government collections for records related to the Sealab program. It became clear that instead of visiting neatly catalogued archives, I would have to rely more on happenstance and a kind of door-to-door research to get my hands on materials stowed away in places like the Bonds' basement. But Osborne did succeed in retrieving hundreds of publicly available pages that would prove critical to documenting the tragedy of Sealab III and the subsequent investigation. His fruitful efforts were above and beyond the call, and I thank him.

Other librarians became essential allies, too, in my hunt for documentation. Bonnie Davis, at the Navy Experimental Diving Unit, promptly responded to my sporadic requests, usually for arcane, half-century-old Navy reports. Beth Kilmarx, who handles the Link Collections at Binghamton University of the State University of New York, made my days spent digging through the library holdings pleasant and productive. With the help of Anne Magill and Merrie Monteagudo, librarians and keepers of the *San Diego Union-Tribune* archives, I was able to rifle through a rough draft of history, mostly about Sealab II, that had been published in the rival newspapers of the time, the *San Diego Union* and the *Evening Tribune*. Douglas Hough at the Man in the Sea Museum in Panama City Beach answered some early questions. The Scripps Institution of Oceanography Library in La Jolla, California, and the Submarine Force Library and Museum in Groton, Connecticut, were both worthwhile destinations for obtaining additional background material. I was able to view the pertinent televised record at the wonderful Museum of Television and Radio, now the Paley Center for Media, at both the Beverly Hills and New York City locations.

Over the course of researching this book, I came to rely on the friendly staff of Pelletier Library at Allegheny College, notably Cynthia Burton and Linda Ernst, whose seemingly magical search powers made even the most obscure old diving books and articles appear. Others at the liberal arts college in Meadville, Pennsylvania—where I filled in as a journalism instructor for a couple of years—would share their knowledge with me, including Professor Glenn Holland, in the Department of Philosophy and Religious Studies, who took the time to help me better appreciate George Bond's biblical references. Lee Coates, associate professor of neuroscience and biology, offered guidance as I got up to speed on physiology and the mechanics of respiration. The expertise I seemed to need most often came from computing services guru Phil Reinhart, who talked me through more than a few technical difficulties. When it came time to digitize the old photographs I had collected for publication, I got first-class lessons from Dennis O'Laughlin, a gifted and indefatigable teacher of commercial art at the Crawford County Career and Technical Center in Meadville.

With respect to the technical, physiological, and even historical side of diving, I found a marvelous teacher in Peter Bennett, whose roots in the world of diving truly run deep. He is professor emeritus of anesthesiology, a longtime director of the F. G. Hall Hyperbaric Center at Duke University Medical Center, founder and first president of the Divers Alert Network, author of innumerable scientific papers, and coeditor with fellow researcher David Elliott of *The Physiology and Medicine of Diving*, a definitive textbook since Bennett and Elliott edited the first edition in 1969. Dr. Bennett cheerfully gave me more than a few outstanding crash courses, and his wealth and breadth of experience, including his familiarity with many key figures in the field of diving research, made him an exceptionally valuable resource. He was also among those who reviewed drafts of my manuscript to weed out mistakes and factually dubious material.

Where diving history was concerned, or almost any question about diving, I could count on Leslie Leaney to point me in the right direction. It was Leaney, after all, who first told me I should enlist the assistance of Bob Barth if I wanted to write anything meaningful about Sealab. Leaney's long-standing interest in diving led him to become a cofounder of the Historical Diving Society, and then founder of *Historical Diver* magazine, a highly informative publication recently renamed *The Journal of Diving History*, which Leaney has served as publisher, editor, writer, and enthusiastic researcher. Leslie's wife, Jill, is herself a great resource, as are other knowledgeable HDS members who came to my aid, like Kent "Rocky" Rockwell. Leslie Leaney was also among those who read and commented on my manuscript. Others from whom I sought comments were Bob Barth, Walt Mazzone, Richard Blackburn, and Robert Bornmann, whose names will be familiar from the Sealab story and from these acknowledgments. A couple of other important manuscript reviewers merit some introduction, although both are well known among their Sealab peers. Don Risk, a former Navy diver who later worked in the offshore oil industry, was involved throughout the program and personally knew many of the participants. James Vorosmarti, in addition to having been a designated medical officer during Sealab III, is an avid researcher of diving history. All of these readers' comments were invaluable in achieving the greatest possible degree of fairness and accuracy. Any remaining deficiencies in the text are nobody's fault but mine.

Much has been written about the legendary Jacques-Yves Cousteau, much of it written or edited by Cousteau himself. To get some further perspective on the documentary record, both written and cinematic, I set off for France in late 2004 to meet with some key members of the Cousteau team, men who began working with JYC early on, even before the *Calypso* first set sail in the early 1950s, and who were with him for years to come. Jean Alinat, a former French naval officer, diver, and longtime leader on the Cousteau team, could not have been more kind and candid when interviewed at his home in Nice, from where he commuted for years to nearby Monaco and his office at the Oceanographic Museum. When my French faltered, he summoned his latent English. I had much the same experience with another Cousteau associate, André Laban, with whom I was able to meet several times during the thirty-first annual Underwater Film Festival in the little seaside city of Juan-les-Pins, west of Nice. Laban, a gentle and creative soul whose clean-shaven head has long been a trademark, is, like so many others I met, a story in himself.

Like the Sealab reunions, the film festival gave me an opportunity to meet and compare notes with a cross section of divers from all over, including Jean-Michel Cousteau, whom I had previously interviewed by phone about his father's work.

In Marseille, at the world headquarters of Comex, I got a warm welcome from Henri Delauze and an educational introduction to the company's Hyperbaric Centre from scientists Bernard Gardette and Claude Gortan. Nicolas Boichot, then Delauze's chief assistant and a specialist in underwater archaeology, took care of business in the most efficient way. Delauze himself, an active septuagenarian who exuded a palpable *joie de vivre*, took time from a busy schedule for several interviews, responded to follow-up questions by e-mail, and provided important documentary materials. Through Delauze I was able to meet Claude Wesly, also of Marseille. A former oceanaut on Conshelf One and Two, Wesly later worked with Delauze and Comex. He shared mementos and memories from his years working with his good friend and mentor, Captain Cousteau, and offered personal insights into Cousteau and Delauze, having known and admired both men. Thanks, too, to former Cousteau associate Dominique Sumian, for helping me make my initial French connections.

The journalists with whom I've worked over the years, many of them still friends, and most still working as journalists, have been my teachers, mentors, and continuing sources of inspiration. There are too many former colleagues to mention, but they know who they are, especially those with whom I worked at the *Santa Barbara News-Press*, where I was a staff writer for most of the 1990s, when the paper was still owned by the New York Times Co. It was a job, and a place, that afforded greater opportunities than I could have imagined. It was in Santa Barbara, too, that the seed of the idea for this book was planted. I owe thanks to a number of editors, including Allen Parsons and Melinda Johnson, whose years at the paper overlapped with most of mine, and whose enthusiasm for my reporting and writing helped put me on the road to tackling a book-length work of nonfiction. Special thanks, too, to R. B. Brenner, whose time as an editor at the paper in the mid-1990s was relatively brief, but long enough for him to become a friend and a great influence on me and my development as a journalist. Jeff Gordinier, yet another *News-Press* alum, has remained a friend and journalistic Yoda ever since our paths crossed in the paper's lifestyle section in the early 1990s. Melissa Grace became a colleague all over again by volunteering to spend some of her rare off-hours at the New York Public Library to aid in my frantic search for a ridiculously elusive article.

Scott Waxman, my agent, immediately embraced my idea for a book about Sealab, took me under his capable wing, and showed me how to make my book proposal fly. Fly it did, into the experienced hands of my editor at Simon & Schuster, Bob Bender, a true veteran of the publishing business, responsible for an extraordinary body of nonfiction work. I was thrilled that someone of his stature saw promise in the Sealab story, and I was grateful that he and the good people at Simon & Schuster, including Bob's assistant, Johanna Li, were willing to take a chance on a first-time author.

Friends and family, of course, make all things possible, and a complete who's who would be impossible. Lucky me. But it took a village to get this book written and a highly abridged list of the villagers must include Rick, Carol, David, and Ellie Holmgren, who often provided a second home to my daughter, among other friendly deeds. My son has had a second home in his public school music programs thanks to an energetic cast of teachers, namely Jill Manning, Irene Kipp, Sandra Corbett, Mary Peters, Duane Banks, and Scott Cresley. Many thanks to villagers Dan Crozier, Roberta Levine, and their kids, Lucy and David, for all the gourmet meals and generous helpings of laughter. Roberta, a wise wordsmith and straight shooter, also read my manuscript and gave me the benefit of her keen literary eye.

And where would I be without my mother, Abby Hellwarth, queen of the village, who's always been an incomparable bastion of love and moral support. My father, Robert Hellwarth, along with my stepmother, Theresia DeVroom, and half-brother, Will, make a formidable cheering section, and I have long relied on my sister, Margaret, and my brother, Tom, for the best in sibling wisdom—which only improved when my sister-in-law, Fiona, joined the family. To Steve and Beverley Dorfman, in-laws extraordinaire, consummate supporters of my family, and voracious readers of nonfiction to boot—including drafts of my manuscript—I say a heartfelt thank you.

To my wife, Jennifer, I owe thanks that run deeper than any dive chronicled in this book. She, more than anyone, has had to endure the prolonged and unpredictable adventure that writing this book became, all while pursuing her own career as a teacher and scholar, and all while we both rode the roller coaster of parenthood. Our amazing kids, Sutter and Camryn, who were in elementary school when this project began, are both easily old enough now to read this book. In between the many lines, I hope they can begin to understand why their dad spent all those hours holed up in his basement office. But they should remember, too, that my greatest fringe benefit was working from home, where I could readily watch (and hear!) the two of them grow up. We've been able to share many moments, mundane and magnificent, and that's been a priceless story in itself.

NOTES

All personal interviews were conducted by the author. Tape-recorded interviews are so noted, as is the location of an interview, when held face-to-face rather than by telephone. For virtually every interview the author has extensive notes, whether they were transcribed from tapes, taken during the interview, or written up shortly afterward. After a first full citation, interviews are cited by the subject's last name and the date of the interview only; in a few instances a first name or initial is needed to distinguish people with the same last name. When information was available from a documentary source, such as a book, magazine, journal, or an official document, that source is often the one cited, even if the same information was obtained through interviews. This was done to direct the interested reader or researcher to the most readily accessible source. Written records, when available, and occasionally audio or visual records, provided valuable corroboration for information obtained from interviews, and often included additional details— and vice versa, making memory a two-way street. But interviews formed the bedrock of this story, whether specifically cited or not. In this regard, the author was fortunate to have been able to attend four reunions of Sealab personnel and others close to the program as part of the research for this book. Two of those reunions were held in San Diego, Calif., March 14–17, 2002, and May 9–11, 2004. Another two were held in Panama City, Fla., March 6–9, 2003, and March 10–12, 2005. Several dozen people typically attended each gathering, thereby making many interviews possible, both on the spot and, in many cases, later, in more formal, one-on-one interviews. Equally important was the opportunity to listen to old friends and shipmates talk and reminisce.

CHAPTER 1: A DEEP ESCAPE

Page

1 *submarine* Archerfish *prepared to release:* Cmdr. George F. Bond as told to Max Gunther, "We Conquered the Deep 300," *True The Man's Magazine*, November 1960, p. 48; Cyril Tuckfield, telephone interview, Aug. 21, 2002, and in Preston, Conn., Sept. 11, 2002.

1 *George Bond was forty-three:* Based on numerous interviews with family, including Bond's older son, George Bond Jr., close friends, and associates; Bond's edited journals, published as *Papa Topside: The Sealab Chronicles of Capt. George F. Bond, USN,* Helen A. Siiteri, ed. (Annapolis, Md.: Naval Institute Press, 1993); unpublished letters and journals; U.S. Navy personnel files, photographs, audio and video recordings. These sources and others are cited below as they pertain to specific information in the text.

1 *resonant baritone:* Cmdr. George F. Bond of the Naval Medical Research Laboratory, New London, Conn., tape recording, keynote speech at the Alpha Omega Alpha initiation ceremonies, Huyck Auditorium, Albany Medical College, Albany, N.Y., April 23, 1959 (in author's possession).

2 *primitive at best:* Carl LaVO, *Back from the Deep: The Strange Story of the Sister Subs* Squalus *and* Sculpin (Annapolis, Md.: Naval Institute Press, 1994), p. 31.

2 *three submarines sank:* A. J. Hill, *Under Pressure: The Final Voyage of the Submarine S-Five* (New York: Free Press, 2002), pp. 61, 206.

2 *escape trunks:* Sir Robert H. Davis, *Deep Diving and Submarine Operations: A Manual for Deep Sea Divers and Compressed Air Workers,* 9th ed. (Cwmbran, Gwent: Siebe, Gorman, 1995), pp. 257, 650; Tuckfield, interviews, Aug. 21, 2002, and Sept. 11, 2002; *Submarine Escape,* Navy-produced training film (video copy in author's possession).

2 *McCann Rescue Chamber:* Davis, *Deep Diving and Submarine Operations,* pp. 655, 670; Peter Maas, *The Terrible Hours: The Man Behind the Greatest Submarine Rescue in History* (New York: HarperCollins, 1999), p. 63.

2 *Several methods evolved to enable:* George F. Bond, Robert D. Workman, and Walter F. Mazzone, "Deep Submarine Escape," U.S. Naval Medical Research Laboratory Report, no. 346, Dec. 30, 1960, p. 1.

2 *Boyle's law:* U.S. Navy Diving Manual, Part I, Department of the Navy, July 1963, p. 18.

2 *The gravest danger:* U.S. Navy Diving Manual, Part III, p. 11; Bond, recorded keynote speech at Albany Medical College.

3 *escape training tank:* Cultural Resource Group, Louis Berger & Associates, "Historic Structure Documentation for Submarine Escape Training Tank, Naval Submarine Base, New London, Groton, Connecticut," prepared for Northern Division Naval Facilities Engineering Command, Philadelphia, Pa., March 1988 (held in the collection of the U.S. Navy Submarine Force Library and Museum, Groton, Conn.); *Submarine Escape.*

3 *Esther Williams:* Tuckfield, interviews.

3 *rite of passage:* Cultural Resource Group, "Historic Structure Documentation for Submarine Escape Training Tank," p. 1.

3 *resembled a silo:* Archival photographs at U.S. Navy Submarine Force Library and Museum, Groton, Conn., viewed by author Sept. 12, 2002.

3 *yes, in theory:* Bond et al., "Deep Submarine Escape," p. 1.

4 *Tuck never swore:* Tuckfield, interview, Sept. 11, 2002.

4 *prior calculations:* Bond et al., "Deep Submarine Escape," pp. 5, 13.

4 *maximum of three minutes:* Bond and Gunther, "We Conquered the Deep 300," p. 105.

5 *breathing nitrogen under pressure: U.S. Navy Diving Manual,* Part I, p. 70; Bond, recorded keynote speech at Albany Medical College.

5 *instructors would routinely jab:* Bob Barth, interviews, Sept. 24 and 26, 2002; Tuckfield, interviews, Aug. 21, 2002, and Sept. 11, 2002; interviews with other former training tank instructors; *Submarine Escape.*

6 *Charlie Aquadro took his turn:* Dr. Charles Aquadro, interview, Beaufort, N.C., May 16–18, 2003.

6 *Bond sweated it out:* Memorandum from Arline Hibbard, secretary of the American Medical Association Council on Rural Health, "Sweetness and Light as a Navy Doctor Sees It," Nov. 10, 1954 (copy in author's possession); Aquadro, interview, May 16–18, 2003.

7 *"After you, Courageous Leader":* Bond and Gunther, "We Conquered the Deep 300," p. 107.

7 *the trunk fell silent:* Ibid., p. 108.

7 *quizzed himself:* Ibid.

8 *raised his left arm:* Bond et al., "Deep Submarine Escape," p. 15.

8 *passed through a thermocline:* Bond and Gunther, "We Conquered the Deep 300," p. 109.

8 *he had nothing left to blow:* Tuckfield, interviews, Aug. 21, 2002, and Sept. 11, 2002.

8 *Fifty-three seconds:* Bond et al., "Deep Submarine Escape," p. 13.

9 *produced his pipe:* Bond and Gunther, "We Conquered the Deep 300," p. 109; official U.S. Navy photograph (in author's possession).

9 *feat was noted in:* "Navy Man Escapes Sea from 300 Feet Deep," *New York Times,* Oct. 4, 1959, p. 35; "Up from the Bottom," *Time,* Oct. 19, 1959, p. 59.

9 Today *show:* A Program Analysis card from the NBC News Archives says Bond was on the show Oct. 9, 1959 (in author's possession).

9 *on* Conquest: Craig Gilbert, "Escape from Danger," *Conquest* shooting script, Feb. 19, 1960 (copy in author's possession).

9 *tests to analyze possible causes:* George F. Bond, "Escapes from Sinking Jet Aircraft Cockpits," U.S. Naval Medical Research Laboratory, New London, Conn., Memorandum Report 59-2, May 25, 1959.

9 *A lengthy story:* Gene Witmer, "Cmdr. Bond Escapes from 'Aircraft' Underwater to Come Up with Answers," *New London* (Conn.) *Evening Day,* Aug. 28, 1959.

9 *born on November 14, 1915:* The date appears on Bond's gravestone at the Church of the Transfiguration, Bat Cave, North Carolina; that date, along with his birthplace, Willoughby, Ohio, is noted in numerous documents.

9   *when his father died:* Robert Malby Bond died Sept. 15, 1925, according to Treasury Department, Office of Commissioner of Internal Revenue, Miscellaneous Tax Unit to Louise F. Bond, Administratrix, Estate of R. M. Bond, DeLand, Florida, Oct. 10, 1928.

9   *worked on the school's monthly: The Karux*, Mercersburg Academy yearbook (1933?), G. F. Bond pictured and listed as one of six associate editors of *The Lit* magazine.

9   *known as the class poet: The Karux*, 1933, p. 57.

10 *lifelong fondness for bourbon:* Interviews; "fondness" may be an understatement, at least once Bond was older, according to many who knew him. Bourbon, while his favorite, was not the only liquor he drank. Bond's habitual drinking tended to take place after hours and in social settings and did not generally interfere with his work. He frequently acknowledged his vice, usually in a lighthearted way, as in his paper presented to the Association of American Medical Colleges, cited later, in which he mentions writing an important proposal "after a nice round of three or four potions of bourbon and water and a good meal." Another typical example is found in Bond's "Sealab I Chronicle," also cited later, in which he describes mixing "a very dry martini of about six ounces" upon arriving home from a difficult day at work. He then adds, "Three hours and several martinis later, the course was crystal clear" (p. 5).

10 *"Rebel": The Karux*, 1933, p. 57.

10 *course list revealed:* Office of the Registrar, scholastic record of George Foote Bond, University of Florida, Gainesville, Fla. (in author's possession).

10 *for his thesis:* George Foote Bond, "A Study of an Appalachian Dialect," a thesis presented to the Graduate Council of the University of Florida in partial fulfillment of the requirements for the degree of master of arts, August 1939.

10 *In those formative years:* George Foote Bond Jr., interviews, Hendersonville, N.C., Oct. 10–13, 2003; Ellen Moorehead (née Barrino), younger sister of Marjorie Bond (née Barrino), and Lynn Moorehead, Ellen's daughter, interviews, Bat Cave, N.C., Oct. 10, 2003; William A. Burch, longtime Bond friend, interview, Bat Cave, N.C., Oct. 11, 2003.

10 *Fate himself served:* Bond, "A Study of an Appalachian Dialect," p. 88.

10 *"take no back-sass":* Ibid., p. 104.

10 *once lived with Ben and Dooge:* Ibid., p. 85.

10 *"George, why don't you go":* George Bond, "Doctor on the Mountain," unpublished manuscript (1975), pp. 13, 30. Edited excerpts from this typed manuscript, including a variation on this quotation, can be found in Chapter 1 of *Papa Topside*. Bond was seeking a publisher for this memoir about his rural practice, as noted in a personal letter to Bond from Judith Joye, a New York literary agent, June 2, 1975, and in a feature article by Lewis W. Green, "Mountains Keep Calling Him Back," *Asheville* (N.C.) *Citizen*, Feb. 1, 1970.

11 *married Marjorie Barrino:* Ellen and Lynn Moorehead, interview, Oct. 10, 2003; Bond Jr., interviews, Oct. 10–13, 2003.

11 *had seen two people die:* Ben M. Patrick and Emma Carr Bivins, "Pioneer for Rural Health," *Country Gentleman,* February 1949, p. 31.

11 *enrolled at the University of North Carolina:* UNC Chapel Hill transcript, 1940–41.

11 *started medical school at McGill:* Letter of certification from Donald Fleming, M.D., Secretary, Faculty of Medicine, McGill University, Oct. 14, 1953 (in author's possession).

11 *applied for a reserve commission:* Bond, *Papa Topside,* p. 11; transcript of Naval Service, Capt. George Foote Bond, Medical Corps, U.S. Navy, Pers-E24-JFR, 582417/2100, Jan. 10, 1966 (in author's possession).

11 *made average grades:* Fleming letter of certification.

11 *"from grammar school up":* Bond to Louise F. Bond, Feb. 14, 1945, from Homeopathic Hospital of Montreal.

11 *lavishly appointed houseboat:* "The 61-Footer Cyrene, Latest in House Boats," *Yachting,* January 1921.

11 *a log cabin:* Bond Jr., interviews, Oct. 10–13, 2003.

11 *"family practice in the home":* Bond, "Doctor on the Mountain," p. 104.

12 *Granny Hill:* Ibid., p. 196; Patrick and Bivins, "Pioneer for Rural Health," p. 111.

12 *learned to perform delicate procedures:* Bond, "Doctor on the Mountain," pp. 105, 110, 182, 187.

12 *power of prayer:* Ibid., p. 111; Bond, *Papa Topside,* p. 15.

12 *religion had not been central:* Bond Jr., interviews, Oct. 10–13, 2003.

12 *served as a lay preacher:* Bond, *Papa Topside,* p. 21; Bond Jr., interviews, Oct. 10–13, 2003.

12 *outbreak of influenza:* Bond, "Doctor on the Mountain," p. 112.

12 *Valley Clinic and Hospital:* Ibid., p. 115.

12 *model for improving medical care:* George F. Bond, M.D., "Plan for Rural Medical Services," *Journal of the American Medical Association* 147 (Dec. 15, 1951): 1543.

12 *written up:* Joseph Phillips, "The Revolution in Hickory Nut Gorge," *Reader's Digest,* November 1952, p. 117.

12 *knew him as "Doctor George":* Bond, *Papa Topside,* p. 28.

13 *eligible for recall to the Army:* Ibid., p. 22.

13 *chose the submarine service:* Ibid.

13 *began his professional conversion:* Ibid., pp. 24, 29.

13 *young family to Pearl Harbor:* Ibid., p. 23.

13 *to revive Charlie Aquadro:* Memorandum from C. C. Cole, Commander Submarine Squadron ONE to Commander Submarine Force, U.S. Pacific Fleet, Subject: Letter of Commendation; recommendation for, Jan. 16, 1956 (copy in author's possession).

13 *adviser for a training film:* Bond, *Papa Topside*, p. 26.

13 *guest of honor on* This Is Your Life: Ibid.; videocassette recording, undated, but a Program Analysis card from the NBC News Archives gives the air date as June 22, 1955 (in author's possession). Among those who appeared with Bond and corroborated highlights of his life were his wife, his mother, his sister, his four children, several former patients, Dooge Conner, and, as noted in the text, Lonnie Hill (copy in author's possession).

13 *planned to go into a partnership:* Bond, *Papa Topside*, p. 28.

14 *Lonnie had hanged himself:* Ibid., p. 24.

14 *volunteered to return to duty:* Ibid., p. 29.

14 *assistant officer-in-charge:* Bond résumé included in Navy press packet for Sealab I, 1964; Bond transcript of Naval Service, Jan. 10, 1966; "Capt. Vogel Leaving Navy; Cmdr. Bond Takes Lab Post," *New London* (Conn.) *Evening Day,* June 4, 1959.

14 *"diving game":* Bond, recorded keynote speech at Albany Medical College; Bond, *Papa Topside*, p. 43.

CHAPTER 2: DIVING

Page

15 *"exploration and exploitation":* Capt. George F. Bond, MC, USN, "Undersea Living—A New Capability," paper presented to the Congress of Petroleum and the Sea, Oceanographic Institute, Monaco, May 13, 1965 (in author's possession).

15 *Bond had in mind a whole range:* Bond, untitled paper presented at Pharmaceutical Wholesalers Convention, Las Vegas, Nev., March 15–18, 1961. This speech is among the earliest and most complete written records found of Bond's oft-stated belief in the need "to explore and exploit the continental shelves," as he says again here, and to use divers to do it.

15 *a speech to an audience:* Bond, recorded keynote speech at Albany Medical College; this represents another early example of Bond's thinking and attitude, with respect to the need to dive on the continental shelves, but with an emphasis on physiological issues and research.

16 *Beebe famously used:* William Beebe, *Half Mile Down* (Rahway, N.J.: Quinn & Boden, 1934), p. 176.

16 *man had to be able to swim freely:* Pharmaceutical Wholesalers speech, p. 6; Bond, "Prolonged Exposure to High Ambient Pressure," in *Second Symposium on Underwater Physiology*, Christian J. Lambertsen, M.D., and Leon J. Greenbaum, Jr., M.D., eds. (Washington, D.C.: National Academy of Sciences—National Research Council, Publication 1181, Feb. 25–26, 1963), p. 29.

16 *a specially designed sea dwelling:* Pharmaceutical Wholesalers speech, p. 8; Neil Hickey, "Your Future Home Under the Sea," *The American Weekly*, Nov. 9, 1958, p. 5.

16 *No one spent hours or days:* Kenneth W. Wallace, Navy master diver, Experimental Diving Unit, 1962–1967, interview, June 1, 2003.

16 *depth limit for most Navy divers: U.S. Navy Diving Manual,* Part I, pp. 96, 98, 109, 110; Wallace, interview, June 1, 2003.

17 *"Dominion over the seas":* Capt. Walter Mazzone, taped interview, Jan. 2, 2002. This wide-ranging interview of several hours was the first of more than a dozen interviews with Mazzone, both in person and by telephone; there were also numerous follow-up e-mails, some of which are cited, as are specific interviews, in the notes that follow. See also Bond, "Man-in-the-Sea Program," paper presented to the Association of American Medical Colleges, 17th annual session, San Diego, Calif., Sept. 29, 1966 (in author's possession), p. 11.

17 *"A Proposal for Underwater Research":* Bond, "Man-in-the-Sea Program," p. 5, is one of the many speeches and articles in which Bond says he submitted this proposal in 1957, which would have been during his first year as assistant officer-in-charge of the Medical Research Lab. This time frame appears to be correct based on other documentary sources and interviews. However, a copy of the proposal, a five-page, single-spaced typed essay that sounds Bond's familiar themes, is undated (in author's possession). Also undated is an official Bureau of Medicine and Surgery research proposal form that may have accompanied the original narrative proposal. On the form, which lists Bond as principal investigator, the "experimental design" outlines a series of animal and ultimately human experiments as part of study no. 3100.03, "Effects of Prolonged Exposure to High Ambient Pressures of Synthetic Gas Mixtures."

17 *needed a breakthrough:* Bond, "Proposal for Underwater Research," p. 1.

17 *"In man's history":* Ibid.

17 *struck him as the essential questions:* Bond, recorded keynote speech at Albany Medical College.

17 *Mark V had been in use:* Robert C. Martin, *The Deep-Sea Diver: Yesterday, Today and Tomorrow* (Cambridge, Md.: Cornell Maritime Press, 1978), p. 5.

17 *Navy became serious:* James Dugan, *Man Under the Sea* (New York: Harper & Bros., 1956), p. 37; *U.S. Navy Diving Manual,* Part I, p. 2.

18 *since Augustus Siebe:* Martin, *The Deep-Sea Diver,* p. 139; Davis, *Deep Diving and Submarine Operations,* p. 645.

18 *breath-holding divers:* Davis, *Deep Diving and Submarine Operations,* p. 546.

18 *Aristotle:* Ibid., p. 550.

18 *Pliny:* Ibid.

18 *da Vinci:* Ibid., p. 551.

18 *variety of apparatuses:* Ibid., pp. 554–67.

18 *the* Royal George: Ibid., pp. 404, 565.

18 *no more than ninety feet:* Ibid., p. 559.

18 *weighed some fifty pounds:* Martin, *The Deep-Sea Diver,* p. 151.

18 *twenty pounds each:* Ibid., p. 202.

18 *two hundred pounds:* Ibid., p. 43.

18 *Tenders had to help:* Ibid., p. 21; *U.S. Navy Diving Manual,* Part II, p. 89.

18 *Improvements had been made:* Martin, *The Deep-Sea Diver,* pp. 60, 139; Davis, *Deep Diving and Submarine Operations,* pp. 71, 645.

18 *Once a hardhat diver was lowered:* Robert C. Martin, "Diving with a Mark V Outfit," in *The Deep-Sea Diver,* pp. 43–50; *U.S. Navy Diving Manual,* Part II, pp. 71–77; Wallace, interview, June 1, 2003; interviews with former hardhat divers.

19 *The pressure of the air:* "Underwater Physics" and "Underwater Physiology," in *U.S. Navy Diving Manual,* Part I, pp. 13–53; Davis, "The Physics and Physiology of Deep Diving," in *Deep Diving and Submarine Operations,* pp. 26–40.

19 *it doesn't take much:* U.S. Navy Diving Manual, Part I, pp. 54, 59, 62.

19 *"lung squeeze":* Ibid., pp. 61, 149.

19 *measure called an atmosphere:* Ibid., p. 16.

19 *enveloping pressures cancel:* Ibid., p. 17; Martin, *The Deep-Sea Diver,* p. 11.

19 *air in the ears, sinuses:* U.S. Navy Diving Manual, Part I, p. 59.

20 *the experience of hardhat diving:* Former hardhat divers, interviews, including Jack Reedy and Charles E. Johnson, Ph.D., during a tour at Naval Sea Systems Command, Navy Experimental Diving Unit, March 11, 2005, part of the Sealab reunion that year in Panama City, Fla.; *U.S. Navy Diving Manual,* Part II, p. 7; Tom Eadie, "What a Dive Is Like," in *I Like Diving: A Professional's Story* (Boston: Houghton Mifflin, 1931), pp. 23–37.

20 *getting "fouled":* U.S. Navy Diving Manual, Part II, p. 80.

20 *shut off the air flow:* Ibid., p. 81.

20 *"blowup":* U.S. Navy Diving Manual, Part I, p. 152; Davis, *Deep Diving and Submarine Operations,* p. 34.

20 *"squeeze":* U.S. Navy Diving Manual, Part I, pp. 21, 80; Davis, *Deep Diving and Submarine Operations,* p. 34; Martin, *The Deep-Sea Diver,* p. 47.

21 *major topic was the bends:* U.S. Navy Diving Manual, Part I, p. 128.

21 *a muddy tomb:* David McCullough, *The Great Bridge* (New York: Simon & Schuster, 1972), p. 197.

21 *"Grecian bend":* Ibid., p. 186; Torrance R. Parker, *20,000 Jobs Under the Sea: A History of Diving and Underwater Engineering,* Don Walsh, ed. (Palos Verdes Peninsula, Calif.: Sub-Sea Archives, 1997), p. 255.

21 *Bert's experiments showed:* Davis, *Deep Diving and Submarine Operations,* p. 4.

22 *carbonated soda pop:* Ibid., pp. 4, 38; *U.S. Navy Diving Manual,* Part I, p. 24.

22 *wreaking corporeal havoc:* Davis, *Deep Diving and Submarine Operations,* p. 126; *U.S. Navy Diving Manual,* Part I, p. 128.

22 *"recompression":* U.S. Navy Diving Manual, Part I, pp. 129, 178.

22 *Navy Yard had chambers:* Ibid., p. 5.

22 *John Scott Haldane:* Davis, *Deep Diving and Submarine Operations,* p. 6.

22 *Haldane's tables:* Ibid., pp. 94, 100–108.

22 *deeper than ninety feet:* U.S. Navy Diving Manual, Part I, p. 3.

23 *nitrogen narcosis:* Ibid., p. 70.

23 l'ivresse des grandes profondeurs: Capt. J. Y. Cousteau with Frédéric Dumas, *The Silent World* (New York: Harper & Brothers, 1953), p. 31.

23 *"Swede" Momsen:* Maas, *The Terrible Hours*, p. 29.

23 *found it in helium:* Ibid., p. 190; Davis, *Deep Diving and Submarine Operations*, p. 180.

23 *"helium hat":* Nat A. Barrows, *Blow All Ballast!: The Story of the* Squalus (New York: Dodd, Mead, 1941), p. 190; *U.S. Navy Diving Manual*, Part II, p. 30.

23 *three hundred pounds:* Martin, *The Deep-Sea Diver*, p. 111.

23 USS Squalus *sank:* LaVO, *Back from the Deep*, p. 34; Maas, *The Terrible Hours*, p. 33.

24 *One diver passed out:* LaVO, *Back from the Deep*, p. 64.

24 *Swede Momsen, who ran:* Maas, *The Terrible Hours*, p. 172.

24 *inaugural use of the McCann:* Ibid., p. 159; Davis, *Deep Diving and Submarine Operations*, p. 670.

24 *recommissioned as the* Sailfish: LaVO, *Back from the Deep*, p. 74.

24 *Rough-hewn scuba designs:* Peter Jackson, "Development of Self-Contained Diving Prior to Cousteau-Gagnan," *Historical Diver* 13 (Fall 1997): 34.

24 *Cousteau and Gagnan would patent:* Leslie Leaney, "Jacques Yves Cousteau: The Pioneering Years," *Historical Diver* 13 (Fall 1997): 18.

24 *growing interest in diving:* Ibid., p. 28; Chuck Blakeslee, "The Early History of Recreational Diving as Recorded in *Skin Diver* Magazine," *Historical Diver* 35, (2002): 37.

24 *Salvage operations such as:* Edward C. Raymer, *Descent into Darkness: Pearl Harbor, 1941: A Navy Diver's Memoir* (Novato, Calif.: Presidio Press, 1996), pp. 25, 28.

25 *The book sold briskly:* Leaney, "Jacques Yves Cousteau: The Pioneering Years," p. 27.

25 *National Geographic Society signed on:* Capt. Jacques-Yves Cousteau, "Fish Men Discover a 2,200-Year-Old Greek Ship," *National Geographic*, January 1954; Robert M. Poole, *Explorers House: National Geographic and the World It Made* (New York: Penguin, 2004), p. 111.

25 *debut on American network:* Omnibus, no. 16, Part 1, on CBS, Jan. 17, 1954, viewed at Museum of Television and Radio, New York City, March 28, 2003.

25 *"You get the impression":* Ibid.

25 *diving school at Key West:* U.S. Navy Diving Manual, Part I, p. 7.

26 *unsuitable beyond 130:* Ibid., Part III, p. 9.

26 *the standard Navy limit:* Ibid., Part I, pp. 96, 98, 109, 110; Wallace, interview, June 1, 2003.

26 *earlier researchers had suggested:* James Vorosmarti Jr., M.D., "A Very Short History of Saturation Diving," *Historical Diving Times* 20 (Winter 1997): 5.

26 How long can a man stay down?: Bond, recorded keynote speech at Albany Medical College.

26 *six hundred hardhat dives:* Maas, *The Terrible Hours*, p. 236.

26 *typical bottom time:* LaVO, *Back from the Deep*, p. 66.

26 *"You are wasting time":* Bond, "Man in the Sea" (undated paper, apparently for an audience of industrial researchers in author's possession; references in the text indicate that this paper must have been written circa 1967; it is similar to Bond's previously cited paper presented to Association of American Medical Colleges, Sept. 29, 1966), p. 4.

26 *considered his BuMed boss:* Ibid.; Gerald J. Duffner, former director of submarine medicine, Bureau of Medicine and Surgery, interview, April 2, 2003.

CHAPTER 3: GENESIS

Page

28 *renowned illustrator:* Robert E. Weinberg, *A Biographical Dictionary of Science Fiction and Fantasy Artists* (Westport, Conn.: Greenwood, 1988), p. 178.

28 *"exclusive first look":* Neil Hickey, "Your Future Home Under the Sea," *The American Weekly*, Nov. 9, 1958, p. 5.

29 *maintained a peripatetic schedule:* Interviews, corroborated by an inch-thick file from Bond's early years in New London containing numerous travel-related memos, such as one from Commander Submarine Force, U.S. Atlantic Fleet to Bond, dated Aug. 19, 1958, pertaining to a three-day trip to Washington, D.C. A letter from Gerald J. Duffner, dated Sept. 23, 1959, also alludes to Bond's frequent travels (in author's possession).

29 *Bond first met Walt Mazzone:* Mazzone, interview, Jan. 2, 2002.

29 *serving as a valued mentor:* Robert C. Bornmann, interview, Feb. 13, 2003.

29 *catalyst for a conversation:* Mazzone, interview, Jan. 2, 2002.

30 *came from a working-class family:* Walter Mazzone, interview, San Diego, Calif., March 17, 2002.

30 *aboard the submarine* Puffer: Ibid.; Theodore Roscoe, *True Tales of Bold Escapes* (Englewood Cliffs, N.J.: Prentice Hall, 1965), pp. 57, 77, 79; Capt. William J. Ruhe, USN (ret.), *War in the Boats: My World War II Submarine Battles* (Washington, D.C.: Brassey's, 1994), p. 196.

31 *on the USS* Crevalle: Mazzone, interview, March 17, 2002; Ruhe, *War in the Boats*, pp. 195, 277, 293.

31 *wearing only the Jack Browne:* Mazzone, interview, March 17, 2002.

32 *something that Mazzone admired:* Mazzone, interview, Jan. 2, 2002.

32 *a favorite maxim:* Ibid.; Mazzone, interview, Jan. 17, 2003.

32 *escapes from 150:* Bond et al., "Deep Submarine Escape," p. 11.

32 *main physiological issues:* Robert D. Workman, George F. Bond, and Walter F. Mazzone, "Prolonged Exposure of Animals to Pressurized Normal and Synthetic Atmospheres," U.S. Naval Medical Research Laboratory Report

no. 374, Bureau of Medicine and Surgery, Navy Department Research Project MR005.14-3100.02, Jan. 26, 1962, p. 1; Mazzone, taped interview, San Diego, Calif., Aug. 7, 2002.

32 *Oxygen toxicity: U.S. Navy Diving Manual,* Part I, p. 70.

33 *Jacques Cousteau found this out:* Cousteau, *The Silent World,* p. 16.

33 *experimented with other gases:* P. B. Bennett, "Inert Gas Narcosis," in *The Physiology and Medicine of Diving and Compressed Air Work,* P. B. Bennett and D. H. Elliott, eds., 1st ed. (Baltimore: Williams & Williams, 1969), p. 162.

33 *boundary was untested: U.S. Navy Diving Manual,* Part I, pp. 102, 110; Wallace, interview, June 1, 2003.

33 *more time decompressing than working:* A. J. Bachrach, "A Short History of Man in the Sea," in *The Physiology and Medicine of Diving and Compressed Air Work,* P. B. Bennett and D. H. Elliott, eds., 2nd ed. (London: Ballière Tindall, 1975), p. 6.

33 *a half-hour of bottom time: U.S. Navy Diving Manual,* Part I, p. 113.

33 *a full hour on the bottom:* Ibid.

33 *Some researchers had suggested:* Vorosmarti, "A Very Short History of Saturation Diving," p. 5; Vorosmarti letter to author, Dec. 21, 2003, with enclosed copy of letter from G. C. C. Damant to Dr. John Scott Haldane dated July 16, 1935; Vorosmarti e-mail to author, Jan. 26, 2004; Mazzone, interview, Aug. 7, 2002; e-mail to author, April 20, 2005.

33 *discovered that nitrogen was responsible:* P. B. Bennett, "Inert Gas Narcosis," in *The Physiology and Medicine of Diving and Compressed Air Work,* 1st ed., p. 156; Peter Bennett, taped interview, Feb. 8, 2008.

33 *pioneered the use of helium:* A. R. Behnke, "Some Early Studies of Decompression," in *The Physiology and Medicine of Diving and Compressed Air Work,* 1st ed., p. 233; Bennett, interview, Feb. 8, 2008.

34 *Behnke retired from the Navy: Who's Who in America,* 41st ed. (Chicago: Marquis Who's Who, 1980–81), p. 235.

34 *a supportive elder and friend:* Helen Siiteri, editor of *Papa Topside,* who interviewed Behnke in the early 1990s before his death, telephone interview by author, May 16, 2006; assorted personal letters from Bond to Behnke (in author's possession), including one that alludes to prior correspondence, Nov. 15, 1971; a cheerful letter from Behnke to Bond, April 20, 1973.

34 *preliminary and relatively shallow steps:* A. R. Behnke, "Some Early Studies of Decompression," in *The Physiology and Medicine of Diving and Compressed Air Work,* 1st ed., p. 248; Vorosmarti, "A Very Short History of Saturation Diving," p. 5.

34 *Dr. Edgar End:* "Diver Tests Human Endurance: Max Gene Nohl Collapses with Bends After 27 Hours in Pressure Tank," *Look,* March 28, 1939, p. 28.

34 *a concept known as "saturation diving":* Bond, *Papa Topside,* pp. 3, 32; Bachrach, "A Short History of Man in the Sea," p. 6.

34 *bandied about in the mid-1940s:* Vorosmarti, "A Very Short History of Saturation Diving," p. 5.

34 *had been understood for years:* A. E. Boycott, D.M., G.C.C. Damant, Lieut. and Inspector of Diving, R.N., and J. S. Haldane, M.D., F.R.S., "The Prevention of Compressed-Air Illness," *Journal of Hygiene* 8 (1908): 345; Eadie, *I Like Diving,* pp. 60, 62.

34 *Decompression and its underlying theories:* R. D. Workman, "Decompression Theory: American Practice," in *The Physiology and Medicine of Diving and Compressed Air Work,* 1st ed., p. 252.

35 *bends are especially dangerous:* D. J. Kidd and D. H. Elliott, "Clinical Manifestations and Treatment of Decompression Sickness in Divers," in *The Physiology and Medicine of Diving and Compressed Air Work,* 1st ed., p. 471.

35 *"fast" tissues:* Workman, "Decompression Theory," in *The Physiology and Medicine of Diving and Compressed Air Work,* 1st ed., p. 256; *U.S. Navy Diving Manual,* Part I, p. 67.

35 *for causing severe aches:* Kidd and Elliott, "Treatment of Decompression Sickness," in *The Physiology and Medicine of Diving and Compressed Air Work,* 1st ed., p. 466.

35 *Mathematical models:* Workman, "Decompression Theory," p. 252.

35 *Dr. Workman succeeded Bond:* Lois Birkner Workman, Dr. Workman's wife of fifty-five years, interview, May 7, 2003. Robert Dean Workman died April 4, 1998, at his farm in Picayune, Miss. He was seventy-six.

35 *genial, but a bit reserved:* Interviews with former colleagues, including Wallace, interview, June 1, 2003, and Barth, interview, Sept. 27, 2002.

36 *always wanted to be a doctor:* Lois Workman, interview, May 7, 2003.

36 *Workman went scuba diving:* Ibid.

36 *changed career course:* Ibid.

36 *keen feel for the mathematical:* Robert C. Bornmann, "Robert Dean Workman, M.D., Past President and Founding Member of the Undersea and Hyperbaric Medical Society, Inc.: A Remembrance," *Pressure* 27, no. 5 (Sept./Oct. 1998) (newsletter of the U&HMS): 2; Mazzone, taped interview, May 8, 2003; Wallace, interview, June 1, 2003.

36 *postgraduate fellowship:* Bornmann, "A Remembrance," p. 2; Lois Workman, interview.

36 *Dr. Christian Lambertsen:* "2000 *Historical Diver* Magazine Diving Pioneer Award," *Historical Diver* 8, no. 4 (Fall 2000): 5.

36 *confirmed a critical point:* Mazzone, interview, May 8, 2003; Mazzone, e-mail to author, April 19, 2005; Bond, *Papa Topside,* pp. 32, 88.

36 *The decompression curve flattened out:* Mazzone, e-mail to author, April 20, 2005.

37 *called these experiments Genesis:* Mazzone, interview, Jan. 2, 2002; Bond, *Papa Topside,* pp. 4, 33.

37 *building adjacent to the bottom:* Barth, interview, June 9, 2003.

37 *much in the way of instrumentation:* Mazzone, interview, Aug. 7, 2002.

37 *struck Bond as a good starting point:* Ibid.

38 *rats were dead:* Workman, Bond, and Mazzone, "Prolonged Exposure of Animals," p. 5.

38 *some two hundred:* Bond, Pharmaceutical Wholesalers speech, p. 10.

38 *a heightened friskiness:* Mazzone, interview, Aug. 7, 2002.

38 *the lab evoked:* Workman, Bond, and Mazzone, "Prolonged Exposure of Animals," p. 25; Mazzone, interview, Aug. 7, 2002.

38 *money out of their own pockets:* Mazzone, interviews, Jan. 2, 2002, and Aug. 7, 2002; Bond, Pharmaceutical Wholesalers speech, p. 9.

38 *costly helium:* Mazzone, taped interview, Feb. 5, 2003.

38 *some scientists had expressed doubt:* Bond, "Prolonged Exposure to High Ambient Pressure," p. 31.

39 *goats enlisted:* Barth, interview, Sept. 26, 2002; Mazzone, interview, Jan. 2, 2002.

39 *a number of goat pairs:* Workman, Bond, and Mazzone, "Prolonged Exposure of Animals," p. 19; Mazzone, interview, Aug. 7, 2002.

39 *nonverbal warning signs:* Workman, Bond, and Mazzone, "Prolonged Exposure of Animals," p. 22; Mazzone, interview, Aug. 7, 2002.

39 *stint as a training tank instructor:* Bob Barth, *Sea Dwellers: The Humor, Drama and Tragedy of the U.S. Navy Sealab Programs* (Houston, Tex.: Doyle Publishing, 2000), p. 9; Bob Barth, interview, Sept. 26, 2002.

39 *enlisted in the Navy at seventeen:* Barth, interview, Sept. 30, 2002.

39 *came time for shore duty:* Barth, interview, Sept. 26, 2002.

39 *"bottom drops":* Barth, *Sea Dwellers,* p. 9; Barth, interview, Sept. 26, 2002.

40 *One favorite stunt:* Barth, interview, Sept. 26, 2002.

40 *pet Kampuchean monkey:* Barth, *Sea Dwellers,* p. 16; Aquadro, interview, May 16–18, 2003.

40 *put the goats on leashes:* Mazzone, interview, Aug. 7, 2002; Barth, interview, Sept. 24, 2002; Barth, *Sea Dwellers,* p. 11.

41 *stopped by a base patrolman:* Barth, interview, Sept. 24, 2002; Barth, *Sea Dwellers,* p. 11.

41 *Able and Baker:* Richard Witkin, "2 Monkeys Survive Flight Into Space in U.S. Rocket and Are Retrieved At Sea," *New York Times,* May 29, 1959, p. 1.

41 *cover of* Life *magazine: Life,* June 15, 1959.

41 *good news for humans:* Workman, Bond, and Mazzone, "Prolonged Exposure of Animals," pp. 32, 35.

41 *a more conservative decompression:* Ibid., p. 32.

42 *report's message was clear:* Peter Bennett, Ph.D., D.Sc., professor of anesthesiology and senior director, F. G. Hall Hypo-Hyperbaric Center, Duke University Medical Center, and coeditor of *The Physiology and Medicine of Diving and Compressed Air Work,* interview, Oct. 27, 2003.

CHAPTER 4: FRIENDLY RIVALS

Page

43 *Arne Zetterström:* Dugan, *Man Under the Sea*, p. 51.

43 *Maurice Fargues:* Leaney, "Cousteau: The Pioneering Years," p. 23.

43 *Wilfred Bollard:* Dugan, *Man Under the Sea*, p. 43.

43 *George Wookey:* "Obituary of Lieutenant George Wookey," (London) *Daily Telegraph*, April 6, 2007, p. 29.

44 *speech to the Boston Sea Rovers:* "The History of the Boston Sea Rovers, Vol. 1, 1953–1974" (photocopy, Boston Sea Rovers, Boston, Mass., [2000]), p. 37.

44 *retired from the French navy:* Jacques-Yves Cousteau with James Dugan, *The Living Sea* (New York: Harper & Row, 1963), p. 327.

44 *the* Calypso: Ibid., p. 23.

44 *captivated by Bond's ideas:* Ibid.

44 *"Your message was important":* Clifford P. Wells Jr. to Bond, Jan. 25, 1960, from Park Ridge, Ill. (in author's possession).

44 *Diver of the Year:* "The History of the Boston Sea Rovers," p. 50.

44 *introduced Bond and Cousteau:* Ibid., p. 51.

44 *his first significant meeting with Cousteau:* Bond, *Papa Topside*, p. 3; "An Evening of Undersea Adventure with Captain Jacques-Yves Cousteau, in person, and Commander George F. Bond, U.S. Navy," program for the Richard Ferg Memorial Fund Benefit at Town Hall, 123 West 43rd Street, New York, sponsored by the Long Island Dolphins Inc., Sept. 10, 1959 (in author's possession).

44 *James Dugan:* Leaney, "Cousteau: The Pioneering Years," p. 24.

45 *Dugan's role:* Ibid., p. 26; "The History of the Boston Sea Rovers," p. 120.

45 *deep into the night:* Bond, *Papa Topside*, p. 3.

45 *son of a businessman:* Axel Madsen, *Cousteau: An Unauthorized Biography* (New York: Beaufort, 1986), p. 6.

45 *lived on 95th Street:* Jean-Michel Cousteau, *Mon père le commandant* (Paris: L'Archipel, 2004), p. 17.

45 *hated riding horses:* Ibid.; Madsen, *Cousteau*, p. 8.

45 *must have been inspiring:* "The History of the Boston Sea Rovers," p. 52; Cousteau, *The Living Sea*, p. 315.

45 *"Jacques Cousteau and me":* Mazzone, interviews, Jan. 2, 2002, and Jan. 23, 2003.

45 *Austrian Hans Hass:* Dugan, *Man Under the Sea*, p. 152; Michael Jung, "Hans Hass: Pioneer Underwater Photographer and Swim Diver," Santa Barbara Underwater Film Festival program (published with Historical Diving Society USA), Sept. 11, 1998, p. 13 (in author's possession).

45 *cover of* Time: *Time*, March 28, 1960.

45 *"one of the great explorers":* National Geographic, July 1961, p. 146.

45 *notably U.S. Divers:* Sara Davidson, "Cousteau Searches for His Whale," *New York Times Magazine,* Sept. 10, 1972, pp. 87–88; Leaney, "Cousteau: The Pioneering Years," p. 28.

46 *Russians also had been engaged in: The History of Russian Diving* (Historical Diving Society of Russia, St. Petersburg), no. 1 (2002), pp. 65, 69, 72; "Diving Boom in the Soviet Union," in *Men Under Water,* James Dugan and Richard Vahan, eds., for the Underwater Society of America (Philadelphia: Chilton, 1965), pp. 50–52.

46 *Edwin Albert Link Jr.:* Martha Clark and Jeanne Eichelberger, "Edwin A. Link, 1904–1981," in *The Link Collections* (Binghamton: State University of New York, Special Collections, 1981, revised 1999), p. 7.

46 *Like Edison:* Robert Conot, *Thomas A. Edison: A Streak of Luck* (New York: Da Capo, 1979), pp. 7–9.

46 *working full-time at his father's business:* Susan van Hoek with Marion Clayton Link, *From Sky to Sea: A Story of Edwin A. Link* (Flagstaff, Ariz.: Best Publishing, 2003), p. 14.

46 *fascinated with gadgets:* Ibid., p. 2.

46 *hang around dusty airfields:* Ibid., p. 14.

46 *threatened to disown:* Ibid., p. 10.

46 *solo flight in 1926:* Ibid., p. 16; Clark and Eichelberger, "Edwin A. Link," p. 7.

46 *killed in training:* Van Hoek with M. C. Link, *From Sky to Sea,* p. 15.

47 *without leaving the ground:* Ibid.

47 *the Link Trainer:* Ibid., pp. 19, 24, 25, 39–60; Clark and Eichelberger, "Edwin A. Link," pp. 7–9.

47 *met Marion Clayton:* Clark and Eichelberger, "Edwin A. Link," p. 73.

47 *set up on a blind date:* Van Hoek with M. C. Link, *From Sky to Sea,* pp. 28–29.

47 *married her best story:* Clark and Eichelberger, "Edwin A. Link," p. 73.

47 *took an active role:* Ibid.; Van Hoek with M. C. Link, *From Sky to Sea,* pp. 30, 33.

47 *carnival ride:* Van Hoek with M. C. Link, *From Sky to Sea,* pp. 34, 170.

47 *the military finally bought in:* Ibid., pp. 34, 42–44, 60.

47 *two million airmen:* Thomas W. Ennis, "Edwin A. Link, 77, Invented Instrument Flight Simulator," *New York Times,* Sept. 9, 1981, Section B, p. 7; Clark and Eichelberger, "Edwin A. Link," p. 73.

47 *ease out of the trainer business:* Van Hoek with M. C. Link, *From Sky to Sea,* pp. 62, 181; Clark and Eichelberger, "Edwin A. Link," p. 9.

47 *self-made millionaire:* Van Hoek with M. C. Link, *From Sky to Sea,* pp. xvi, 70.

47 *holder of twenty patents:* Link Biographical Statement, Sept. 15, 1981, including a chronology of patents, Folder 777, the Link Collections, Special Collections, State University of New York at Binghamton (hereafter abbreviated as LC).

47 *restless and taken up sailing:* Van Hoek with M. C. Link, *From Sky to Sea,* p. 62.

47 *tried the Aqualung:* Ibid., p. 80.

47 *first of many expeditions:* Ibid., pp. 78, 83.

47 *reputable organizations:* Ibid., pp. 79, 158, 236, 238, 251.

47 *enthralled with the history:* Ibid., p. 85.

47 *routes of Christopher Columbus:* Ibid., pp. 92, 116.

48 Santa María: Ibid., p. 95.

48 *designing needed tools:* Ibid., pp. 85, 89.

48 *caught wind of Bond's Genesis experiments:* Memorandum from LCDR Robert
    C. Bornmann, MC, USN, 614527/2100 to Chief of Naval Operations, "The
    E. A. Link Diving Expedition and the 'Man in Sea' Project," Enclosure 1:
    "Background and Narrative Summary of Dives," Oct. 31, 1962, p. 1 (in author's
    possession).

48 *to convince the Navy brass:* Link to Dr. Charles Aquadro, July 9, 1962, from
    aboard *Sea Diver* en route to Monaco, Folder 236; Link to M. M. Payne,
    National Geographic Society, Feb. 11, 1962, Folder 229, LC.

48 *project he called Man-in-Sea:* Ibid., p. 32; Edwin A. Link, "Our Man-in-Sea
    Project," *National Geographic,* May 1963, p. 713.

48 *society's backing:* Marion Clayton Link, *Windows in the Sea* (Washington,
    D.C.: Smithsonian Institution Press, 1973), p. 34; Link, "Our Man-in-Sea
    Project," p. 715.

48 *little interest in money:* Van Hoek with M. C. Link, *From Sky to Sea,* p. 181.

48 *battered Volkswagen Bug:* Link to Mr. and Mrs. William Link, Aug. 31, 1962,
    from aboard *Sea Diver,* Monaco, Folder 236, LC.

48 *put up $500,000:* Van Hoek with M. C. Link, *From Sky to Sea,* p. 181.

48 Sea Diver: Ibid., pp. 160–64, 172, 182; M. C. Link, *Windows in the Sea,* p. 20.

48 *Link met with Jacques Cousteau:* Van Hoek with M. C. Link, *From Sky to Sea,*
    p. 255.

48 *office at the Oceanographic Museum:* Cousteau, *The Living Sea,* p. 300.

49 *a short walk past:* Author's visit to the museum, Nov. 3, 2004.

49 *home to JYC's friends:* Madsen, *Cousteau,* p. 113.

49 *Link kept* Sea Diver *berthed:* Van Hoek with M. C. Link, *From Sky to Sea,*
    p. 254.

49 *craft moored nearby:* M. C. Link, *Windows in the Sea,* p. 34.

49 *always dreamed of flying:* Jean-Michel Cousteau, *Mon père le commandant,*
    p. 21.

49 *training to become a Navy pilot:* Madsen, *Cousteau,* p. 13.

49 *a late-night car crash:* Ibid., p. 18; Jean-Michel Cousteau, *Mon père le comman-
    dant,* p. 22.

49 *took up swimming:* Madsen, *Cousteau,* p. 21; Jean-Michel Cousteau, *Mon père
    le commandant,* p. 29.

49 *Bobby, a pilot, who was killed:* Ellen and Lynn Moorehead (Bond relatives, as
    cited in notes to Chapter 1), e-mail to author, April 6, 2005.

49 *lunch with Cousteau:* Link to Cousteau at Oceanographic Institute, Dec. 14,
    1961, Folder 227, LC.

49 *possibility of teaming up:* Ibid.

49 *met again several times:* Van Hoek with M. C. Link, *From Sky to Sea*, p. 255.

49 *Navy was in no hurry:* Link to Aquadro, July 9, 1962, Folder 236; Link to M. M. Payne, Feb. 11, 1962, Folder 229, LC.

49 *a "house" big enough:* Van Hoek with M. C. Link, *From Sky to Sea*, p. 255.

50 *an aluminum cylinder:* M. C. Link, *Windows in the Sea*, p. 45; Lord Kilbracken, "The Long, Deep Dive," *National Geographic*, May 1963, p. 719.

50 *backing from the Smithsonian:* Link, "Our Man-in-Sea Project," p. 715; Van Hoek with M. C. Link, *From Sky to Sea*, p. 257.

50 *diving bell:* Davis, *Deep Diving and Submarine Operations*, pp. 197, 602.

50 *siege of Tyre:* Ibid., p. 603.

50 *Submersible Decompression Chamber:* Ibid., pp. 19, 137; this citation uses the term "submerged," but this may be a typographical error, as such chambers were commonly known as "submersible," according to Peter Bennett, coeditor of *The Physiology and Medicine of Diving and Compressed Air Work*, interview, May 15, 2008.

50 *gave Link some ideas:* M. C. Link, *Windows in the Sea*, p. 15.

50 *tedious and exhausting:* Ibid., p. 136.

50 *pressurized elevator to shuttle divers:* Bornmann to CNO, "Summary of Dives," Oct. 31, 1962, p. 2.

50 *new cylinder on board:* Van Hoek with M. C. Link, *From Sky to Sea*, pp. 257, 259.

50 *tons of Byzantine artifacts:* Robert Sténuit, *The Deepest Days*, translated by Morris Kemp (New York: Coward-McCann, 1966), p. 50.

51 *near the Lipari Islands:* Memorandum from Link to Our friends and relatives interested in our underwater archeological and diving activities, "Report No. 2," Oct. 5, 1962, p. 2, Folder 237, LC.

51 *a convenient excuse:* Ibid.

51 *John T. Hayward:* Link to Captain Jacques Cousteau, Jan. 8, 1962; Link to Vice Admiral John T. Hayward, Jan. 26, 1962, Folder 228, LC.

51 *money to be made in selling:* Link to Mendel Peterson, Aug. 31, 1962, from aboard *Sea Diver*, Monaco, Folder 236, LC.

51 *"this double-crossing by Cousteau":* Ibid.

51 *Mendel Peterson:* Van Hoek with M. C. Link, *From Sky to Sea*, pp. 79, 85, 91.

51 *Bond sent a letter:* Bond to Lt. Robert Bornmann, MC, U.S. Navy, Aug. 31, 1962, from U.S. Naval Medical Research Laboratory, New London, Folder 358, LC.

52 *Navy approval to proceed:* Ibid.

52 *selfish interest on Bond's part:* Link to Aquadro, Sept. 19, 1962, Folder 237, LC.

52 *his concept:* Bond, *Papa Topside*, p. 32; Bond, "Man in the Sea" (1967), p. 3; Bond, "Tomorrow the Seas," p. 6.

52 *nor even the cautious U.S. Navy brass:* Bornmann to Link, Nov. 15, 1962, Folder 360, LC.

52 *steal Bond's thunder:* Link to Dr. Charles Aquadro, July 9, 1962, Folder 236, LC.

52 *Bond deserved all due credit:* Ibid.; M. C. Link, *Windows in the Sea*, p. 41.

52 *prod the Navy into action:* Link to Aquadro, July 9, 1962.

52 *gladly drop his own:* Ibid.

52 *any serious mishap:* Bornmann to Link, Nov. 15, 1962, Folder 360, LC.

52 *would have heeded Bond's warning:* Link to Aquadro, Sept. 19, 1962, Folder 237, LC.

53 *just thirty-three feet:* Cousteau, *The Living Sea*, p. 315; Link to Mendel Peterson, from aboard *Sea Diver*, Monaco, Folder 236, LC.

53 *great publicity value:* Link to Our Friends, "Report No. 2," p. 3.

53 *live in a swimming pool:* Ibid.; M. C. Link, *Windows in the Sea*, p. 39.

53 *a variety of possible depths:* M. C. Link, *Windows in the Sea*, p. 38.

53 *Navy's Sixth Fleet:* Ibid.

53 *delivering Link's cylinder:* Van Hoek with M. C. Link, *From Sky to Sea*, p. 262.

53 *provide him with the costly helium:* Ibid.; Link to Cmdr. Robert J. Alexander, Office of Chief of Naval Operations, Nov. 17, 1962, Folder 238; John Ellis, Smithsonian Institution, to Link, July 13, 1962, Folder 357, LC.

53 *have a submarine rescue ship:* Robert Bornmann, interview, June 22, 2003; Link to Vice Adm. D. L. MacDonald, Oct. 3, 1962, Folder 237, LC; Bornmann to CNO, "Summary of Dives," Oct. 31, 1962, p. 4.

53 *requested that Charlie Aquadro:* Link to Adm. George C. Anderson, Chief of Naval Operations, Feb. 11, 1962, Folder 229; Link to Aquadro, July 9, 1962, Folder 236, LC.

53 *befriended Link while stationed:* Aquadro, interview, July 12, 2003.

53 *underwater archaeological expeditions:* Van Hoek with M. C. Link, *From Sky to Sea*, pp. 165, 213; M. C. Link, *Windows in the Sea*, p. 197.

53 *Aquadro would have been happy:* Aquadro to Link, Oct. 20, 1962, from Arlington, Va., Folder 359, LC.

53 *Link's request for medical assistance:* Ibid.; Van Hoek with M. C. Link, *From Sky to Sea*, p. 220.

53 *Dr. Robert Bornmann:* Bornmann, interviews, Feb. 13, 2003, and June 22, 2003.

54 *rarely had direct exchanges:* Bob Barth, interview, Dec. 15, 2003; Walt Mazzone, e-mails to author, June 28, 2003, and Jan. 25, 2005; also confirmed by a lack of correspondence between Link and Bond in the Link Collections.

54 *Among the items Bornmann packed:* Ibid.; Bornmann to CNO, "Summary of Dives," Oct. 31, 1962, p. 1.

54 *a bit of a temper:* Van Hoek with M. C. Link, *From Sky to Sea*, p. xviii.

54 *little time to diversions:* Ibid.

54 *At fifty-eight:* Bornmann, interview, June 22, 2003; Kilbracken, "The Long, Deep Dive," pp. 718, 731; Sténuit, *The Deepest Days*, p. 48.

54 *a series of test dives:* Van Hoek with M. C. Link, *From Sky to Sea*, p. 260.

54 *anything he wouldn't do himself:* Ibid., p. 262.

54 *up to his neck:* Robert Bornmann, interview, Aug. 7, 2003.

54 *weight of more than two tons:* M. C. Link, *Windows in the Sea*, p. 36.

55 *shot out of the water:* Bornmann, interview, Aug. 7, 2003.

55 *they had done enough:* Ibid.

55 *On August 28, Link attempted:* Kilbracken, "The Long, Deep Dive," p. 725; Bornmann to CNO, "Summary of Dives," Oct. 31, 1962, p. 4; Sténuit, *The Deepest Days*, p. 72.

55 *mostly by caisson workers:* Behnke, "Early Studies of Decompression," in *The Physiology and Medicine of Diving and Compressed Air Work*, 1st ed., pp. 248–49.

55 *diving tables maxed out:* U.S. Navy Diving Manual, Part I, pp. 98, 110.

55 *donned scuba gear:* M. C. Link, *Windows in the Sea*, p. 41.

55 *spent much of the afternoon:* Kilbracken, "The Long, Deep Dive," p. 725.

55 *twenty-one-year-old Clayton:* Ibid., pp. 723, 725.

55 *find time for a nap:* Ibid.

55 *hunched over:* Ibid., p. 726.

55 *hauled the sealed-up cylinder onto the deck:* Ibid.

56 *crunched some decompression figures:* Bornmann to CNO, "Summary of Dives," Oct. 31, 1962, p. 4; Bornmann, interview, June 22, 2003.

56 *"exceptional exposures":* U.S. Navy Diving Manual, Part I, p. 102.

56 *Bornmann opened the hatches:* Kilbracken, "The Long, Deep Dive," p. 726; Bornmann, interview.

56 *nearly fifteen hours, including:* Kilbracken, "The Long, Deep Dive," p. 726.

56 *Edmund Halley:* Davis, *Deep Diving and Submarine Operations*, pp. 608–9; Dugan, *Man Under the Sea*, p. 309.

56 *narrowly averted getting bent:* U.S. Navy Diving Manual, Part I, p. 98; James T. Joiner, "Diving Bells; Diver Lockout Systems," in *A Pictorial History of Diving*, Arthur J. Bachrach, Ph.D., Barbara Mowery Desiderati, and Mary Margaret Matzen, eds. (San Pedro, Calif.: Best Publishing, 1988), p. 5.

56 *considered a renaissance man:* Van Hoek with M. C. Link, *From Sky to Sea*, p. xi.

56 *getting a little old for this:* Kilbracken, "The Long, Deep Dive," p. 726; Bornmann, interview, Aug. 7, 2003.

57 *met Sténuit at Vigo Bay:* Sténuit, *The Deepest Days*, p. 33.

57 *impressed Link as skillful:* Edwin A. Link, "Introduction," in *The Deepest Days*, p. 12.

57 *received Link's invitation:* Sténuit, *The Deepest Days*, p. 66; M. C. Link, *Windows in the Sea*, p. 40.

57 *first man on the moon:* Robert Sténuit, taped interview, Nov. 26, 2004.

CHAPTER 5: DEPTH AND DURATION

Page

58 *On September 5:* Kilbracken, "The Long, Deep Dive," p. 727; Sténuit, *The Deepest Days,* p. 85; M. C. Link, *Windows in the Sea,* p. 47; Sténuit, interview, Nov. 26, 2004. For another account of the Sténuit dive, see Marion Link to Bill and Clayton Link, Sept. 11, 1962, Folder 237, LC.

58 *fantastic conductor of heat:* U.S. *Navy Diving Manual,* Part I, p. 14.

58 *almost pure helium:* Kilbracken, "The Long, Deep Dive," p. 727.

58 *a good night's rest:* M. C. Link, *Windows in the Sea,* p. 48.

58 *installing an additional heater:* Ibid., p. 47.

59 *itself a recent innovation:* E. R. Cross and D. J. Styer Sr., "The History of the Diving Suit," in *A Pictorial History of Diving,* p. 42.

59 Life *and* National Geographic: Sténuit, *The Deepest Days,* p. 89.

59 *some simple tasks:* Ibid.

59 *broiled Sténuit's head:* Ibid., p. 88.

59 *nearly froze:* Ibid., pp. 91, 93; M. C. Link, *Windows in the Sea,* p. 50.

59 *collect detailed physiological data:* Bornmann to CNO, "Link Diving Expedition," Oct. 31, 1962, p. 3.

59 *to sleep by nine:* Sténuit, *The Deepest Days,* p. 92.

59 *Céline's* D'un château l'autre: Ibid., p. 90.

59 *two ten-minute dives:* Bornmann to CNO, "Summary of Dives," Oct. 31, 1962, p. 5.

59 *called off the experiment:* M. C. Link, *Windows in the Sea,* p. 51.

59 *Sténuit protested:* Kilbracken, "The Long, Deep Dive," p. 730.

59 *leak in the homemade ventilation:* Ibid.; Bornmann to CNO, "Summary of Dives," Oct. 31, 1962, p. 5.

59 *couldn't risk running out:* M. C. Link, *Windows in the Sea,* p. 53.

59 *return of rough weather:* Bornmann to CNO, "Summary of Dives," Oct. 31, 1962, p. 5; M. C. Link, *Windows in the Sea,* p. 51.

59 *a real breakthrough:* Edwin A. Link, "Modern Equipment for Locating Wrecks," in *The Undersea Challenge: A Report of the Proceedings of the Second World Congress of Underwater Activities,* Bernard Eaton, ed. (London: British Sub-Aqua Club, 1963), p. 128.

59 *believed they had proved:* Link to Our friends, "Report No. 2," p. 4.

59 *Mazzone were much less impressed:* Mazzone, e-mail to author, June 28, 2003.

60 *Sunbird came to the rescue:* Bornmann to CNO, "Summary of Dives," Oct. 31, 1962, p. 5; M. C. Link, *Windows in the Sea,* p. 55.

60 *Bornmann factored in both:* Bornmann to CNO, "Summary of Dives," pp. 5–7; M. C. Link, *Windows in the Sea,* p. 55.

60 *tables for helium and the Genesis results:* Link, "Modern Equipment for Locating Wrecks," in *The Undersea Challenge,* p. 131.

60 *felt a groaning ache:* Sténuit, *The Deepest Days*, p. 100.

60 *equivalent of seventy feet:* Bornmann to CNO, "Summary of Dives," p. 5.

60 *rough weather to pass:* Kilbracken, "The Long, Deep Dive," p. 730.

60 *ten feet every few hours:* Bornmann to CNO, "Summary of Dives," pp. 6–7.

60 *twenty minutes past six o'clock:* Kilbracken, "The Long, Deep Dive," p. 730.

60 *For almost twenty-six of those:* Ibid.; Bornmann to CNO, "Summary of Dives," pp. 6–7.

61 *admirer John Godley:* John Kilbracken to Link, July 26, 1961, from London, Folder 349, LC.

61 *spread across fourteen pages:* Kilbracken, "The Long, Deep Dive," pp. 718–31.

61 *noncommittal paragraphs:* "Diver Up After 34 Hours," *New York Times*, Sept. 9, 1962, p. 15.

61 *catchy headline:* "Two Frenchmen to Spend Week Living in House on Floor of Sea," *New York Times*, Sept. 9, 1962, p. 15.

61 *dubbed it Diogenes:* Cousteau, *The Living Sea*, p. 315.

62 *ungainly clumps:* James W. Miller and Ian G. Koblick, *Living and Working in the Sea*, 2nd ed. (Plymouth, Vt.: Five Corners, 1995), photograph, p. 31.

62 *"no-decompression" dives:* U.S. Navy Diving Manual, Part I, pp. 96–99; Jean Alinat, a top Cousteau associate from 1947 to 1989, taped interview, Nice, France, Oct. 26, 2004.

62 *breathe ordinary compressed air:* Dr. X. Fructus, "Operation Pre-Continent No. 1," in *The Undersea Challenge*, pp. 79, 81.

62 Conshelf One: Ibid., pp. 76–86; Cousteau, *The Living Sea*, pp. 314–25; Maurice Dessemond and Claude Wesly, *Les hommes de Cousteau* (France: Le Pré aux Clercs, 1997), pp. 95–106; Claude Wesly, interview, Marseille, France, Nov. 2, 2004.

63 *standard dive principles:* Fructus, "Operation Pre-Continent No. 1," p. 77.

63 *bouts of gastritis:* Ibid., p. 83.

63 *Doctors came down:* Ibid., p. 81.

63 *wanted to be left alone:* Cousteau, *The Living Sea*, p. 321; Wesly, interview, Nov. 2, 2004.

63 *On the seventh day:* Cousteau, *The Living Sea*, p. 324.

64 *Cousteau's "exhibition":* Marion C. Link to Lord Kilbracken, Oct. 1, 1962, Sea Diver, Port of Monaco, Folder 237, LC.

64 *Cousteau would acknowledge:* Cousteau, *The Living Sea*, p. 315.

64 *timid:* Jacques-Yves Cousteau, "The Era of 'Homo Aquaticus,'" in *The Undersea Challenge*, p. 10.

64 *"the first men to occupy":* Cousteau, *The Living Sea*, p. 315; see also p. 317.

64 *Falco and Wesly, too, took pride:* Dessemond and Wesly, *Les hommes de Cousteau*, pp. 97–98; Jean-Michel Cousteau, *Mon père le commandant*, p. 120; Wesly, interview, Nov. 2, 2004.

64 *Second World Congress:* "Preface," in *The Undersea Challenge*, p. 2.

64 *Sténuit's record dive:* Link, "Modern Equipment for Locating Wrecks," in *The Undersea Challenge*, p. 128.

64 *"a healthy specimen":* Ibid., p. 129.

64 *flamboyant keynote speech:* Cousteau, "Era of 'Homo Aquaticus,'" in *The Undersea Challenge*, p. 8.

65 *"Homo aquaticus":* Ibid.

65 *earnestly reported:* "Cousteau Envisions a 'Water Man,'" *New York Times*, Oct. 10, 1962, p. 6; "Aquanauts," *Newsweek*, Oct. 29, 1962, p. 59.

65 *Dugan breathed further life:* James Dugan, "Portrait of Homo Aquaticus," *New York Times Magazine*, April 21, 1963, p. 20.

CHAPTER 6: EXPERIMENTAL DIVERS

Page

66 *necessary seals of approval:* Bond, *Papa Topside*, p. 49; Secretary of the Navy to Officer in Charge, U.S. Naval Medical Research Laboratory, U.S. Naval Submarine Base New London, Groton, Connecticut; Subject: Permission to use naval personnel as subjects in prolonged exposure to pressure; request for; Permission is granted . . . , Aug. 17, 1962. This is the fifth and final endorsement attached to Bond's memo of May 15, 1962, requesting permission (full citation below; copy in author's possession).

66 *Dr. James Wakelin:* J. Y. Smith, "James Wakelin, Pioneer in Oceanography, Dies," *Washington Post*, Dec. 23, 1990, p. B6.

66 *among Ed Link's friends:* James H. Wakelin Jr., *"Thresher:* Lesson and Challenge," *National Geographic*, June 1964, p. 763; Link to Wakelin, Aug. 1, 1964; Link to D. W. Smith, president, General Precision Inc., Aug. 1, 1964, Folder 256, LC.

66 *recalled getting a visit:* Bond, undated handwritten manuscript, for a section of the book in progress "Tomorrow the Seas," pp. 8–10, probably circa 1970, as previously noted.

66 *visitor at the secretary's Pentagon office:* Capt. Steve Anastasion, aide to Assistant Navy Secretary James Wakelin from 1960 to 1963, interview, Nov. 21, 2003.

67 *"Gentleman Jim":* Rear Adm. J. Edward Snyder Jr., assistant to Wakelin from 1963 to 1968, interview, Oct. 27, 2003.

67 *test called Genesis C:* Bond, *Papa Topside*, p. 33.

67 *the president was moved:* "Pride of Place, National Naval Medical Center History," www.bethesda.med.navy.mil/visitor/pride_of_place/pop_committee/history.html.

67 *looked neglected and unworthy:* Bond, "Tomorrow the Seas," pp. 13–14.

67 *Bond and Mazzone intended:* Officer in Charge (Bond), U.S. Naval Medical Research Laboratory, U.S. Naval Submarine Base New London, Groton, Connecticut to the Secretary of the Navy via Commanding Officer, U.S. Naval Submarine Base New London; Chief, Bureau of Medicine and Surgery; Chief

of Naval Personnel; Subject: Permission to use naval personnel as subjects in prolonged exposure to pressure; request for (four "volunteer subjects" listed include Bond and Mazzone), May 15, 1962 (copy in author's possession).

67 *supervisors doubling as test subjects:* Commanding Officer, U.S. Naval Submarine Base New London, Groton, Connecticut to the Secretary of the Navy; this is the first of five endorsements attached to Bond's memo seeking permission to use human subjects, June 18, 1962 (copy in author's possession).

67 *trust and admire Commander Bond:* Barth, *Sea Dwellers,* pp. 10, 40.

67 *played growing up in Manila:* Ibid., pp. 49, 50.

67 *son of an Army officer:* Bob Barth, interview, Sept. 30, 2002.

68 *Pan Am Clipper flying boats:* Barth, *Sea Dwellers,* p. 50.

68 *what was meant by lung squeeze:* Ibid.

68 *ship called the* Westward Ho: Barth, interview, Sept. 30, 2002.

68 *On November 27, 1962:* Lt. John C. Bull Jr., "Project Genesis I Phases C and E: General Consideration of Helium as an Inert Component of a Respiratory Medium with Particular Emphasis on Its Properties of Increased Thermal Conductivity and Effect on Blood Coagulation," paper written as partial fulfillment of requirements for qualification in submarine medicine, December 1962 (copy in author's possession).

68 *John Bull and Albert Fisher:* Barth, *Sea Dwellers,* p. 14; Mazzone, taped interview, June 4, 2003.

68 *housed in a basement:* Barth, interview, July 11, 2003.

68 *examine the heat-sapping:* Bull, "Project Genesis I Phases C and E."

68 *endless physical exam:* Barth, interview, June 9, 2003; Barth, *Sea Dwellers,* p. 14.

69 *slight pressure differential:* Barth, interview, June 9, 2003; Barth, *Sea Dwellers,* pp. 13–14; Bond, *Papa Topside,* p. 33.

69 *gaseous cacophony signaled:* Barth, interview, July 11, 2003.

69 *without breathing the element:* Bond, *Papa Topside,* p. 35; Bond, "Tomorrow the Seas," p. 32.

69 *chamber was leaking:* Mazzone, interview, June 4, 2003; Barth, *Sea Dwellers,* p. 14; Bond, "Tomorrow the Seas," pp. 18–19.

69 *integrity of the experiment:* Bond, "Tomorrow the Seas," p. 20; Mazzone, interview, June 4, 2003.

69 *Bond had recently made captain:* "Captain George Foote Bond, Medical Corps, U.S. Navy Transcript of Naval Service," Pers. E24-JFR: kc, 582417/2100, Jan. 10, 1966 (in author's possession).

69 *night shifts could stretch into the day:* Mazzone, interviews, Jan. 7, 2003, and June 4, 2003.

70 *thrown a typewriter:* Mazzone, interview, Jan. 2, 2002.

70 *made possible the opportunity:* Mazzone, interview, Aug. 7, 2002.

70 *dabbed suspect points:* Mazzone, interview, June 4, 2003.

70 *Barth could see Mazzone:* Barth, *Sea Dwellers*, p. 14.

70 *slap it with monkey shit:* Mazzone, interview, June 4, 2003.

70 *more difficult to contain than air:* Ibid.

70 *using dental cement:* Bond, "Tomorrow the Seas," p. 21.

70 *cured with a blow-dryer:* Ibid., p. 22.

70 *never played cards:* Barth, interview, Sept. 26, 2002.

71 *"helium speech":* U.S. *Navy Diving Manual*, Part I, p. 29; Barth, *Sea Dwellers*, p. 13.

71 *"Hey topside! I think":* *Sea Dwellers*, p. 15; Mazzone, interview, Aug. 7, 2002.

71 *watched with glee:* Barth, *Sea Dwellers*, p. 15.

71 *a mischievous order:* Ibid.

71 *breathing from a scuba tank:* Barth, *Sea Dwellers*, p. 16.

71 *Aquadro wheeled him through:* Ibid.; Aquadro, interview, July 12, 2003.

72 *well-publicized deep diving exercise:* Thomas Tillman, "The 1962 Hannes Keller 1,000 Foot Dive—Chapter 2," *Historical Diver* 33 (Fall 2002): 34; see also "Hannes Keller—Deep Dive Expert," *Skin Diver*, January 1963, p. 22.

72 *Hannes Keller staged:* Al Tillman, "The 1962 Hannes Keller 1,000 Foot Dive—Chapter 3," *Historical Diver* 33 (Fall 2002): 36; see also Al Tillman, "An Experiment in Depth," *Skin Diver*, February 1963, p. 8.

72 *secret breathing gas recipes:* T. Tillman, "1,000 Foot Dive—Chapter 2," p. 34.

72 *Dr. Albert Bühlmann:* Ibid.

72 *as an outsider of sorts:* Ibid.

72 *"make history" by touching:* Hannes Keller, "Towards the Limits of the Continental Shelf," in *The Undersea Challenge*, p. 167.

72 *signed on as an underwriter:* T. Tillman, "1,000 Foot Dive—Chapter 2," p. 34; Stephen Bullock, "Hannes Keller and USN," letter to the editor of *Historical Diver*, no. 38, p. 8.

72 *Navy sent Dr. Workman:* Officer in Charge, U.S. Navy Experimental Diving Unit to Chief, Bureau of Ships, Enclosure 1 of 6, "West Coast Trip Report (CDR Nickerson and CDR Workman) in Connection with BuShips 'Keller' Contract #N62558-3052, Phase III, Open Sea Dives," Dec. 14, 1962 (copy in author's possession); Mazzone, interview, Feb. 5, 2003.

73 *arriving at Catalina in late November:* OIC, EDU to Chief, BuShips, Enclosure 1: "West Coast Trip Report," p. 1.

73 *made available by the Shell Oil Co.:* Ibid., p. 1; A. Tillman, "1,000 Foot Dive—Chapter 3," p. 37.

73 *diving bell called Atlantis:* T. Tillman, "1,000 Foot Dive—Chapter 2," p. 34.

73 *Keller's diving partner, Peter Small:* "Tools for the Job," in *The Undersea Challenge*, p. 120, see also p. iii; A. Tillman, "1,000 Foot Dive—Chapter 3," p. 37.

73 *Atlantis to three hundred feet:* OIC, EDU to Chief, BuShips, Enclosure 1: "West Coast Trip Report," p. 2; T. Tillman, "1,000 Foot Dive—Chapter 2," p. 35.

73 *relatively mild cases of bends:* OIC, EDU to Chief, BuShips, Enclosure 1: "West Coast Trip Report," p. 2.

73 *rather than running further tests:* Ibid., pp. 2, 3.

73 *boatload of journalists:* A. Tillman, "1,000 Foot Dive—Chapter 3," p. 38; Dugan and Vahan, eds., *Men Under Water*, p. 206.

73 *adding pressure to go ahead:* "The 1962 Hannes Keller 1,000 Foot Dive—Chapter 5 (U.S. Findings on the Fatal Dive)," *Historical Diver* 33 (Fall 2002): 43.

73 *breathing hoses got tangled up:* Ibid., p. 44; T. Tillman, "1,000 Foot Dive—Chapter 2," p. 35; A. Tillman, "1,000 Foot Dive—Chapter 3," p. 38.

73 *he passed out:* OIC, EDU to Chief, BuShips, Enclosure 6 of 6, "Presentation by Hannes Keller on 11 December 1962 at EDU Concerning Diving Operations off Catalina Island," p. 1.

73 *gas mixture had gone bad:* "1,000 Foot Dive—Chapter 5 (U.S. Findings)," p. 42.

73 *Peter Small blacked out:* OIC, EDU to Chief, BuShips, Enclosure 6: "Presentation by Hannes Keller," p. 1.

73 *Christopher Whittaker:* A. Tillman, "1,000 Foot Dive—Chapter 3," pp. 38–39.

74 *Keller regained consciousness:* OIC, EDU to Chief, BuShips, Enclosure 6: "Presentation by Hannes Keller," p. 1.

74 *Small came around:* Ibid.

74 *Small felt cold:* Ibid.

74 *eight hours locked in Atlantis:* OIC, EDU to Chief, BuShips, Enclosure 3 of 6, "300 Meter Dive of Keller-Small on 3 Dec 1962," p. 2.

74 *coroner said the cause of death:* OIC, EDU to Chief, BuShips, Enclosure 6: "Presentation by Hannes Keller," p. 2.

74 *Keller insisted that other factors:* Ibid.; "1,000 Foot Dive—Chapter 5 (U.S. Findings)," pp. 42–43.

74 *Genesis D, as they called this:* Bond, *Papa Topside*, p. 35; Barth, *Sea Dwellers*, p. 19.

74 *Unit shared Building 214:* Interviews with former EDU personnel about the Unit, its chambers, and its old building, demolished years ago, including Kenneth W. Wallace, Navy master diver at EDU, 1962–1967, interviews, June 1, 2003, and July 22, 2003.

75 *down into the "wet pot":* Bond, *Papa Topside*, pp. 35–36; Barth, interview, June 9, 2003; *U.S. Navy Diving Manual*, Part I, p. 5.

76 *Manning and Lavoie were hospital:* Barth, *Sea Dwellers*, p. 23.

76 *also sent in a canary:* Ibid.

76 *"They're gonna hurt you":* Ibid., p. 22.

76 *palpable air of skepticism:* Barth, *Sea Dwellers*, p. 22; Mazzone, interview, Aug. 7, 2002, and taped interview, June 2, 2003.

76 *Unit personnel couldn't help but:* Wallace, interview, July 22, 2003.

76 *recently played host to a Swiss:* Keller, "Limits of the Continental Shelf," in *The Undersea Challenge*, p. 165; Bullock, letter to *Historical Diver*, no. 38, p. 8; Mazzone, interview, Feb. 5, 2003.

76 *a few people around the Unit:* Barth, *Sea Dwellers*, p. 22.

76 *Dr. Workman, who had recently:* Ibid., p. 25; Workman résumé (copy in author's possession).

76 *historic first for saturation diving:* Bond, *Papa Topside*, p. 38.

76 *wintry and downright steamy:* Barth, *Sea Dwellers*, p. 26.

77 *flimsy folding seat:* Barth, interview, July 11, 2003; Barth, *Sea Dwellers*, p. 21; Mazzone, interview in San Diego, Aug. 7, 2002.

77 *counterweighted trapeze:* Wallace, interviews, June 1 and July 22, 2003.

77 *blood pressure shot up:* Tape recording of Genesis D activity at the Experimental Diving Unit, with commentary dictated by Dr. George Bond, made at various points during the six-day test, April 25–30, 1963 (in author's possession).

77 *Atlantis dive provided a chilling case:* "1,000 Foot Dive—Chapter 5 (U.S. Findings)," p. 42.

77 *take tranquilizers:* Tape recording of Genesis D.

77 *consult a Ouija board:* Ibid.

78 *fire was on Bond's mind:* Ibid.

78 *"Well, I hope you all feel":* Ibid.

78 *"Notice how your voices":* Ibid.

79 *smoke in the roof of the igloo:* Ibid.; Barth, *Sea Dwellers*, p. 27; Barth, interview, June 9, 2003.

79 *faintly acrid smell:* Tape recording of Genesis D.

79 *"The subjects are being held":* Ibid.

79 *To celebrate:* Bond, *Papa Topside*, pp. 38–39.

CHAPTER 7: DEEP LOSS, DEEPER THINKING

Page

80 *tragedy of USS* Thresher: Wakelin, "*Thresher*: Lesson and Challenge," p. 759.

80 *diving to thirteen hundred feet:* Norman Polmar, *The Death of the USS* Thresher (Guilford, Conn.: Lyons, 1964, 2001), pp. 3, 117.

80 *Bond received a call asking:* Bond, *Papa Topside*, p. 1; Barth, *Sea Dwellers*, p. 32.

80 *close friends and shipmates:* Barth, *Sea Dwellers*, p. 32.

80 *orders to join the* Thresher: Barth, interview, Jan. 17, 2003.

81 *bathyscaphe* Trieste: Lt. Cmdr. Donald L. Keach, USN, "Down to *Thresher* by Bathyscaphe," *National Geographic*, June 1964, pp. 764–77.

81 *Navy bought the* Trieste: Edward C. Cargile, "*Trieste*—The Deepest Dive Ever," *Asian Diver* 8, Dec./Jan. 2000, p. 79.

81 *record drop to the deepest:* Ibid., p. 77.

81 Thresher *remains scattered:* Keach, "Down to *Thresher* by Bathyscaphe," pp. 766–67.

81 *best tool the Navy had:* Wakelin, "*Thresher:* Lesson and Challenge," p. 761; "United States Navy Briefing to Industry on the Deep Submergence Systems Project," sponsored by Special Projects Office, Department of the Navy, Nov. 24, 1964 (published and distributed by the National Security Industrial Association), p. 10.

81 *impossibility of a rescue:* Edwin A. Link, "Tomorrow on the Deep Frontier," *National Geographic,* June 1964, p. 786.

81 *established the Deep Submergence:* Memorandum from Secretary of the Navy Fred Korth to Distribution List; SecNav Notice 3100; Subj: Deep Submergence Systems Review Group, establishment of, April 24, 1963 (Edwin A. Link Special Collection, Evans Library, Florida Institute of Technology, online at edwin.lib.fit.edu/cdm4/about.php; hereafter abbreviated as FIT).

82 *submarine rescue—while crucial:* Rear Adm. E. C. Stephan, USN (ret.), "National Level Implications of the Deep Submergence Systems Research Groups, Deep Submergence Systems Project and Later Navy Ocean Engineering Evolutions; and Current Non-Military and Non-Government Interests in Ocean Engineering," lecture to ED Salvor/Ocean Engineering Class at Naval Ship Systems Command Headquarters, Washington, D.C., July 5, 1967, p. 3 (copy in author's possession).

82 *summoned by Rear Admiral E. C. Stephan:* Ibid., p. 2; Link, "Tomorrow on the Deep Frontier," p. 778.

82 *late at Stephan's home:* Stephan, "National Level Implications," p. 2.

82 *Link agreed to participate:* Link, "Tomorrow on the Deep Frontier," p. 780; Van Hoek with M. C. Link, *From Sky to Sea,* p. 275.

82 *Link headed the industry division:* Van Hoek with M. C. Link, *From Sky to Sea,* p. 274.

82 *tapped Charlie Aquadro:* Aquadro, interview, May 16–18, 2003.

82 *sailed* Sea Diver *to Washington:* Van Hoek with M. C. Link, *From Sky to Sea,* p. 276.

82 *provided with an office:* Ibid., p. 278.

83 *frequently be seen around the Unit:* Wallace, interview, July 22, 2003; Sténuit, *The Deepest Days,* p. 117.

83 *assistance of a medical student:* Joan Membery to Marion and Ed Link, Oct. 28, 1962, Folder 359, LC; Ed Link to Membery, Nov. 6, 1962, Folder 238, LC.

83 *locking mice into:* M. C. Link, *Windows in the Sea,* p. 63; Joan H. Membery and Edwin A. Link, "Hyperbaric Exposure of Mice to Pressures of 60 to 90 Atmospheres," *Science* 5 (June 4, 1964): 1241.

83 *verify a decompression schedule:* Sténuit, *The Deepest Days,* p. 135; C. H. Hedgepeth, U.S. Navy Experimental Diving Unit to Link, March 27, 1964, FIT; M. C. Link, *Windows in the Sea,* p. 63.

83 *"the first human colony":* "Two Frenchmen to Spend Week Living in House on Floor of Sea," *New York Times,* Sept. 9, 1962, p. 15; see also Cousteau, "The Era of 'Homo Aquaticus,'" in *The Undersea Challenge,* p. 10.

83 *French national petroleum office:* Capt. Jacques-Yves Cousteau, "At Home in the Sea," *National Geographic,* April 1964, p. 506.

83 *team of some forty-five men:* Ibid., p. 468.

83 *Conshelf Two's main sanctuary:* Jacques-Yves Cousteau, *World Without Sun,* James Dugan, ed. (New York: Harper & Row, 1964), p. 19.

83 *chef, Pierre Guibert:* Ibid., p. 79.

84 Bifteck sauté: Ibid., p. 93.

84 *depth of about two atmospheres:* Cousteau, "At Home in the Sea," p. 473. The habitat's precise depth is difficult to pinpoint in Cousteau's writing, but, as in the case of Conshelf One, the oceanauts appear to have been living close to or within the no-decompression limit at that time of two atmospheres, or thirty-three feet. On p. 466, Cousteau writes that Starfish House is "on a coral ledge 36 feet beneath the surface." But once perched on its telescopic legs (see pp. 476, 481, and the photograph of Starfish House on p. 475), its actual depth would have been several feet less. In *World Without Sun,* p. 70, Cousteau says Starfish House "stands five feet above the coral sand"; in the book's Foreword on p. 5 he gives the depth as thirty-three feet. In Dessemond and Wesly, *Les hommes de Cousteau,* p. 108, the ledge's depth is given as ten meters, or thirty-three feet; if that was the case, and the habitat stood five feet above the ledge, then the entry in the habitat floor would have been at about twenty-eight feet.

84 *"only at peril of their lives":* Cousteau, "At Home in the Sea," p. 482.

84 *"would die there in the shallows":* Cousteau, *World Without Sun,* p. 73.

84 *chamber on board* Rosaldo: Cousteau, "At Home in the Sea," p. 482; Cousteau, *World Without Sun,* p. 198.

84 *"the most beautiful":* Dessemond and Wesly, *Les hommes de Cousteau,* p. 107.

84 *"a diver's paradise":* Ibid.

84 *in the mid-eighties Fahrenheit:* Ibid., p. 501.

84 *Five "oceanauts":* Cousteau, "At Home in the Sea," pp. 468, 470, 482.

84 *smoke their Gitanes:* Cousteau, *World Without Sun,* pp. 90–91.

84 *Tuscan cigar:* Cousteau, "At Home in the Sea," p. 465.

84 *a handsome parrot:* Cousteau, *World Without Sun,* pp. 77, 79.

84 *onion-shaped dome:* Cousteau, "At Home in the Sea," p. 468.

85 *silver-hued wet suits:* Ibid., p. 482; Cousteau, *World Without Sun,* p. 72; Wesly, interview, Nov. 2, 2004.

85 *about eighty-five feet deep:* Cousteau, *World Without Sun,* pp. 48, 90, 102; Cousteau, "At Home in the Sea," pp. 498, 501. As with Starfish House, there is some ambiguity in Cousteau's writing about the actual depth at which the oceanauts lived in Deep Cabin, a twenty-foot-tall structure said to be anchored at ninety feet. The depths of 3.5 atmospheres, or 82.5 feet, and eighty-five feet, both of which appear on the pages cited here, seem to be fair estimates, since the entry hatch was in the floor of the midsection, about five to seven feet above the sea floor.

85 *an uncomfortable seven days:* Cousteau, "At Home in the Sea," p. 501.

85 *perspired "like fountains":* Ibid.

85 *a persistent gas leak:* Ibid., p. 503.

85 *depth of 330 feet:* Ibid., p. 505.

86 *to 363 feet:* Ibid., p. 506.

86 *Aqualungs were quickly depleted:* Ibid.; Cousteau, *World Without Sun,* p. 161.

86 *surfaced with the other oceanauts:* Cousteau, "At Home in the Sea," p. 506.

86 *suggested that sea floor habitats be placed:* Bond, "Proposal for Underwater Research," pp. 4–5.

86 *Cousteau his second Oscar:* Ephraim Katz, *The Film Encyclopedia,* revised by Fred Klein and Ronald Dean Nolen, 3rd ed. (New York: HarperPerennial, 1998) p. 302; Internet Movie Database, online at www.imdb.com.

86 *story for* National Geographic: Cousteau, "At Home in the Sea," pp. 465–507.

86 *on August 26, 1963:* Lt. John A. Lynch, MC, USN, "Exercise Tolerance Studies in an Artificial Atmosphere Under Increased Barometric Pressure," U.S. Naval Medical Research Laboratory Report No. 419, Bureau of Medicine and Surgery, Research Project MR005.14.3002-4.11. Although undated itself, this report includes more precise dates of Genesis E than seem to be recorded anywhere else. These dates correspond to the handwritten ones displayed by the test subjects in Navy photographs from the experiment (in author's possession); Barth, *Sea Dwellers,* p. 33, gives the general time period as "mid August of 1963"; Bond, *Papa Topside,* pp. 40–41, says the experiment would have a "total 'bottom time' of twelve days," but gives an erroneous start date of Aug. 1, 1962.

86 *Bond liked the coincidence that E:* Bond, *Papa Topside,* p. 39.

87 *able to do useful work:* Wakelin, "*Thresher:* Lesson and Challenge," p. 762; Mazzone, interview, Jan. 2, 2002; Lynch, "Exercise Tolerance Studies," p. 1.

87 *from experimental chambers into:* Barth, *Sea Dwellers,* p. 31; Bond, *Papa Topside,* p. 39.

87 *chamber had finally been installed:* Bond, *Papa Topside,* pp. 39, 41; Barth, *Sea Dwellers,* p. 31.

87 *control console along one side:* Navy photographs from Genesis E (in author's possession).

87 *sported a rectal probe:* Barth, *Sea Dwellers,* p. 38.

88 *"Before crossing emergency jumper":* Bond, *Papa Topside,* p. 45.

88 *Christmas tree decorations:* Ibid., p. 47.

88 *slightly elevated oxygen content:* Mazzone, interview, Aug. 7, 2002.

88 *passed a cake through:* Barth, *Sea Dwellers,* p. 39.

88 *showed some minor changes:* Lynch, "Exercise Tolerance Studies," p. 9; tape-recorded dictation of Genesis E results by Dr. George Lord, who was on Bond's staff at the Medical Research Lab (in author's possession).

88 *"shattering glass in the room":* Bond, *Papa Topside,* p. 42.

89   *"sonic boom"*: Ibid., p. 45.

89   *"all hell broke loose!"*: Ibid.

89   *apparently outright fiction*: Neither Mazzone, in interviews on March 17, 2002, and June 2, 2003, nor Barth, in an undated interview, recall these dramatic incidents, which Bond wrote that they, too, witnessed. Embellishment and exaggeration in Bond's writing were familiar to them, and to others interviewed, but not usually outright fiction, as appears to be the case in these particular accounts.

89   *After nearly twelve days living*: Lynch, "Exercise Tolerance Studies," pp. 1, 2, 9.

89   *schedule lasting twenty-six hours*: Bond, *Papa Topside*, p. 47.

89   *Barth, Bull, and Manning showed no*: Ibid.; Dictation of Genesis E results by Dr. George Lord.

CHAPTER 8: TRIANGLE TRIALS

Page

90   *build a prototype sea dwelling*: Barth, *Sea Dwellers*, p. 43.

90   *crews of a Texas Tower*: Ibid.

90   *improvised hacking and welding*: Barth, interview, June 9, 2003.

90   *"Bond's Folly—Sealab I"*: Mazzone, interview, Jan. 29, 2003.

90   *got a call from Captain Bond*: Barth, interview, June 9, 2003.

91   *old minesweeping floats*: Ibid.; Barth, *Sea Dwellers*, p. 46; Bond, *Papa Topside*, p. 49.

91   *Formal Navy approval*: Bond, *Papa Topside*, pp. 48–49.

91   *happy to try living on the*: Barth, interviews.

91   *"aquanaut," as Captain Bond liked to call*: Bond frequently used this term in his writings and in audiotape recordings made during Sealab.

91   *considered the French explorer a hero*: Scott Carpenter and Chris Stoever, *For Spacious Skies: The Uncommon Journey of a Mercury Astronaut* (Orlando, Fla.: Harcourt, 2002), p. 313.

91   *met Cousteau at the Massachusetts*: Ibid., p. 314; Scott Carpenter, taped interview, Jan. 30, 2002.

91   *fear of the deep ocean*: Carpenter and Stoever, *For Spacious Skies*, p. 314; Carpenter, interview, April 6, 2004.

91   *Cousteau said that*: Carpenter, interview, Jan. 30, 2002.

91   *Carpenter had never heard of*: Ibid.

91   *attained official status*: No one interviewed knew for sure, and no documentation as to the origin of the name could be found, but Mazzone, in the Jan. 29, 2003, interview cited above, was certain that the name originated with him, although he did not know how it became official. Lewis Melson, in a telephone interview on Jan. 24, 2003, said his engineer, Al O'Neal, came up with the name. O'Neal was not yet involved when Mazzone said he made up the name and wrote it on the Texas Tower pod. But perhaps O'Neal heard it, liked

it, and helped make it official once the project was taken on by the Office of Naval Research. Bond would seem most likely to elaborate on such a point; his reason for choosing to call his early experiments Genesis was well known. But Bond's only documented comment on the Sealab name is a passing reference in *Papa Topside*, p. 49.

92 *lent back to the Navy:* Carpenter and Stoever, *For Spacious Skies*, p. 315.

92 *they would train him:* Barth, *Sea Dwellers*, p. 56; Barth, interview, Jan. 16, 2004.

92 *knew he would have to prove himself:* Carpenter, taped interview, April 6, 2004.

92 *lifetime hobbies and achievements:* Carpenter and Stoever, *For Spacious Skies*, pp. 61, 70, 72.

92 *hazing of their new recruit:* Carpenter, interview, April 6, 2004; Barth, interview, Jan. 16, 2004.

92 *underfunded the Sealab program:* Carpenter, interviews, Jan. 25, 2002, and Oct. 20, 2006.

92 *formed after World War II:* A fiftieth anniversary history of ONR, originally published in *Naval Research Reviews* 48, no. 1 (1996), online at www.onr. navy.mil/about/history.

92 *$200,000:* Lewis B. Melson, Captain, U.S. Navy, Retired, "Remarks Concerning the Inception and Operations of the Sealab I and Sealab II Projects," Feb. 15, 1989, p. 5 (copy in author's possession, obtained during interview with Melson, Annapolis, Md., Nov. 11, 2003).

92 *about $35,000:* Bill Culpepper, former senior engineer at the Mine Defense Laboratory assigned to Sealab, interview, Seagrove Beach, Fla., March 11, 2003.

92 *few specific guidelines:* Ibid.; Barth, *Sea Dwellers*, p. 56.

92 *earning a master's degree:* Mazzone, e-mail to author, June 4, 2003; Mazzone, interview, Oct. 17, 2002.

92 *Bond could see Sealab:* Bill Culpepper, interview, June 13, 2003.

92 *cigar-shaped capsule:* Barth, *Sea Dwellers*, pp. 46, 147.

93 *fixtures from Sears:* Melson, "Remarks," p. 7; Culpepper, interview, March 11, 2003.

93 *avoid an ordinary fridge:* Culpepper, interview, March 11, 2003.

93 *Sears garden hose:* Ibid.; Lewis Melson, taped interview, Jan. 28, 2003.

93 *"Knowledge of the oceans":* Public Papers of the Presidents of the United States, John F. Kennedy, 1961 (Washington, D.C.: U.S. Government Printing Office, 1962), p. 241.

93 *"We know less of the oceans":* Public Papers of the Presidents of the United States, John F. Kennedy, 1963 (Washington, D.C.: U.S. Government Printing Office, 1964), p. 805.

94 *shipyard work he preferred:* Melson, e-mail to author, Jan. 10, 2004.

94 *late August 1963:* Melson, e-mail to author, June 14, 2003.

94 *decided to act on an idea:* Melson, "Remarks," pp. 3–4.

94 *Typhoon Karen's trashing:* Ibid., p. 2.

94 *Melson saw firsthand:* Ibid.

94 *ought to have better ways:* Ibid., p. 3.

94 *put his top engineer:* Ibid., p. 4.

94 *"diving establishment":* Ibid., p. 5.

94 *Neither Melson nor O'Neal was a diver:* Melson, interview, Nov. 11, 2003.

94 *found out about Captain Bond:* Melson, e-mail to author, Feb. 1, 2004; Melson, "Remarks," p. 6.

94 *invited Bond to Washington:* Ibid.

94 *Melson still had doubts:* Ibid.

94 *tried to recruit Dr. Workman:* Ibid., p. 5; Melson, interview, June 13, 2003.

94 *lent his support and asked:* Melson, "Remarks," p. 5; Melson, e-mail to author, Feb. 2, 2004.

94 *pooled funds from other projects:* Melson, "Remarks," p. 5.

95 *put Sealab I on view:* Barth, *Sea Dwellers*, p. 58.

95 *Bond stood on the tarmac:* Ibid.

95 *A few thousand people:* Ibid., p. 146; George F. Bond, Capt., USN, "Sealab I Chronicle," pp. 44–45 (copy in author's possession of this unedited journal on which the Sealab I chapter in *Papa Topside* is based), p. 1.

95 *its first tests at sea:* Barth, *Sea Dwellers*, p. 61.

95 *orange whale chasing after: Sealab I*, a documentary film produced by the U.S. Naval Photographic Center; in addition to providing a valuable visual record of many activities, the film includes corroborating information in its voice-over narration (video copy in author's possession, but also available in libraries).

95 *support ship that would serve:* H. A. O'Neal, G. F. Bond, R. E. Lanphear, and T. Odum, "Project Sealab Summary Report: An Experimental Eleven-Day Undersea Saturation Dive at 193 Feet," Office of Naval Research, Washington, D.C., June 14, 1965, p. 31; "Special Projects Barge YFNB-12," Sealab I press release, undated (in author's possession).

95 *"Your Friendly Navy Barge":* Mazzone, Barth, and others, interviews.

95 *attached Sealab to the barge:* Culpepper, interview, March 11, 2003.

95 *effect of the stretching line:* Ibid.; Barth, *Sea Dwellers*, p. 62.

95 *lab flooded and crash-landed:* Barth, *Sea Dwellers*, p. 62; Bond, "Sealab I Chronicle," p. 6.

96 *threatened with termination:* Bond, "Sealab I Chronicle," p. 8.

96 *hanging by a hair:* LCDR Robert E. Thompson, MC, USN, "Sealab I: A Personal Documentary Account," U.S. Naval Submarine Medical Center, Groton, Conn., Memorandum Report No. 66-9, Bureau of Medicine and Surgery, Navy Department Research Work Unit MF011.99-9003.05, March 30, 1966, p. 5.

96 *left on the bottom overnight:* Ibid., p. 6.

96 *twenty-seven miles southwest:* O'Neal et al., "Project Sealab Summary Report," p. 3.

96 *relatively warm and clear water:* Ibid., p. 2; *Sealab I* film.

96 *known as the Bermuda Triangle:* Charles Berlitz, with the collaboration of J. Manson Valentine, *The Bermuda Triangle* (Garden City, N.Y.: Doubleday, 1974), pp. 11, 19.

96 *"They vanished as completely":* Ibid., photograph section, p. 2.

96 *deepened the mystery:* Ibid., p. 35.

97 *Link wrapped up:* M. C. Link, *Windows in the Sea*, p. 63.

97 *habitat he called SPID:* Ibid.

97 *envisioned as an elevator:* Sténuit, *The Deepest Days*, pp. 163–65.

97 Nahant *used its echo:* Ibid., p. 162.

97 *extinct volcano:* Ibid., p. 148; Hans W. Hannau, *Bermuda* (Munich: Wilhelm Andermann Verlag, 1962), p. 5.

97 *an undersea plateau:* M. C. Link, *Windows in the Sea*, p. 66.

97 *seas were calm:* Sténuit, *The Deepest Days*, p. 163.

97 *leaks had to be plugged:* Ibid.

97 *Lindbergh, the eldest son:* Ibid., p. 156; M. C. Link, *Windows in the Sea*, p. 66; Jon M. Lindbergh, "What Direction for Men in the Sea?," *Undercurrents*, defunct magazine about commercial diving), June 1970, pp. 8–9.

98 *affable but considerably more taciturn:* Sténuit, interview, Nov. 26, 2004.

98 *By early afternoon:* Sténuit, *The Deepest Days*, p. 167.

98 *collided in midair:* "Colony Shocked; Seven Survive Plane Collision: Five Bodies Are Recovered," *Bermuda Sun*, June 29, 1964, p. 1; "Moment of an Awful Plunge," *Life*, July 10, 1964, p. 32; "Seven Dead, 10 Missing as Planes Collide," *New York Times*, June 30, 1964, p. 67.

98 *Sealab crew got an unexpected:* Barth, *Sea Dwellers*, p. 65.

98 *depths down to 250:* Bond, *Papa Topside*, p. 51.

98 *considered putting Sealab down near:* Ibid., p. 53; Thompson, "Sealab I: A Personal Documentary Account," p. 14.

98 *pipefuls of Sir Walter:* Bond, *Papa Topside*, p. 53; George Bond Jr., interview, Oct. 11, 2003.

98 *stick with their original plan:* Bond, *Papa Topside*, p. 54.

98 *observer and emissary:* Aquadro, interviews, including Oct. 8, 2002, and May 16–18, 2003.

99 *perilous dives to three hundred feet:* Ibid.; Colin Simpson, *The Lusitania* (Little Brown & Co., 1973), p. 4; Kenneth MacLeish, "Was There A Gun?," *Sports Illustrated*, Dec. 24, 1962, p. 46.

99 *for a television documentary:* Charles Aquadro, interview, June 30, 2006.

99 *a golden age for diving:* Vorosmarti, "History of Saturation Diving," p. 4.

99 *surprised him with an offer:* Aquadro, interview, Oct. 8, 2002.

99   *arrangement with the Office of Naval Research:* Aquadro, interview, May 16–18, 2003; Commander F. J. Kelley, ONR Branch Office, London to Commandant Jacques-Yves Cousteau, Director, Institut Oceanographique, June 24, 1964 (copy in author's possession).

99   *jet remains were widely scattered:* Thompson, "Sealab I: A Personal Documentary Account," p. 14; Robert Thompson, interview, Feb. 19, 2004; Bond, "Sealab I Chronicle," p. 19; Aquadro, interview, May 16–18, 2003; Cyril Tuckfield, interview, March 24, 2004.

99   *unhappy and unrewarding:* Bond, *Papa Topside,* p. 52.

99   *Tuck came up with:* Tuckfield, interview, March 24, 2004; Thompson, "Sealab I: A Personal Documentary Account," p. 15.

100   *Aquadro had managed to extricate:* Aquadro, interview, May 16–18, 2003; Thompson, "Sealab I: A Personal Documentary Account," p. 15.

100   *seemed the least fazed:* Carpenter, interview, Oct. 20, 2006; Carpenter and Stoever, *For Spacious Skies,* p. 317.

100   *panting like winded dogs:* Sténuit, *The Deepest Days,* p. 172.

100   *their heater failed:* Ibid., p. 171.

100   *lightbulb imploded:* Ibid.

100   *divers scribbled notes:* Ibid., pp. 172, 178.

100   *distance of fifty feet:* Ibid., p. 175.

101   *emptied within minutes:* Edwin A. Link, "Outpost Under the Ocean," *National Geographic,* April 1965, p. 532.

101   *dive for several hours:* Sténuit, *The Deepest Days,* p. 178.

101   *seventy-two degrees:* Ibid., p. 169.

101   *Sténuit nearly froze:* Ibid., pp. 179, 180.

101   *deepest photographs ever shot:* Ibid.

101   *unnerving thuds:* Ibid., p. 178.

101   *Forty-nine hours after:* Ibid., pp. 168, 181.

101   *Link called down:* Ibid., p. 180; M. C. Link, *Windows in the Sea,* p. 78.

101   *divers thought they might as well:* Sténuit, *The Deepest Days,* p. 180.

101   *thought they had been fortunate:* M. C. Link, *Windows in the Sea,* p. 81; Link to Jon Lindbergh, Aug. 13, 1964, Folder 256, LC.

101   *nothing more to be gained:* Sténuit, *The Deepest Days,* p. 180; Sténuit, interview, Nov. 26, 2004.

101   *afternoon on July 2:* Sténuit, *The Deepest Days,* p. 181.

101   *made an airtight transfer:* M. C. Link, *Windows in the Sea,* pp. 78–79.

101   *sailed back to Miami:* Ibid., p. 79.

101   *Link's three medical specialists:* Ibid., p. 66.

101   *schedule that Dr. Workman had devised:* Sténuit, *The Deepest Days,* p. 181; Link to Cmdr. C. H. Hedgepeth, Officer in Charge, U.S. Navy Experimental Diving Unit, July 21, 1964, Folder 255, LC.

102   *inside of his leg:* Sténuit, *The Deepest Days,* p. 184.

102 *schedule was extended:* Ibid.

102 *Bond considered a scientific sin:* Bond, "Sealab I Chronicle," p. 21.

102 *NASA could continue to assess:* Richard Witkin, "Clouds Threaten Astronaut Flight; Officials Hoping to Launch Cooper on a 22-Orbit Trip This Morning in Florida," *New York Times,* May 14, 1963, p. 1.

102 *explosion to rock* Sea Diver: Diary of Marion C. Link, pp. 16–17, Folder 107, LC; Link to Dr. C. J. Lambertsen, University of Pennsylvania School of Medicine, Aug. 13, 1964; Link to Dr. Joan Membery, Mayo Clinic, Rochester, Minn., Aug. 14, 1964; Link to Kenneth MacLeish, assistant editor, National Geographic Society, Aug. 14, 1964, Folder 256, LC; Sténuit, interview, Nov. 26, 2004.

103 *paralysis around his ankles:* Sténuit, interview, Nov. 26, 2004.

CHAPTER 9: "BREATHE!"

Page

104 *observation platform called Argus Island:* O'Neal et al., "Project Sealab Summary Report," p. 3; *Sealab I* film.

104 *tricky process of lowering:* O'Neal et al., "Project Sealab Summary Report," p. 5.

104 *Bond's heart sank:* Bond, "Sealab I Chronicle," p. 41.

104 *like Panama City:* Barth, *Sea Dwellers,* p. 64.

104 *had good reason to worry:* Bond, "Sealab I Chronicle," p. 40.

104 *called into a conference:* Ibid., p. 41.

104 *monstrous handling difficulties:* Ibid.

104 *the job was hopeless:* Ibid.

105 *source of tension:* Melson to Helen Siiteri, editor of *Papa Topside,* Aug. 1, 1992 (copy in author's possession); Melson, "Remarks," p. 9; Mazzone, interviews, including Jan. 2, 2002.

105 *impressive credentials:* Sealab I press releases (in author's possession).

105 *ignored his commands:* Melson to Siiteri, Aug. 1, 1992; Melson, interview, Jan. 28, 2003.

105 cannot accede to command: Mazzone, interviews, including Jan. 2, 2002.

105 *worthy of court-martial:* Melson to Siiteri, Aug. 1, 1992; Melson e-mail to author, June 26, 2003.

105 *an ingenious approach:* Bond, "Sealab I Chronicle," p. 41.

105 *suggested relocating the test:* Ibid.

105 *just seventy feet:* Ibid.

105 *exchanged hopeful glances:* Ibid., p. 43.

105 *use the big crane on Argus:* Ibid; O'Neal et al., "Project Sealab Summary Report," p. 5.

105 *Bond was relieved:* Bond, "Sealab I Chronicle," p. 43.

106 *front-page headline: Bermuda Sun,* July 17, 1964.

106 *newly acquainted colleagues rode:* Carpenter, interview, April 6, 2004; Aquadro, interview, May 16–18, 2003; *Royal Gazette* (Bermuda), July 17, 1964.

106 *called to the dispensary:* Aquadro, interview, May 16–18, 2003.

106 *Bond was summoned:* Ibid.; Bond, *Papa Topside*, p. 54; Bond, "Sealab I Chronicle," p. 44.

106 *thirty-nine-year-old spaceman:* Carpenter and Stoever, *For Spacious Skies*, p. 11.

106 *dozed off at the handlebars:* Carpenter, interview, April 6, 2004.

106 *compound fracture:* Ibid., p. 317.

106 *looking badly bruised:* Aquadro, interview, May 16–18, 2003.

106 *totaled half a dozen cars:* Carpenter and Stoever, *For Spacious Skies*, p. 60.

106 *beloved '34 Ford coupe:* Ibid., p. 88.

106 *off the roster:* Bond, "Sealab I Chronicle," p. 44.

106 *sorely disappointed:* Ibid.; Carpenter, interview, April 6, 2004.

106 *Thompson, a younger submarine medical officer:* Robert Thompson, taped interview, Fallbrook, Calif., Dec. 29, 2003; résumé from a Sealab I press packet (in author's possession).

107 *Anderson had been enthusiastic:* Barth, interview, June 9, 2003.

107 *the two hit it off:* Barth, *Sea Dwellers*, p. 23.

107 *had swarthy features:* Navy photographs (in author's possession); *Sealab I* film.

107 *an inveterate prankster:* "Lester Everett 'Andy' Anderson, GMI (DV), Aug. 20, 1932–Aug. 31, 1999," biographical sketches by family and friends compiled by Sherry Anderson, Lester's eldest daughter, who sent a copy to author on May 19, 2003, p. 34; interviews with Anderson's former shipmates.

107 *"midnight requisitioning":* Barth, *Sea Dwellers*, p. 57; Barth, interview, Panama City, Fla., including a tour of the Navy Experimental Diving Unit and the surrounding base, formerly the Mine Defense Laboratory, Dec. 17, 2001.

107 *Andy Anderson was just fifteen:* "Lester Everett 'Andy' Anderson," p. 9; résumé of Lester Everett Anderson, 1984 (in author's possession).

107 *and all were fathers:* O'Neal et al., "Project Sealab Summary Report," pp. 57–58.

107 *structure with respect to its divers:* Barth, interview about Anderson's specific case; interviews with other Navy divers; see also *The Bluejackets' Manual*, revised by Bill Wedertz and Bill Beardon, 20th ed. (Annapolis, Md.: United States Naval Institute, 1978), pp. 16–18.

107 *jobs that happened to be underwater:* Owen Lee, "The Master of the Master Divers," *Skin Diver*, September 1965, p. 53.

108 *ran scuba training sessions:* Barth, interview, Sept. 24, 2002.

108 *seas were blessedly calm:* Bond, *Papa Topside*, pp. 54, 56; *Sealab I* film.

108 *secured by a bridle:* Barth, *Sea Dwellers*, p. 150; *Sealab I* film.

108 *issue of gradually raising the lab's:* Barth, *Sea Dwellers*, p. 62.

108 *"ballast bins":* Ibid., p. 61; O'Neal et al., "Project Sealab Summary Report," p. 25; *Sealab I* film.

108 *axles were another product:* Melson, "Remarks," p. 8.

108 *Although somewhat unwieldy:* Bond, *Papa Topside*, p. 56; Cyril Tuckfield, interview, March 24, 2003; *Sealab I* film.

108 *like steel pellets:* Culpepper, interview, March 11, 2003; Link, "Outpost Under the Ocean," p. 533.

108 *wrangled them into the ballast:* Tuckfield, interview, March 24, 2004.

108 *Barth and Dr. Thompson made:* Thompson, "Sealab I: A Personal Documentary Account," p. 21.

109 *the "air space":* Ibid.; Bill Culpepper, interview, Dec. 12, 2003.

109 *Shortly before two o'clock:* Bond, "Sealab I Chronicle," p. 48.

109 *SDC was another hand-me-down:* O'Neal et al., "Project Sealab Summary Report," p. 32; *Sealab I* film.

109 *Mazzone had to pick up:* Mazzone, interview, Aug. 7, 2002.

109 *he gave it a trial run:* Walter Mazzone, taped interview, San Diego, Calif., Dec. 28, 2003.

109 *Bond and Mazzone each took:* Bond, *Papa Topside*, p. 57.

109 *They swam the distance:* Ibid.

109 *deepest breath-holding dive:* Ibid.

109 *similar swim from Link's cylinder:* Sténuit, *The Deepest Days*, p. 169.

109 *coral sand on the sea floor:* Bond, "Sealab I Chronicle," p. 50; James Atwater and Roger Vaughan, "Room at the Bottom of the Sea," *Saturday Evening Post*, Sept. 5, 1964, pp. 20, 21, 23.

109 *the shark cage:* Bond, *Papa Topside*, p. 57; Barth, *Sea Dwellers*, p. 77.

110 *maintained at nearly seven times:* O'Neal et al., "Project Sealab Summary Report," pp. 29, 34, 39.

110 *potentially great chapter:* Bond, *Papa Topside*, p. 62.

110 *the interior was a cross between:* O'Neal et al., "Project Sealab Summary Report," pp. 4, 14, 15, 25, 26; *Sealab I* film.

110 *exhaled purposefully:* Bond, *Papa Topside*, p. 57.

110 *SDC was cramped:* Barth, *Sea Dwellers*, p. 81.

111 *teemed with marine life:* Thompson, "Sealab I: A Personal Documentary Account," pp. 21, 22; Barth, *Sea Dwellers*, p. 68.

111 *version of "O Sole Mio":* O'Neal et al., "Project Sealab Summary Report," p. 39.

111 *"Sealab I, this is":* Tape recording of Sealab I aquanauts upon arrival in the habitat (in author's possession).

111 *a favorite of Andy's:* Sherry Anderson, e-mail to author, Dec. 13, 2003.

111 *"Well, Lester Anderson":* Tape recording of Sealab I aquanauts.

111 *muttering over the intercom:* Ibid.

111 *"As I take it now":* Ibid.

111 *about seventy-four Fahrenheit:* O'Neal et al., "Project Sealab Summary Report," pp. 19, 39.

111 *set up the cameras:* Thompson, "Sealab I: A Personal Documentary Account," p. 22.

111 *"Smile!":* Tape recording of Sealab I aquanauts.

111 *Anderson tried again:* Ibid.

112 *"Well, I guess that's okay":* Ibid.

112 *"Papa Topside":* Bond, *Papa Topside,* p. 62.

112 "my *aquanauts":* Bond, *Papa Topside,* p. 54; O'Neal et al., "Project Sealab Summary Report," p. 43.

112 *variety of setup procedures:* O'Neal et al., "Project Sealab Summary Report," p. 39; Barth, *Sea Dwellers,* p. 68.

112 *sparkling Mateus rosé:* Ibid., p. 63; Thompson, "Sealab I: A Personal Documentary Account," p. 24.

112 *wrong with the thermoelectric:* O'Neal et al., "Project Sealab Summary Report," pp. 39, 50.

112 *had the use of their electrowriter:* Ibid., p. 32.

112 *"HELP!":* Bond, "Sealab I Chronicle," p. 51.

112 *just a lungful of air:* Barth, *Sea Dwellers,* p. 79.

112 *they'd go out on their own:* Thompson, "Sealab I: A Personal Documentary Account," makes this clear, as does a handwritten Sealab I logbook kept by the aquanauts, with extemporaneous notes on their daily activities (in author's possession).

112 *crystalline conditions:* O'Neal et al., "Project Sealab Summary Report," pp. 40, 41; Barth, *Sea Dwellers,* p. 67; *Sealab I* film.

112 *crisscrossed the sandy sea floor:* Barth, *Sea Dwellers,* p. 66; *Sealab I* film.

112 *water was about seventy degrees:* O'Neal et al., "Project Sealab Summary Report," p. 44; Thompson, "Sealab I: A Personal Documentary Account," p. 29.

113 *"hookah":* O'Neal et al., "Project Sealab Summary Report," pp. 21, 30.

113 *killing conversation:* Barth, interview, Jan. 16, 2004.

113 *the new Mark VI:* "Service Manual for Mark VI Underwater Breathing Apparatus," *U.S. Navy Diving Manual,* Section D-E, Department of the Navy, Bureau of Ships, NAVSHIPS 393-0653, pp. 594–631; interviews with numerous Navy divers who used the rig, including Barth, interview, Jan. 11, 2002, and Richard Blackburn, a Sealab III aquanaut and team member for the Mark VI field evaluation conducted by Explosive Ordnance Disposal Unit Two, U.S. Atlantic Fleet Mine Force, Charleston, S.C., December 1962 to March 1963, several interviews, including Jan. 27, 2005; another helpful source was Kent Rockwell, scuba historian and former editor of *Historical Diver,* including his e-mails to author of May 17, 2004, and June 13, 2005.

113 *less likely to kill a man:* Blackburn, interview, Jan. 27, 2005; "Mark VI Semi-Closed Circuit Breathing Apparatus—Field Evaluation, December 1962–March 1963," Evaluation Report 1–63, conducted by Explosive Ordnance Disposal Unit Two, submitted by J. C. Bladh, project officer, p. 1; Jim Bladh, interview during Sealab reunion, Panama City, Fla., March 12, 2005.

114 *allowed to rise about one atmosphere:* Walter Mazzone, interview, Jan. 19, 2005; Reynold T. Larsen and Walter F. Mazzone, "Excursion Diving from Saturation Exposures at Depth," in *Underwater Physiology—Proceedings of the Third Symposium on Underwater Physiology,* sponsored by the Committee on Undersea Warfare of the National Academy of Sciences—National Research Council and the Office of Naval Research, in Washington, D.C., March 23–25, 1966, C. J. Lambertsen, ed. (Baltimore: Williams & Wilkins, 1967), p. 251; Charles Hillinger, "Aquanauts Under Sea Talk to Press Above," *Los Angeles Times,* Sept. 10, 1965, p. A8.

114 *"explosive decompression":* Peter Bennett, coeditor of *The Physiology and Medicine of Diving and Compressed Air Work,* taped interview, April 5, 2006.

114 *a few technical difficulties:* O'Neal et al., "Project Sealab Summary Report," p. 39.

114 *their canned sardines:* Barth, *Sea Dwellers,* p. 74; *Sealab I* film.

114 *George and Wally:* Command Information Bureau, Project Sealab I press releases, July 22 and July 27, 1964 (in author's possession).

114 *two sharks that appeared:* Command Information Bureau, Project Sealab I press release, July 23, 1964.

115 *behavior that irritated:* Bond, "Sealab I Chronicle," p. 56; Mazzone, interview, Jan. 2, 2002; O'Neal et al., "Project Sealab Summary Report," p. 38.

115 *upset some topside sensibilities:* Thompson, "Sealab I: A Personal Documentary Account," p. 27; Barth, *Sea Dwellers,* p. 79.

115 *"aquanaut breakaway phenomenon":* Bond, *Papa Topside,* p. 35, with further elaboration in an undated, typed paper by Bond, titled "Sealab I," p. 5; this paper appears to be a later version of his "Sealab I Chronicle" (copy in author's possession); Mazzone, interview, Jan. 2, 2002.

115 *studied marine biology:* Thompson, interview, Dec. 29, 2003.

115 *informal demonstration:* Thompson, "Sealab I: A Personal Documentary Account," pp. 1, 26, 32.

115 *fishy aroma:* Barth, interview; O'Neal et al., "Project Sealab Summary Report," p. 43.

115 *divers shuttled supplies:* Barth, *Sea Dwellers,* p. 79.

115 *kept in the mid-eighties:* O'Neal et al., "Project Sealab Summary Report," pp. 39–40, 42, 49.

115 *variety of physical discomforts:* Thompson, "Sealab I: A Personal Documentary Account," p. 23; O'Neal et al., "Project Sealab Summary Report," p. 42.

115 *everyone's joints ached:* Thompson, "Sealab I: A Personal Documentary Account," p. 23.

115 *pain in Barth's left shoulder:* Ibid., p. 24.

115 *Thompson and Manning had headaches:* Ibid., p. 23.

115 *spiked fevers and remained feverish:* O'Neal et al., "Project Sealab Summary Report," pp. 47, 49.

116 *didn't write or think clearly:* Thompson, "Sealab I: A Personal Documentary Account," p. 24.

116 *complained about rapid breathing:* Ibid., p. 31; O'Neal et al., "Project Sealab Summary Report," p. 47.

116 *nearly burned his lips:* Thompson, interview, Dec. 29, 2003; Thompson, "Sealab I: A Personal Documentary Account," p. 30.

116 *with a thermometer:* Robert Thompson, interview, Sept. 20, 2002.

116 *a nude blonde:* Aquadro, interview, May 16–18, 2003 (he had kept the original as a Sealab I souvenir); Atwater and Vaughan, "Room at the Bottom of the Sea," p. 21.

116 *books were propped up:* Atwater and Vaughan, "Room at the Bottom of the Sea," p. 22 (photo); *Sealab I* film.

116 *recited prose and poetry:* Bond, "Sealab I Chronicle," p. 57; George Bond Jr., e-mail to author, Feb. 24, 2003.

116 *On the first Sunday:* Handwritten Sealab I logbook, entry for July 26, 1964; Command Information Bureau, Project Sealab I press release, July 27, 1964.

116 *"Sealab Prayer":* Retirement ceremony program for Capt. George F. Bond, Dec. 1, 1975 (in author's possession); recited from memory by Walt Mazzone at Sealab reunions.

117 *reached several ham operators:* Barth, *Sea Dwellers*, p. 73; tape recording of Sealab I activity, dated July 28, 1964 (in author's possession); Thompson, "Sealab I: A Personal Documentary Account," p. 33.

117 *like undersea summer camp:* Barth, *Sea Dwellers*, pp. 73–74, 79; O'Neal et al., "Project Sealab Summary Report," p. 41.

117 *work Melson found lacking:* Melson, "Remarks," p. 10; Melson, interview, Jan. 28, 2003.

117 *end of the first week:* Thompson, "Sealab I: A Personal Documentary Account," p. 31.

117 *arrival of Star I:* Ibid.; Bond, "Sealab I Chronicle," p. 62.

117 *Albert "Smoky" Stover:* Barth, interview, Jan. 16, 2004; Barth, *Sea Dwellers*, p. 75.

117 *also set up a TV camera:* Barth, *Sea Dwellers*, p. 73.

117 *asked Manning to take:* "Inside Log Sealab I, L. Anderson, GM1, Diary," p. 1 (in author's possession).

117 *sub wasn't much bigger than:* Navy photograph No. XAB-18077, U.S. Naval Station, Bermuda, July 1964 (in author's possession); Barth, *Sea Dwellers*, p. 152.

117 *Barth had grabbed on to a piece:* Barth, *Sea Dwellers*, p. 76.

118 *Manning ran out of film:* "Anderson, GM1, Diary," p. 1; Manning was said to be quite ill when research for this book began, could not be reached for interviews, and is said to have died in late March 2004 (Bob Barth, e-mail to author and others, April 9, 2004).

118 *gave the camera to Anderson:* "Anderson, GM1, Diary," p. 1.

118 *swam the twenty yards or so:* O'Neal et al., "Project Sealab Summary Report," p. 55.

118 *bubbles had ceased to burp:* Ibid.; Bond, "Sealab I Chronicle," p. 63.

118 *Bond could see him scurry:* Bond, "Sealab I Chronicle," p. 63.

118 *activate the bypass:* O'Neal et al., "Project Sealab Summary Report," p. 55.

118 *heard a clang:* Ibid.; "Anderson, GM1, Diary," p. 2; Barth, *Sea Dwellers,* p. 76; Atwater and Vaughan, "Room at the Bottom of the Sea," p. 24.

118 *figured it must be Manning:* "Anderson, GM1, Diary," p. 2.

118 *Anderson looked down and:* O'Neal et al., "Project Sealab Summary Report," p. 56.

118 *managing to get:* "Anderson, GM1, Diary," p. 2.

118 *stuck his index finger:* Ibid.

118 *a hundred pounds of diving gear:* Atwater and Vaughan, "Room at the Bottom of the Sea," p. 24.

118 *shouted for help and rapped:* Ibid.; O'Neal et al., "Project Sealab Summary Report," p. 56.

118 *"Breathe!":* "Anderson, GM1, Diary," p. 2.

118 *when he heard the SOS:* Thompson, interview, Feb. 19, 2004; Thompson, "Sealab I: A Personal Documentary Account," p. 32.

119 *not making much sense:* "Anderson, GM1, Diary," p. 2.

119 *insisted nothing was wrong:* Thompson, interview, Sept. 20, 2002.

119 *Dr. Thompson's checkup:* O'Neal et al., "Project Sealab Summary Report," p. 56; Thompson, "Sealab I: A Personal Documentary Account," p. 32.

119 *beat Anderson at cribbage:* Atwater and Vaughan, "Room at the Bottom of the Sea," p. 24.

119 *Manning's eyes:* Ibid.; Thompson, "Sealab I: A Personal Documentary Account," p. 32.

119 *"face mask squeeze":* U.S. Navy Diving Manual, Part I, p. 150; Thompson, interview, Dec. 29, 2003.

119 *"I've seen eyes like":* Atwater and Vaughan, "Room at the Bottom of the Sea," p. 24.

119 *received a special supper:* Thompson, "Sealab I: A Personal Documentary Account," p. 32.

119 *talking-to from Captain Mazzone:* O'Neal et al., "Project Sealab Summary Report," p. 48.

119 *hurricane conditions:* Bond, "Sealab I Chronicle," pp. 55, 68; *Sealab I* film.

119 *could tear the support barge:* Bond, "Sealab I Chronicle," p. 58.

119 *temporarily self-sufficient:* O'Neal et al., "Project Sealab Summary Report," pp. 26, 46.

119 *hunker down and take their chances:* Atwater and Vaughan, "Room at the Bottom of the Sea," p. 24; Thompson, interview, Sept. 20, 2002; *Sealab I* film.

119 *Bond was satisfied:* Bond, *Papa Topside,* p. 62; Melson, "Remarks," p. 10.

120 *settled on a plan:* Bond, "Sealab I Chronicle," pp. 70–72, 74.

120 *crashing to the bottom:* Barth, *Sea Dwellers,* p. 80.

120 *began its slow ascent:* O'Neal et al., "Project Sealab Summary Report," p. 19; *Sealab I* film.

120 *three-quarter moon:* Bond, "Sealab I Chronicle," p. 69.

120 *The ascent rate:* Ibid., pp. 72–73.

120 *air space for a smoke:* Thompson, "Sealab I: A Personal Documentary Account," p. 35.

120 *first of many holds:* O'Neal et al., "Project Sealab Summary Report," p. 53.

120 *up to ninety feet:* Bond, "Sealab I Chronicle," p. 74.

120 *hardly brush his teeth:* Ibid.

120 *slow-motion roller coaster:* Barth, *Sea Dwellers,* p. 80.

120 *Bond had to concede:* Bond, "Sealab I Chronicle," p. 74.

121 *took a last breath:* Barth, *Sea Dwellers,* p. 80; O'Neal et al., "Project Sealab Summary Report," p. 53.

121 *Anderson avoided looking:* Thompson, "Sealab I: A Personal Documentary Account," p. 36.

121 *did not like the sensation:* Ibid.

121 *cable caught on an:* Bond, "Sealab I Chronicle," p. 76.

121 *heard a* pop: Thompson, interview, Dec. 29, 2003.

121 *stuck his thumb over:* Ibid.; Frank Hogan, "Thumb in Leak Saves 4 Aquanauts," (San Diego) *Evening Tribune,* Feb. 6, 1965.

121 *fitted with a box-shaped frame:* Barth, *Sea Dwellers,* pp. 66, 150–51; *Sealab I* film.

121 *sloshed in several inches:* Barth, *Sea Dwellers,* p. 81.

121 *makeshift urinals:* Ibid., p. 82.

121 *reporters were due to arrive:* Ibid., p. 83; O'Neal et al., "Project Sealab Summary Report," p. 54.

121 *wouldn't let his aquanauts out:* Barth, *Sea Dwellers,* p. 83.

122 *more than just frustration:* Ibid.; Barth, interview, June 9, 2003.

122 *almost fifty-five hours:* O'Neal et al., "Project Sealab Summary Report," pp. 53–54.

122 *informal press conference:* Barth, *Sea Dwellers,* p. 83; Thompson, "Sealab I: A Personal Documentary Account," p. 36.

122 *wore sunglasses:* Navy photograph No. XAB-18102, U.S. Naval Station, Bermuda, July 1964 (in author's possession); Barth, *Sea Dwellers,* p. 155.

122 *jock itch and air embolism:* Barth, *Sea Dwellers,* p. 83.

122 *formal press conference:* Bond, *Papa Topside,* pp. 64–65; Bond, "Sealab I Chronicle," pp. 78–80.

122 *"Double Eagle":* Bond, "Sealab I Chronicle," p. 79.

122 *sat on a temporary stage:* Navy photograph No. N24627.203, U.S. Navy Underwater Sound Laboratory, July 1964 (in author's possession).

122 *Manning again wore:* Ibid.; *Sealab I* film.

122 *which his wife detested:* Rose Anderson, interview, April 13, 2004.

122 *longest printable statement:* Bond, *Papa Topside*, p. 65.

122 *triple-decker banner headlines:* Richard Witkin, "Ranger Takes Close-Up Moon Photos," *New York Times*, Aug. 1, 1964, p. 1.

122 *ran sixteen paragraphs by:* Hanson W. Baldwin, "Navy Men Set Up 'House' Under Atlantic and Find Biggest Problem Is Communication," *New York Times*, Aug. 11, 1964, p. 35.

123 *fired off a letter the next day:* Link to Turner Catledge, managing editor, *New York Times*, Aug. 12, 1964; see also Link to Dr. C. J. Lambertsen, University of Pennsylvania School of Medicine, Aug. 24, 1964, Folder 256, LC.

123 *A few of Link's friends:* H. G. Place, former chairman and president, General Precision Equipment Corp., to Hanson W. Baldwin, Aug. 28, 1964, Folder 381, LC; L. F. Weidman, president, Royal Tours, to Catledge, Aug. 24, 1964, Folder 381, LC.

123 *Baldwin responded:* Baldwin to Capt. Paul Hammond, USNR (ret.), 230 Park Ave., New York, Aug. 13, 1964, Folder 381, LC.

123 *"Simply because everyone":* George Palmer, assistant to the managing editor, *New York Times*, to Link, Aug. 18, 1964, Folder 381, LC.

123 *failed completely to recognize:* C. H. Hedgepeth, Officer in Charge, Navy Experimental Diving Unit, to Link, July 9, 1964, Folder 380, LC.

123 *smattering of magazine articles:* Atwater and Vaughan, "Room at the Bottom of the Sea," p. 18; "A Home Under the Sea," *Life*, Sept. 4, 1964, p. 74; "Men Beneath the Sea," *Newsweek*, July 20, 1964, p. 56; "Men Under Pressure," *Newsweek*, Aug. 24, 1964, p. 56; Coles Phinzy, "Settlers at the Bottom of the Sea," *Sports Illustrated*, June 29, 1964, p. 22; Bill Barada, "U.S. Navy Aquanauts; The Men of Sealab I," *Skin Diver*, August 1964, p. 12.

123 *on* To Tell the Truth: Barth, interview, June 9, 2003.

CHAPTER 10: THE TILTIN' HILTON

Page

124 *planning began in earnest:* D. C. Pauli and G. P. Clapper, Office of Naval Research, eds., "Project Sealab Report: An Experimental 45-Day Undersea Saturation Dive at 205 Feet," Sealab II Project Group, ONR Report ACR-124, Department of the Navy, Washington, D.C., March 8, 1967, p. 2. Two Navy-sanctioned documentary films provided a valuable visual record of Sealab II activities and added corroborating information in their voice-over narrations. *To Sink a House*, produced by the Naval Undersea Research and Development Center, Motion Picture Productions Technical Information Staff, gives an overview of the project but with emphasis on the engineering involved with lowering the habitat; *Man in the Sea—The Story of Sealab II*, produced by John J. Hennessy and W. A. Palmer Films, contains some of the

same material but covers the project more broadly (video copies in author's possession, but also available in libraries). A valuable audio record came from hours of assorted tape recordings made during the Sealab II project, including two press conferences (in author's possession). Also helpful was the Navy's informally produced "Navy compilation of Sealab II press coverage" (in author's possession). It contains many copies of a wide variety of articles published during the project, which complemented those that the author researched and obtained individually, from key publications such as the two daily San Diego, Calif., newspapers of the time, the *San Diego Union* and the *Evening Tribune*, and also the *Los Angeles Times, The New York Times*, and assorted magazines. Along with the available documentation, interviews with Sealab II participants were indispensable, including those cited below.

124 *three teams:* Pauli and Clapper, "Project Sealab Report," pp. 167–76; Barth, *Sea Dwellers*, pp. 90, 102, 181.

124 *close to $2 million:* Stated by Bond in tape recording of Sealab II press conference for Team 3, Oct. 12, 1965 (in author's possession); Bond, *Papa Topside*, p. 145.

124 *edge of an undersea canyon:* Pauli and Clapper, "Project Sealab Report," p. 134.

124 *around fifty degrees Fahrenheit:* Ibid., p. 131.

124 *"hostile environment":* Ibid., pp. 1, 16; Bond, "Sealab II Chronicle," pp. 75, 82, 91, 97 (copy in author's possession of this unedited journal on which the Sealab II chapters in *Papa Topside* are based).

124 *Scripps Institution:* Pauli and Clapper, "Project Sealab Report," p. 131.

124 *mainland support base:* Ibid., p. 12.

125 *thirty consecutive days:* Barth, *Sea Dwellers*, p. 102.

125 *do most of the setting up:* Bond, *Papa Topside*, p. 107.

125 *methods for military operations:* Pauli and Clapper, "Project Sealab Report," pp. 18–19.

125 *Medical monitoring of all:* Ibid., pp. 14, 16, 203; Raymond J. Hock, Ph.D., Northrop Space Laboratories, Captain George F. Bond, MC, USN, and Captain Walter F. Mazzone, MSC, USN, "Physiological Evaluation of Sealab II: Effects of Two Weeks Exposure to an Undersea 7-Atmosphere Helium-Oxygen Environment," printed by Nortronics, a Division of Northrop Corporation, Anaheim, Calif., for the Deep Submergence Systems Project, U.S. Navy, December 1966, pp. iii, 9, 98 (copy in author's possession).

125 *most important mission:* Pauli and Clapper, "Project Sealab Report," p. 13.

125 *christening ceremony:* Tape recording of Sealab II christening; copy of ceremony program (in author's possession); *To Sink a House.*

125 *Morse and a dozen others:* Tape recording of Sealab II christening.

125 *"successful beyond expectation":* Ibid.

126 *"as commonplace as jet":* Ibid.

126 *band launched into:* Ibid.; *To Sink a House.*

126 *Bond wept, unashamedly:* Bond, "Sealab II Chronicle," p. 15.

126 *their names read aloud:* Tape recording of Sealab II christening.

126 *favored divers he knew:* Bond, "Sealab II Chronicle," p. 130.

126 *Many more had applied:* Judith Morgan, "Achievements of Sealab II Are Revealed," *San Diego Union*, Oct. 13, 1965, p. A13.

126 *still in Monaco:* Aquadro, interview, May 16–18, 2003; Dessemond and Wesly, *Les hommes de Cousteau*, p. 126.

126 *spat with Scott Carpenter:* Bond, "Sealab II Chronicle," p. 16.

126 *seemed to get along fine:* Carpenter, interview, April 6, 2004.

126 *autograph some pictures:* Rose Anderson, interview, April 13, 2004.

126 *sensed a souring:* Billie Coffman, interview, Sept. 29, 2004.

126 *witnessed a horrific accident:* Ibid.

126 *Fred Jackson and John Youmans:* Geoffrey A. Wolff, "2 Experimental Navy Divers Killed in Decompression Chamber Fire," *Washington Post*, Feb. 17, 1965, p. 1.

126 *to test a decompression schedule:* John V. Harter, "Fire at High Pressure," in *Underwater Physiology—Proceedings of the Third Symposium on Underwater Physiology*, sponsored by the Committee on Undersea Warfare of the National Academy of Sciences—National Research Council and the Office of Naval Research, in Washington, D.C., March 23–25, 1966, C. J. Lambertsen, ed. (Baltimore: Williams & Wilkins, 1967), p. 56.

126 *had switched:* Coffman, interview, Sept. 29, 2004.

127 *"We have a fire":* Ibid.; Harter, "Fire at High Pressure," in *Underwater Physiology*—proceedings of the Third Symposium, p. 58.

127 *Coffman saw a flame:* Coffman, interview, Sept. 29, 2004.

127 *staring into a blast furnace:* Ibid.

127 *"thermal explosion":* Harter, "Fire at High Pressure," in *Underwater Physiology*, p. 63.

127 *joined a dozen others:* Wolff, "Experimental Navy Divers Killed," *Washington Post*.

127 *blackened like the inside:* Ibid., pp. 58–61; Ken Wallace, master diver at EDU and a witness to the fire, interview, June 1, 2003.

127 *everyone was stunned:* John Harter, interview, July 19, 2003; Harter, the previously cited author of "Fire at High Pressure," arrived at the EDU shortly after the fire; others interviewed also recalled the accident's aftermath, including Lester Anderson's wife, Rose Anderson, interview, April 30, 2004; and Lois Workman, the wife of Robert Workman, interview, May 23, 2003.

127 *take the accident especially hard:* Coffman, interview, Sept. 29, 2004.

127 *to the Green Derby:* Ibid.

127 *Al Vogel:* Ibid.; Wallace, Harter, and other former EDU divers and personnel interviewed recalled the bygone days of Vogel and the Green Derby.

127 *noticed a rebelliousness:* Bond, *Papa Topside*, p. 76; Anderson is referred to only

as "A"; a similar account appears in Bond's unedited "Sealab II Chronicle," in which Bond discreetly refers to Anderson by his initials, "L.A."

127 *and had goaded Carpenter:* Bond, *Papa Topside,* p. 76.

127 *didn't show up for the christening:* Ibid.

127 *They would miss him:* Coffman, interview, Sept. 29, 2004; Bond, *Papa Topside,* p. 77.

127 *tried to find the reasoning:* Bond, *Papa Topside,* p. 77.

128 *"He's going to kill somebody":* Rose Anderson, interview, April 2, 2004.

128 *agreed that Anderson would have to go:* Bond, *Papa Topside,* p. 77.

128 *revolts against authority:* Ibid.

128 *accepted his fate:* Ibid.

128 *formalized as the Man-in-the-Sea:* Pauli and Clapper, "Project Sealab Report," p. 2.

128 *a new Deep Submergence Systems Project:* Edwin A. Link and Philip D. Gallery, Rear Admiral, U.S. Navy (ret.), "Deep Submergence and the Navy," *U.S. Naval Institute Proceedings* (1966?), Folder 1094, LC. Among those acknowledged for assistance in preparation of this eleven-page article are Capt. Lewis Melson and Rear Adm. E. C. Stephan.

128 *A busy agenda:* Pauli and Clapper, "Project Sealab Report," pp. 3–4.

128 *in the organizational chart:* Ibid., p. 13.

128 *its ultimate destination:* Ibid., pp. 131, 134.

128 *handful of mentors:* Ibid., pp. 67–68, 87; Culpepper, interview, March 11, 2003.

128 *like an extra-large tank car:* Ibid., p. 2.

128 *new Sealab insignia:* Ibid., p. 5.

129 *pair of barges:* Ibid., p. 121.

129 *like a floating construction site:* Official U.S. Navy photograph, no. VC-3-3480-9-65 (in author's possession).

129 *Mazzone coined one day:* Mazzone, e-mail to author, July 1 and 2, 2004; Bond, *Papa Topside,* p. 71.

129 *almost a mile offshore:* Pauli and Clapper, "Project Sealab Report," p. 135.

129 *oceanographers had mapped:* Ibid., p. 131.

129 *Finding a suitable site:* Ibid., pp. 131–34; Bond, *Papa Topside,* p. 88.

129 *underwater cameras to assess: Story of Sealab II* (film).

129 *lying unconscious:* Bond, *Papa Topside,* pp. 85–86; Walter Mazzone, taped interview, May 5, 2004.

129 *Bunton had to abandon:* Bill Bunton, interview, San Diego, Calif., Aug. 4, 2002.

129 *swam at top speed:* Scott Carpenter, taped interview, May 27, 2004.

129 *tried to slow him:* Barth, *Sea Dwellers,* p. 99.

130 *Disoriented, he hung upside down:* Carpenter, interview, May 27, 2004.

130 *joined the Navy after high school:* Mary Cannon, Berry's widow, interview, Jan. 23, 2002.

130 *several hundred others:* Morgan, "Achievements of Sealab II Are Revealed," *San Diego Union*, p. A13.

130 *as a Western pioneer:* M. Cannon, interview, Jan. 23, 2002.

130 *start of something historic:* Numerous aquanauts and others close to the project expressed this in interviews.

130 *Sealab II touched down:* "Sealab II Log," a handwritten log with extemporaneous daily entries mostly made by Bond and Mazzone during their respective watches, pp. 6–7 (copy in author's possession).

130 *a few other technical difficulties:* Ibid.; Pauli and Clapper, "Project Sealab Report," p. 99; Bond, "Sealab II Chronicle," pp. 56–57; *Story of Sealab II.*

130 *fitted with a counterweight:* Pauli and Clapper, "Project Sealab Report," p. 124; *To Sink a House.*

130 *two-hundred-ton habitat:* Pauli and Clapper, "Project Sealab Report," p. 84.

130 *much as a submarine operates:* Ibid., pp. 39, 42; Bond, *Papa Topside*, p. 69.

130 *within a couple of hours:* "Sealab II Log," p. 7.

130 *sizable press corps:* Bond, *Papa Topside*, p. 92.

131 *Two dozen photographers:* Personal diary of Berry Cannon from Sealab II, entry for Aug. 28, 1965 (copy in author's possession).

131 *down to the lab in pairs:* Diary of Berry Cannon, entry for Aug. 28, 1965; *Story of Sealab II.*

131 *crowded onto a lower deck: Story of Sealab II.*

131 *took a last puff:* Ibid.

131 *Wilbur Eaton:* Barth, *Sea Dwellers*, pp. 79, 116; Barth, e-mail to author, Dec. 15, 2003.

131 *bluff that would have afforded:* Bond, "Sealab II Chronicle," p. 55.

131 *through the four-foot:* Pauli and Clapper, "Project Sealab Report," p. 40.

131 *fits of laughter:* Scott Carpenter, "200 Feet Down, the Next U.S. Frontier," *Life*, Oct. 15, 1965, p. 100B; Carpenter, interview, April 6, 2004.

131 *few days behind schedule:* Pauli and Clapper, "Project Sealab Report," p. 30.

131 *lead the first two teams:* Ibid., p. 168; Carpenter, "200 Feet Down," p. 100B.

131 *preparatory tasks to do:* Pauli and Clapper, "Project Sealab Report," p. 182; Diary of Berry Cannon, entries for Aug. 28 and 29, 1965.

131 *PTC was a pressurized:* Pauli and Clapper, "Project Sealab Report," pp. 112–13.

132 *fish had already congregated:* Diary of Berry Cannon, entry for Aug. 28, 1965.

132 *sensed the warmth:* Ibid.

132 *light reassured him:* Ibid.

132 *a safe haven:* Pauli and Clapper, "Project Sealab Report," pp. 112, 128.

132 *porous latex material:* Ibid., p. 278.

132 *fit more snugly:* Ibid., p. 279; interviews with aquanauts, including Cyril Tuckfield, interviews, Aug. 15 and 21, 2002.

132 *sporting electrodes:* Pauli and Clapper, "Project Sealab Report," pp. 15, 237; Diary of Berry Cannon, entry for Aug. 28, 1965; *To Sink a House.*

132 *galley and workspace:* Pauli and Clapper, "Project Sealab Report," pp. 26, 41; *Story of Sealab II.*

132 *lab was on a slant:* Pauli and Clapper, "Project Sealab Report," p. 326; Diary of Berry Cannon, entry for Aug. 28, 1965.

132 *made Cannon a little nervous:* Diary of Berry Cannon, entry for Aug. 28, 1965.

132 *six-inch-thick line was dropped:* "Sealab II Log," pp. 66, 68, 70, 71, 73, 79; Bond, *Papa Topside,* p. 106; Bryant Evans, "Aquanauts Eat Hearty, Don't Gain," *San Diego Union,* Sept. 10, 1965.

132 *mule-haul the line:* Coffman, interview, Sept. 29, 2004; Cliff Smith, "Aqua-nauts Report on Sealab Life; Vast Future Seen," *San Diego Union,* Sept. 15, 1965, p. A1.

132 *Robert Sonnenburg:* Leo Bowler, "Craft Is Readied for Lowering Task," (San Diego) *Evening Tribune,* Aug. 24, 1965, p. A15; "No More the Voiceless Deep," *Roche Medical Image,* December 1965, p. 6.

133 *outrageous prank:* Barth, e-mail to author, June 29, 2004; several others described it, too.

133 *leave with Team 1 and come back:* Pauli and Clapper, "Project Sealab Report," p. 174; *Story of Sealab II.*

133 *earned the Bronze Star:* Coffman, interview, Sept. 29, 2004.

133 *Tom Clarke:* Tom Clarke, taped interview, April 30, 2004.

133 *lanky and bespectacled:* Pauli and Clapper, "Project Sealab Report," p. 169; Smith, "Aquanauts Report on Sealab Life," *San Diego Union* (with photo).

133 *goldfish in the claw-foot:* Clarke, e-mail to author, June 8, 2005.

133 *average age was thirty-six:* Pauli and Clapper, "Project Sealab Report," pp. 167–76.

133 *fathers of eight boys:* Ibid.

133 *Jay Skidmore:* Ibid., p. 174.

133 *observe the captive aquanauts:* Ibid., p. 245; Roland Radloff and Robert Helm-reich, *Groups Under Stress: Psychological Research in Sealab II* (New York: Appleton-Century-Crofts, 1968), pp. 31, 51–90.

134 *value of decent food:* Pauli and Clapper, "Project Sealab Report," p. 412.

134 *frying was verboten:* Ibid.

134 *two thousand cans:* Ibid., p. 414.

134 *tendency to slide:* Diary of Berry Cannon, entry for Aug. 31, 1965; Radloff and Helmreich, *Groups Under Stress,* pp. 72, 74.

134 *Tiltin' Hilton:* Bond, *Papa Topside,* pp. 93, 94.

134 *Cannon listened as:* Diary of Berry Cannon, entry for Aug. 28, 1965.

134 *in his bathrobe:* Barth, *Sea Dwellers,* p. 167; *Story of Sealab II.*

134 *seemed unable to comprehend:* Tape recording of Sealab–Gemini communica-tions, dated Aug. 28, 1965 (in author's possession).

134 *eleven-thirty that night:* "Sealab II Log," p. 12.

134 *117th orbit:* "It Was Smooth Sailing for Pair in Final Orbit," *Washington Post,* Aug. 30, 1965, p. A9.

134 *link fizzled:* Tape recording of Sealab–Gemini; Bryant Evans, "Sealab Man, Astronaut Have a Talk," *San Diego Union,* Aug. 29, 1965.

134 *filled with stories about* Gemini 5: Evert Clark, "2 Astronauts End 8-Day Flight Tired but in 'Wonderful Shape'; Johnson Hails U.S. Space Gains," *New York Times,* Aug. 30, 1965, p. 1.

135 *NBC current events program:* Audio tape recording of *Survey '65* (in author's possession); Bond, *Papa Topside,* pp. 81–83.

135 *aired on a Saturday night:* Recording of *Survey '65* is labeled Aug. 14, 1965, which fits with dates mentioned during the program.

135 *exploded in riots:* Valerie Reitman and Mitchell Landsberg, "Watts Riots, 40 Years Later," *Los Angeles Times,* Aug. 11, 2005, p. 1.

135 *final American shows:* San Diego Historical Society, San Diego timeline, online at www.sandiegohistory.org/timeline; see also www.rarebeatles.com/photopg7/photopg7.htm#1965.

135 *kinship with the spaceship:* "Sealab II Log," p. 16.

135 *pulled out his harmonica:* Ibid.; *Story of Sealab II.*

135 *"What a life":* "Sealab II Log," p. 16.

135 *first full working day:* Ibid., p. 15.

135 *recited the Sealab prayer:* Ibid., p. 16; Bond, *Papa Topside,* p. 93.

CHAPTER 11: LESSONS IN SURVIVAL

Page

136 *dual Arawak umbilical:* Pauli and Clapper, "Project Sealab Report," p. 94.

136 *a half-hour to free them:* Diary of Berry Cannon, entry for Sept. 4, 1965.

137 *gave the Arawak a try:* Carpenter, interview, April 6, 2004; Carpenter, "200 Feet Down," p. 103.

137 *had seen his team leader:* Coffman, interview, Sept. 29, 2004.

137 *straightening out their hoses:* Ibid.; Pauli and Clapper, "Project Sealab Report," p. 182.

137 *Aquasonic:* Pauli and Clapper, "Project Sealab Report," p. 143.

138 *Even old-style hardhat:* Davis, *Deep Diving and Submarine Operations,* p. 73; Martin, *The Deep-Sea Diver,* pp. 5, 141, 149.

138 *mask fitted around the mouth:* G. F. Bond, "Medical Problems of Multi-Day Saturation Diving in Open Water," in *Underwater Physiology—Proceedings of the Third Symposium,* p. 85.

138 *helium speech was incomprehensible:* Pauli and Clapper, "Project Sealab Report," p. 144.

138 *Cannon's verdict:* Diary of Berry Cannon, entry for Sept. 7, 1965.

138 *"working with mail-order"*: Story of Sealab II.

138 *Tending to the "pots"*: Diary of Berry Cannon, entry for Aug. 31, 1965; Pauli and Clapper, "Project Sealab Report," p. 128; Story of Sealab II.

139 *"way station"*: Commander Thomas Blockwick, "Sealab II (Pictorial)," United States Naval Institute Proceedings, June 1966, p. 108; Bryant Evans, "Trolley to Shift Gear," San Diego Union, Sept. 11, 1965 (photo), p. A17.

139 *tumbled into Scripps*: "Sealab II Log," pp. 45, 121; Leo Bowler, "Aquanauts Try Deep Plunge Below Sealab," (San Diego) Evening Tribune, Sept. 21, 1965, p. A8; Bryant Evans, "2nd Sealab Unit Aboard Mother Ship," San Diego Union, Sept. 28, 1965, p. A17.

139 *"psychomotor" machines*: Pauli and Clapper, "Project Sealab Report," pp. 246–48.

139 *was a weather station*: Ibid., p. 369.

139 *like a teepee frame*: Ibid., p. 376.

139 *eventually found the contraption*: "Sealab II Log," pp. 46–74.

139 *dragged the weather station*: Ibid., pp. 62, 72; Pauli and Clapper, "Project Sealab Report," p. 375.

139 *Scripps scientists had decided*: Pauli and Clapper, "Project Sealab Report," p. 375.

139 *too heavy to swim*: Diary of Berry Cannon, entry for Sept. 7, 1965.

139 *huffing and puffing*: Ibid.

140 *Headaches, fatigue, ear infections*: "Sealab II Log," pp. 96, 99, 128; Bond, Papa Topside, p. 143; Leo Bowler, "9 Aquanauts' Ears Infected, but Activities Not Curtailed," (San Diego) Evening Tribune, Sept. 8, 1965, p. A2; Radloff and Helmreich, Groups Under Stress, p. 74; "No More the Voiceless Deep," Roche Medical Image, pp. 7–8.

140 *hook up a battery charger*: Diary of Berry Cannon, entry for Sept. 2, 1965.

140 *Anchovies!*: Ibid.; "Sealab II Log," pp. 37, 41, 43, 45.

140 *nowhere to go for shelter*: Bond, Papa Topside, p. 101.

140 *rid the PTC of the decaying fish*: Ibid.; "Sealab II Log," pp. 47–49; Cyril Tuckfield, interview, April 26, 2004.

140 *capsule had to be raised*: "Sealab II Log," pp. 83–85, 87; Barth, Sea Dwellers, p. 109.

140 *another night without*: Diary of Berry Cannon, entries for Sept. 10 and 11, 1965.

140 Scorpaena guttata: Radloff and Helmreich, Groups Under Stress, pp. 63–64; California Department of Fish and Game, Marine Sportfish Indentification, www.dfg.ca.gov/marine/mspcont4.asp.

141 *one every square yard*: Carpenter, interview, April 6, 2004.

141 *usually dissolved into darkness*: Interviews with aquanauts; see also Radloff and Helmreich, Groups Under Stress, pp. 62–63; "Sealab II Log," pp. 23 and 60, for example, contain telling entries on the limited visibility; Diary of Berry Cannon, entry for Sept. 3, 1965, says about the silty bottom, "once

stirred, visibility goes to zero"; *Story of Sealab II* shows just how dark the water often was.

141 *unnerving to lose sight:* Radloff and Helmreich, *Groups Under Stress,* pp. 62–63; "Sealab II Log," p. 59.

141 *practice of carrying tethers:* Pauli and Clapper, "Project Sealab Report," p. 21.

141 *ripe for physiological research:* "Sealab II Log," p. 59; *Story of Sealab II.*

141 *everything seemed to take longer:* Radloff and Helmreich, *Groups Under Stress,* pp. 65–67; Diary of Berry Cannon, entries for Aug. 30 and Sept. 2, 1965; interviews with aquanauts.

141 *Donning a snug-fitting wet suit:* Diary of Berry Cannon, entry for Aug. 30, 1965; Carpenter, "200 Feet Down," p. 103.

141 *frequently burned out:* Pauli and Clapper, "Project Sealab Report," pp. 92, 185.

141 *take an hour to replace:* Diary of Berry Cannon, entry for Sept. 2, 1965; Radloff and Helmreich, *Groups Under Stress,* p. 66.

141 *"This will certainly be":* "Sealab II Log," p. 20.

142 *brief ceremonial visit:* Coffman, interview, Sept. 29, 2004; Leo Bowler, "Torpedoman Re-Enlists 205 Ft. Down," (San Diego) *Evening Tribune,* Aug. 31, 1965, p. A1; *Story of Sealab II.*

142 *For their brief thirteen minutes: Story of Sealab II.*

142 *cake had crumbled:* "Cake Falls—Aquanaut's Wife Crumbles," *San Diego Union,* Sept. 7, 1965, p. A21; *Story of Sealab II.*

142 *Bond led the singing:* Tape recording of assorted Sealab communications, labeled Day 4–12 (in author's possession).

142 *asked to sit for an interview:* Bond, *Papa Topside,* pp. 110–11; a less edited, more colorful description of Bond's thoughts can be found in "Sealab II Chronicle," pp. 77–79.

142 *When the article appeared:* Peter Bart, "Sealab Explores an 'Inner World,'" *New York Times,* Sept. 9, 1965, p. 30.

142 *unwelcome consternation:* Bond, *Papa Topside,* p. 110.

142 *posted the marked-up clipping:* Ibid., p. 111.

142 *Carpenter had just returned:* "Sealab II Log," p. 91.

142 *stuck his bare hand in:* Carpenter, interview, Jan. 25, 2002.

143 *nose and sinuses filled:* Carpenter, "200 Feet Down," p. 104.

143 *assortment of drugs:* "Sealab II Log," pp. 92, 94.

143 *on the intercom with Bond:* Tape recording of assorted Sealab II communications from the day Carpenter was stung (in author's possession).

143 *postponed their pressurized PTC ride:* Ibid.; S. A. Desick, "Carpenter Treated; 9 Men Brought Up," *San Diego Union,* Sept. 13, 1965, p. A17.

143 *find it irresistible to point out:* "Navy compilation of Sealab II press coverage," pp. 100–104.

143 *zapped him on the leg:* Coffman, interview, Sept. 29, 2004.

143 *hardly a word was printed:* "Navy compilation of Sealab II press coverage," combined with research in other publications, made this clear.

143 *ten-day duration mark:* Sealab I was often referred to as an eleven-day dive, as it was in the previously cited full title of O'Neal et al., "Project Sealab Summary Report." But in the log included in that official report, pp. 39 and 53, and as noted in other sources, the actual bottom time, between the aquanauts' initial entry into the lab and its raising, with the aquanauts still inside, was closer to ten days.

143 *team leader's acting deputy:* Bond, "Sealab II Chronicle," p. 85.

144 *the sight of the* Berkone: Barth, *Sea Dwellers,* p. 103.

144 *Barth had been looking forward:* Ibid., p. 106.

144 *Some stingy bureaucrat:* Ibid., p. 104.

144 *out two hundred and forty bucks:* Ibid.

144 *Carpenter's persistent questions:* Bond, *Papa Topside,* p. 125.

144 *functioning of a pneumofathometer:* Ibid., pp. 111–12; "Sealab II Log," pp. 74, 113.

144 *like a patient father:* Tape recordings during Sealab II of dictations by Bond and of assorted topside conversations with aquanauts in the habitat; a dialogue with Carpenter about the pneumofathometer is on a tape recorded Sept. 22, 1965, according to Bond's dictation (in author's possession).

144 *"Our Lord gave us":* Tape recordings during Sealab II; Bond, *Papa Topside,* p. 113.

144 question everything: Tape recordings during Sealab II.

144 *an electrically heated wet suit:* Pauli and Clapper, "Project Sealab Report," pp. 272–305.

145 *getting an ice cube shoved:* Carpenter, "200 Feet Down," p. 103.

145 *chilled within a half-hour:* Diary of Berry Cannon, entry for Aug. 30, 1965.

145 *arms shaking uncontrollably:* Tape recordings of Sealab II press conference for Team 2 (in author's possession); Carpenter, "200 Feet Down," p. 103.

145 *"unbelievably delicious":* Carpenter, "200 Feet Down," p. 103.

145 *"better than sex":* Barth, *Sea Dwellers,* p. 111.

145 *Eight aquanauts:* Pauli and Clapper, "Project Sealab Report," pp. 297–98.

145 *wrapped up a four-year stint:* Wally Jenkins, taped interview, May 3, 2004.

145 *Tolbert, who at thirty-nine:* Pauli and Clapper, "Project Sealab Report," p. 175.

145 *grateful to have been picked:* Tolbert statement in tape recording of Sealab II press conference for Team 2.

145 *crushed by an ice truck:* George Dowling, Tolbert's Mine Defense Lab colleague at the time, interview during a Sealab reunion, Panama City, Fla., March 12, 2005.

146 *watched as it vanished:* Tape recording of Sealab II press conference for Team 2.

146 *release scores of detectors:* Dowling, interview, March 12, 2005; Judith Morgan, "2 Plan to Litter Sea with Papers," *San Diego Union*, Sept. 4, 1965, p. A2.

146 *severe case of rash:* Dowling, interview, March 12, 2005.

146 *"You are completely enclosed":* Morgan, "2 Plan to Litter Sea with Papers," *San Diego Union*, p. A2.

146 *four "targets":* Pauli and Clapper, "Project Sealab Report," p. 251.

146 *Tolbert realized he was drifting:* Radloff and Helmreich, *Groups Under Stress*, p. 56. This first-person account, although anonymous in the cited text, clearly came from Tolbert. See also "Sealab II Log," p. 107. Other aquanauts interviewed, including Cyril Tuckfield, on April 26, 2004, were familiar with this incident. Tolbert died before research on this book began.

146 *slipped toward his knees:* Jenkins, interview, May 3, 2004; this previously cited interview, along with Tolbert's published account, cited above, are the primary sources for the details of this close call.

147 *Ken Conda:* Pauli and Clapper, "Project Sealab Report," pp. 170, 408; Conda, like Tolbert, died before research on this book began.

147 *scared him quite like this:* Radloff and Helmreich, *Groups Under Stress*, p. 56.

147 *cadre of marine mammals:* "Brief History of the Navy's Marine Mammal Program," in *Annotated Bibliography of Publications from the U.S. Navy's Marine Mammal Program*, Technical Document 627, Revision D (San Diego, Calif.: Space and Naval Warfare Systems Center, May 1998), pp. v–vii, www.spawar. navy.mil/sti/publications/pubs/td/627/index.html.

147 *hit TV series:* Tim Brooks and Earle Marsh, *The Complete Directory to Prime Time Network and Cable TV Shows, 1946–Present*, 6th ed. (New York: Ballantine, 1995), p. 1263.

147 *act as a courier:* Cliff Smith, "A Tougher Tuffy Carries Mail," *San Diego Union*, Sept. 18, 1965.

147 *undersea St. Bernard:* Pauli and Clapper, "Project Sealab Report," p. 407; *Story of Sealab II*.

147 *web of tethers:* Pauli and Clapper, "Project Sealab Report," p. 21; Cyril Tuckfield, interview, Aug. 15, 2002.

147 *in about a minute:* Pauli and Clapper, "Project Sealab Report," p. 408.

148 *Tuffy seemed to be wary:* Ibid.

148 *potential to work with future Sealabs:* Ibid., p. 409.

148 *mirthful headline:* Ted Thackery Jr., "Porpoise Tuffy Chickens Out on Sealab Test," *Los Angeles Herald-Examiner*, Sept. 17, 1965.

148 *White's space walk:* www.hq.nasa.gov/office/pao/History/SP-4203/ch11-2. htm; the walk was widely covered in the press but this official NASA Web site provides the essential details.

148 *NASA had to lease:* Ibid.

148 *give or take an atmosphere:* Reynold T. Larsen and Walter F. Mazzone, "Excursion Diving from Saturation Exposures at Depth," in *Underwater*

Physiology—Proceedings of the Third Symposium, p. 251; Charles Hillinger, "Aquanauts Under Sea Talk to Press Above," *Los Angeles Times*, Sept. 10, 1965, p. A8.

148 *"excursion dives":* Larsen and Mazzone, "Excursion Diving from Saturation Exposures at Depth," in *Underwater Physiology—Proceedings of the Third Symposium*, pp. 241–54; Mazzone, interview, Jan. 19, 2005.

148 *Dr. Workman wielded:* Larsen and Mazzone, "Excursion Diving from Saturation Exposures at Depth," in *Underwater Physiology—Proceedings of the Third Symposium*, p. 242; Mazzone, interview, Jan. 19, 2005.

149 *results were promising enough:* Larsen and Mazzone, "Excursion Diving from Saturation Exposures at Depth," in *Underwater Physiology—Proceedings of the Third Symposium*, p. 247; Mazzone, e-mail to author, Jan. 20, 2005.

149 *Mazzone never saw any:* Mazzone, interviews, including Jan. 23 and Feb. 5, 2003; Hock, Bond, and Mazzone, "Physiological Evaluation of Sealab II," p. 4.

149 *say, 430 feet:* Larsen and Mazzone, "Excursion Diving from Saturation Exposures at Depth," in *Underwater Physiology—Proceedings of the Third Symposium*, p. 251.

149 *much more expensive to operate:* Mazzone, interview, Jan. 19, 2005.

149 *could foresee a time:* Stated by Bond in tape recording of Sealab II press conference for Team 3, Oct. 12, 1965.

149 *placed like way stations:* Bond, "Proposal for Underwater Research," p. 3.

149 *yield greater insights:* Bond, *Papa Topside*, p. 128.

150 *others were cut short:* Jenkins, interview, May 3, 2004; "Sealab II Log," pp. 112, 113, 113b, 118, 118b, 119, contain examples of some difficulties, as do a smattering of local news items, such as Bryant Evans, "Two Sealab Divers Reach 253 Feet; Instrument Failure Forces Men into Swift Return," *San Diego Union*, Sept. 24, 1965, p. A17.

150 *edge of the Grand Canyon:* Jenkins, interview, May 3, 2004.

150 *along the rocky canyon rim:* Ibid.

150 *stopped where they were:* Ibid.

150 *reached their depth goal:* Ibid.; Bryant Evans, "Aquanauts Dive Below Sealab to Sheer Cliff," *San Diego Union*, Sept. 25, 1965.

150 *had just lowered Conshelf Three:* Capt. Jacques-Yves Cousteau, "Working for Weeks on the Sea Floor," *National Geographic*, April 1966, p. 508.

150 *after midnight on September 22:* Ibid., p. 509.

150 *almost 330 feet:* Ibid., p. 504.

150 *a steel sphere:* Ibid., pp. 502, 504, 519.

151 *minimize the habitat's reliance:* Ibid., p. 504.

151 *same hostile conditions:* André Laban, leader of Conshelf Three crew and longtime Cousteau associate, interview, Juan-les-Pins, France, Oct. 27 and 31, 2004.

151 *shooting a film while living:* Ibid.; Cousteau, "Working for Weeks on the Sea Floor," p. 506.

151 *prime sponsors:* Cousteau, "Working for Weeks on the Sea Floor," p. 502.

151 *"Christmas tree":* Ibid., p. 500; *World of Jacques Cousteau,* National Geographic Special, April 28, 1966, viewed at Museum of Television and Radio, New York City, March 28, 2003.

151 *lower a mock Christmas tree:* Cousteau, "Working for Weeks on the Sea Floor," pp. 522, 527.

151 *370 feet:* Ibid., p. 527.

151 *Oil engineers would be watching:* Ibid.

151 *at seven-thirty in the morning:* "Sealab II Log," p. 18A.

152 *Barth was the last:* Ibid.

152 *football helmets:* Jenkins, interview, May 3, 2004; Radloff and Helmreich, *Groups Under Stress,* p. 77; *Story of Sealab II.*

152 *could see out a little porthole:* Barth, *Sea Dwellers,* p. 115.

152 *exposed the aquanauts to as much danger:* Bond, *Papa Topside,* p. 146; Radloff and Helmreich, *Groups Under Stress,* pp. 77–78.

152 *had to sweat out its decompression:* Diary of Berry Cannon, entry for Sept. 13, 1965.

152 *six feet an hour:* Pauli and Clapper, "Project Sealab Report," p. 112.

152 *just over twenty feet long:* Ibid., p. 118.

153 *like "Goodnight, Irene":* Carpenter, interview, Jan. 25, 2002; *Story of Sealab II.*

153 *quite a few congratulatory calls:* Barth, *Sea Dwellers,* p. 115.

153 *from President Lyndon Johnson:* S. A. Desick, "President Calls Up Crew in Chamber," *San Diego Union,* Sept. 27, 1965, p. A17.

153 *Alas, it was not:* Tape recording of the conversations with the telephone operators and the president (in author's possession).

153 *operator sounded mortified:* Ibid.

153 *big headlines about Cooper:* Evert Clark, "2 Astronauts End 8-Day Flight Tired but in 'Wonderful Shape'; Johnson Hails U.S. Space Gains," *New York Times,* Aug. 30, 1965, p. 1, was one of many.

154 *"I'm not sure he got much":* Tape recording of the conversations.

154 *ashore for a more formal press conference:* Bryant Evans, "Carpenter Reports on Life for 30 Days Under the Sea," *San Diego Union,* Sept. 29, 1965 (with photo), p. 1.

154 *whose members had sat willy-nilly:* Smith, "Aquanauts Report on Sealab Life," *San Diego Union* (with photo), p. 1.

154 *spoke from a lectern:* Story of Sealab II.

154 *"If we had been seeking records":* Tape recording of Sealab II press conference for Team 2.

155 *"hardships":* Ibid.

155 *never worked harder:* Ibid.

155 *"Commander Carpenter":* Ibid.

155 *"It's going on":* Ibid.

155 *"This is the second":* Ibid.

156 *"We are really in the dawn":* Ibid.

CHAPTER 12: THE THIRD TEAM

Page

157 *Robert Carlton Sheats:* Owen Lee, "The Master of the Master Divers," *Skin Diver,* September 1965, p. 50; this article, a Q&A with Sheats, added to the overall description of Sheats's life and times that came from other sources, including those cited in the notes that follow.

157 *"the finest man":* Tape recording of Sealab II press conference for Team 3; see also Bond, "Sealab II Chronicle," p. 21, for a further expression of his unequivocal respect for Sheats.

157 *compact and muscular:* Barth, *Sea Dwellers,* p. 166 (photo).

157 *supervisor for Sealab I:* Ibid., p. 55.

157 *took part in the macabre recovery:* Lee, "The Master of the Master Divers," p. 52.

157 *dives he was forced to make:* Robert C. Sheats, *One Man's War: Diving as a Guest of the Emperor 1942* (Flagstaff, Ariz.: Best Publishing, 1998), pp. 31–52.

158 *Bond had first met Sheats:* Tape recording of Sealab II press conference for Team 3; Lee, "The Master of the Master Divers," p. 53.

158 *friends and diving buddies:* Lee, "The Master of the Master Divers," p. 53; Carl Sheats, the younger of Sheats's two sons, interview, April 27, 2004.

158 *glad to have Sheats leading:* Bond, "Sealab II Chronicle," p. 21.

158 *his obsession with safety:* Ibid.; Phil Sheats, the older of Sheats's two sons, interviews, July 8 and 21, 2004.

158 *ordered five pairs:* Bob Sheats Daily Log—Sealab II, entry for Sept. 20, 1965 (copy in author's possession).

158 *lapses in safety procedures:* Ibid., entry for Sept. 25, 1965.

158 *Mark VI had a long way to go:* Ibid., entry for Sept. 30, 1965; see also entries for Sept. 2 and 5, and Oct. 3; "Sealab II Log, Part 2," p. 41; Radloff and Helmreich, *Groups Under Stress,* pp. 55–56, 58.

159 *long face:* Pauli and Clapper, "Project Sealab Report," p. 173.

159 *fresh out of diving school:* Bill Meeks, taped interview, April 29, 2004.

159 *paired with Sheats during:* Ibid.

159 *gave him nightmares:* Ibid.

159 *when Bond brought it up:* Ibid.

159 *"foam in salvage":* Pauli and Clapper, "Project Sealab Report," pp. 385–89.

160 *enabled the Team 3 aquanauts to refill:* Ibid., pp. 182, 419.

160 *swam a hundred feet over:* Sheats Daily Log, entry for Sept. 27, 1965.

160 *the FIS gun:* Pauli and Clapper, "Project Sealab Report," p. 387.

160 *to take the first shots:* Meeks, interview, April 29, 2004.

160 *between two and eighteen minutes:* Pauli and Clapper, "Project Sealab Report," p. 386.

160 *foam escaped:* Leo Bowler, "2 Aquanauts Study Coral During Dive," (San Diego) *Evening Tribune*, Oct. 5, p. A3.

160 *becoming overly buoyant:* Meeks, interview, April 29, 2004.

160 *foam was clogging:* Ibid.

160 *signaled to each other:* Ibid.

160 *Sheats let it be known:* Sheats Daily Log, entry for Sept. 30, 1965; "Sealab II Log, Part 2," pp. 37A, 39.

160 *switch to standard scuba rigs:* Bond, *Papa Topside*, p. 137.

161 *problem with the foam:* Pauli and Clapper, "Project Sealab Report," p. 388.

161 *an important step forward:* Sheats Daily Log, entry for Sept. 29, 1965.

161 *blasted studs into steel:* Pauli and Clapper, "Project Sealab Report," p. 389.

161 *Meeks preferred it:* Ibid., p. 390; Meeks, interview, April 29, 2004.

161 *prototype rubber pontoon:* Pauli and Clapper, "Project Sealab Report," p. 391.

161 *bilingual confusion:* Bond, "Sealab II Chronicle," pp. 116–17; "Phone Links Yanks, French 'Oceanauts,'" (San Diego) *Evening Tribune*, Oct. 1, 1965, p. 1.

161 *blithely summed up the exchange:* Bond, "Sealab II Chronicle," pp. 116–17; an abridged and less capricious version of Bond's account appears in *Papa Topside*, p. 139; Leo Bowler, "Sealab's Eye, Maybe Ear Kept on French," (San Diego) *Evening Tribune*, Sept. 9, 1965, p. A3.

161 *were stung by scorpion fish:* "Sealab II Log, Part 2," pp. 45, 52; Sheats Daily Log, entry for Oct. 3, 1965; Meeks, interview, April 29, 2004.

161 *going to lose someone:* Sheats Daily Log, entry for Oct. 7, 1965.

161 *elevated carbon monoxide:* Sealab II Log, Part 2, pp. 32, 33, 34a, 38; Pauli and Clapper, "Project Sealab Report," pp. 423–25; Sheats Daily Log, entry for Oct. 1, 1965.

161 *responsible for some of the aquanauts' headaches:* Leo Bowler, "Pain in Mornings Puzzles Doctors; Carbon Monoxide Detected in Atmosphere of Sealab," (San Diego) *Evening Tribune*, Oct. 1, 1965, p. A17; tape recording of Sealab II press conference for Team 3.

162 *doing most of their own cooking:* Sheats Daily Log, entry for Sept. 18, 1965.

162 *send down a cake:* Sheats Daily Log, entry for Sept. 30, 1965; "Raw Fish, Cake Mark a Birthday," (San Diego) *Evening Tribune*, Oct. 1, 1965, p. A17.

162 *had his team eating sashimi:* "Raw Fish, Cake Mark a Birthday," (San Diego) *Evening Tribune*; "Navy compilation of Sealab II press coverage," p. 168.

162 *sent some up to Captain Mazzone:* Sheats Daily Log, entry for Oct. 3, 1965.

162 *netting hordes of plankton:* "Two Dive 300 Feet, Sample Plankton," (San Diego) *Evening Tribune*, Oct. 6, 1965, p. A17; Judith Morgan, "Plunge to 300 Feet," *San Diego Union*, Oct. 6, 1965, p. A17; "Sealab II Log, Part 2," p. 52.

162 *pleasantly nutty:* Meeks, interview, April 29, 2004.

162 *Not bad with cereal:* Sheats Daily Log, entries for Oct. 4 and 6, 1965.

162 *have a woodsman's knack:* Bond, "Sealab II Chronicle," p. 119.

162 *precocious little pinniped:* Leo Bowler, "Sea Lions Discover Sealab as Fine Spot to Stop and Eat," (San Diego) *Evening Tribune*, Sept. 23, 1965, p. A1; "Sam the Sea Lion Shaped for Sealab," *Tribune*, Sept. 25, 1965, p. A1. Numerous stories about the sea lions appeared, mainly in the San Diego papers; the two cited here cover the main points.

162 *made more excursion dives:* "Sealab II Log, Part 2," pp. 57, 62; Morgan, "Plunge to 300 Feet," *San Diego Union*, p. A17.

162 *another first for Sealab II:* "Sealab II Log, Part 2," p. 54.

162 *around one hundred hours diving:* Press release, Sealab II C.I.B. (Command Information Bureau), Oct. 12, 1965 (in author's possession). The figures in this seven-page release are the only such tallies for dive times found in any of the various source materials. They are: 6,067 min. (Team 1); 5,850 min. (Team 2); 8,678 min. (Team 3). An entry in "Sealab II Log, Part 2," p. 65, which mentions the differences in the team totals but not the total times themselves, helps to corroborate the press release's numbers.

162 *150 hours in the water:* Ibid.

162 *almost eight hundred minutes:* Situation Report (SitRep) no. 50, Oct. 7, 1965 (in author's possession); SitReps, written by Bond or other commanders, gave brief official summaries of a day's activities and were relayed to some two dozen Navy offices. These reports provided further background and also corroboration of other documentation and interviews.

163 *how much more efficient:* "Sealab II Log, Part 2," p. 61.

163 *essentially breaking camp:* Sheats Daily Log, entries for Oct. 7–10, 1965.

163 *television circuit on:* "Sealab II Log, Part 2," pp. 68–69.

163 *Dodgers won that third game:* Baseball Almanac, www.baseball-almanac.com/ws/yr1965ws.shtml.

163 *farewell of a sermon:* Tape recording of assorted Sealab II communications from the final day (in author's possession).

163 *by eleven o'clock:* Sheats Daily Log, entry for Oct. 10, 1965.

163 *approximately thirty hours:* Press release, Sealab II C.I.B., Oct. 12, 1965; Pauli and Clapper, "Project Sealab Report," p. 112.

163 *beautiful, adventurous, rewarding:* Sheats Daily Log, entry for Oct. 6, 1965.

163 *enough, even for him:* Ibid.

163 *he had to force himself:* Ibid.

163 *disappointed him:* Ibid.

163 *planned to rate them all:* Ibid., entry for Oct. 10, 1965; Sheats's initial ratings and brief critiques appear here.

163 *passed the time reading:* Ibid.

163 *listened to the Dodgers:* Ibid., "Sealab II Log, Part 2," pp. 67, 68, 69.

163 *notice a telltale pain:* Pauli and Clapper, "Project Sealab Report," p. 120.

163 *did not want to be the one:* Sheats Daily Log, entry for Oct. 11, 1965.

163 *nagging knee ache had spread:* Ibid.

164 *twice in twenty-seven years:* Leo Bowler, "Sealab Diver Better After Case of Bends" and "Aquanaut's Bends Hit in Same Leg as Before," (San Diego) *Evening Tribune*, Oct. 12, 1965, in "Navy compilation of Sealab II press coverage," p. 166.

164 *verify his friend's diagnosis:* Bond, *Papa Topside*, p. 147.

164 *have the nine others move:* Ibid.

164 *trading places with Bond:* Ibid.

164 *"among the proudest ribbons":* Tape recording of Sealab II press conference for Team 3; for a good print account of the conference, see Morgan, "Achievements of Sealab II," *San Diego Union*, Oct. 13, 1965, p. A13, previously cited.

164 *"the next frontier":* Tape recording of Sealab II press conference for Team 3.

164 *"The deep dive we made":* Ibid.

164 *a heartfelt story:* Ibid.

165 *believed that advice once saved:* Ibid.

165 *another day for observation:* SitRep no. 54, Oct. 12, 1965; Leo Bowler, "Diver Suffers Bends After Sealab Ascent," (San Diego) *Evening Tribune*, Oct. 12, 1965, p. A1.

165 New York Times *dispatched:* Peter Bart, "Success of Sealab Program to Spur Navy Undersea Research," *New York Times*, Oct. 13, 1965, p. 16.

165 *surfaced the following day:* Cousteau, "Working for Weeks on the Sea Floor," pp. 532, 536; "6 French Aquanauts Finish 3-Week Stint in Diving Bell," *New York Times*, Oct. 14, 1965, p. 39; *World of Jacques Cousteau*.

165 *weather and helium leaks:* Cousteau, "Working for Weeks on the Sea Floor," pp. 508, 514, 515.

165 *three and a half man-years:* Tape recording of Sealab II press conference for Team 3 (total bottom time, as stated here by Bond, could be checked this way: 45 days x 30 aquanauts = 1,350 days/365 days = 3.7 "man-years").

165 *sending twenty-eight men down:* Ibid.

165 *spontaneous mini-sermon:* Ibid.

166 *"We have seen a program grow":* Ibid.

CHAPTER 13: THE DAMN HATCH

A day-long, taped interview with Bob Barth in Panama City, Fla., on Dec. 16, 2001, provided significant detail and background on the ill-fated dives to Sealab III, as did an initial interview with Richard Blackburn in San Diego, Calif., on March 16, 2002, followed by telephone interviews, including those on June 6, 2002, Sept. 18, 2002, March 22, 2004, and Sept. 19, 2006. There were also a number of e-mail responses from Blackburn, including those dated Oct. 7, 10, and 28, 2003. John Reaves, the only other surviving aquanaut to have made dives to the lab, could not be located either by Sealab friends or the author during the research for this book. However, Reaves's testimony about the dives, along with a wealth of other information, is included in the "Record of Proceedings," a half-foot-tall stack of paperwork obtained through the Freedom of Information Act. Its full title is: "Record of Proceedings of a Formal Board of Investigation Convened at the Deep Submergence Systems Project Technical Office, 139 Sylvester Road, San Diego, California, 92106, by order of the Chief of Navy Material, Department of the Navy; To inquire into the circumstances of the death of Berry L. Cannon in the vicinity of San Clemente Island, California, on 17 February 1969; Ordered 21 February 1969." This documentation consists of pages numbered 1–320, with some additional pages identified by both a number and letter, e.g., page 50-C. Included with the main document are exhibits numbered 1 through 86; in the notes that follow they are cited by the abbreviation "Ex." and the relevant exhibit number, e.g., Ex. 47, plus an exhibit title or description.

The "Record of Proceedings" provided valuable details and a means of corroborating and fleshing out information obtained in interviews with Barth and Blackburn, and also other Sealab III personnel, including, in alphabetical order: Richard Bird, Robert Bornholdt, Robert Bornmann, William J. Bunton, Scott Carpenter, Richard Cooper, Matthew Eggar, Richard Garrahan, David Martin Harrell, Paul G. Linaweaver Jr., Fernando Lugo, Walter Mazzone, Keith Moore, James H. Osborn, Andres Pruna, Terrel W. "Jack" Reedy, Don C. Risk, Donald J. Schmitt, Jackson M. Tomsky, Cyril J. Tuckfield, James Vorosmarti Jr., and William W. Winters; also interviewed were Royal Navy participants Derek J. "Nobby" Clark and Cyril F. Lafferty. Many were interviewed more than once. Some interviews were done in person, some at the Sealab reunions, some by phone, and there were often supplemental exchanges by e-mail. Bunton, a Sealab II aquanaut who was to take part in Sealab III, wrote a short book, *Death of an Aquanaut* (Flagstaff, Ariz.: Best Publishing, 2000), cited below, which provided a useful perspective and further information.

Additional details were gleaned from an audio tape recording of two press conferences given by Capt. William Nicholson, project manager of the Deep Submergence Systems Project. The first was dated Feb. 12, 1969, and apparently held in Washington, D.C., a few days before Sealab III got started; the second was from Feb. 17, 1969, at San Clemente Island, soon after the accident (in author's possession; cited below as "Capt. Nicholson press conference," plus the appropriate date). A valuable visual record, in addition to Navy photographs, came from *100 Fathoms Deep*, a short documentary film about Sealab III produced by the U.S. Navy's Deep Submergence Systems Project. Similar footage can be seen in another Navy film, this one a brief documentary called *The Aquanauts* (video copies of both in author's possession, but they are also available in libraries). The two daily San Diego newspapers at the time, the *San Diego Union* and the *Evening Tribune*, regularly reported on Sealab III; only the most pertinent articles are cited below, but taken together this collection of stories, obtained from the *San Diego Union-Tribune* news library, also helped to sharpen the picture of what took place. The notes that follow, mainly for Chapters 13 and 14, cite the "Record of Proceedings" whenever possible, even if the same or similar information may also have been obtained from other published sources or personal interviews, because this is an official, accessible documentary record. But this record, an abridged transcript of the proceedings, was itself corroborated and often given flesh on its bones by other sources, especially the aforementioned interviews. The notation "interviews with Sealab III personnel" indicates that a particular point was frequently echoed in interviews. Interviews with specific sources are cited below, sometimes in tandem with citations from the "Record of Proceedings" because those interviews shed significant additional light on the available written record.

Page

167   *stand-alone organization:* Jack Tomsky, taped interview, Escondido, Calif., Dec. 30, 2003; Bond, "Sealab III Chronicle," pp. 1–3.

167   *porpoise and an amiable sea lion:* "Press Handbook, Sealab III Experiment, the U.S. Navy's Man-in-the-Sea Program," September 1968, p. 10–1, p. 5–2 (in author's possession; handbook page numbers are grouped by chapter number; a copy in which the brief aquanaut biographies are redacted is included in "Record of Proceedings" as Ex. 21); Capt. Nicholson press conference, Feb. 12, 1969.

167   *"operational" use:* Tomsky, interviews, Aug. 5, 2002, and Dec. 30, 2003; Capt. Nicholson press conference, Feb. 12, 1969.

167   *expanded and remodeled:* "Press Handbook," p. 4–9.

167   Elk River, *originally built:* "Press Handbook," p. 10–1; *100 Fathoms Deep.*

168   *Some $10 million:* Capt. Nicholson press conference, Feb. 12, 1969; Bond, *Papa Topside,* p. 157, puts the cost at "about $15 million," but Nicholson gives his figure with apparent certainty and, as program manager, was in a good position to know. No cost figure was found in official documentation.

168   *island served as a testing range:* "Press Handbook," p. 2-1; *100 Fathoms Deep.*

168   *half-hour away:* Bond, *Papa Topside,* p. 86.

168   *Jackson M. Tomsky:* Tomsky, interviews, May 17, 2002, and June 7, 2004.

168   *"Black Jack":* Tomsky, interview, May 17, 2002; Bond, "Sealab III Chronicle," p. 2.

168   *had to do some talking:* Tomsky, interview, May 17, 2002; John Piña Craven, *The Silent War* (New York: Simon & Schuster, 2001), pp. 150–51.

168   *de facto leader:* Craven, *The Silent War,* p. 149; Capt. Nicholson press conference, Feb. 12, 1969.

169   *"authority, responsibility":* Craven, *The Silent War,* p. 150; Craven, brief interview during a three-day reunion of Sealab participants, San Diego, Calif., March 15, 2002; Craven was also a featured speaker at the reunion dinner

of March 15 at the San Diego Bayside Holiday Inn; he did not attend any of the three subsequent reunions, which the author did; he did not respond to a May 4, 2004, e-mail request for an interview.

169 *that of project manager:* Tomsky, interviews, May 17, 2002, and Dec. 30, 2003; Craven, *The Silent War,* p. 150; "Press Handbook," p. 3-1.

169 *By 1966:* Tomsky, interview, June 7, 2004.

169 *got along well enough:* Tomsky, interview, Dec. 30, 2003.

169 *principal investigator:* Ibid.; "Press Handbook," p. 3-1; "Record of Proceedings," p. 98-B.

169 *professor emeritus:* Tomsky, interview, Dec. 30, 2003.

169 *key lieutenants:* Ibid.

169 *ten of fifty-four aquanauts:* "Press Handbook," pp. 11-1 to 12-1.

169 *ultimately decided not to participate:* Carl Sheats, the younger of Sheats's two sons, interview, April 27, 2004; Phil Sheats, the older of Sheats's two sons, interview, July 8, 2004.

169 *despite Bond's encouragement:* Bond to Sheats, Aug. 8, 1967; Bond to Sheats, May 21, 1968; both on Deep Submergence Systems Project letterhead (in author's possession).

170 *cooperated in diving training:* "Press Handbook," p. 12-1; Bornholdt, interview, Feb. 22, 2002.

170 *Philippe Cousteau:* Bond, "Sealab III Chronicle," pp. 22, 30; Bond to family, on Deep Submergence Systems Project letterhead, Oct. 26, 1968 (in author's possession); "Record of Proceedings," Ex. 72: A training status memo, which includes team rosters showing Cousteau on Team 3; Barth, *Sea Dwellers,* p. 173 (photo).

170 *ideas of his own for opening scenes:* Bond, *Papa Topside,* p. 158.

170 *sign of bone necrosis:* Ibid., pp. 154, 167; Carpenter, interview, April 6, 2004; David Perlman, "Scott Carpenter and the Great Sealab Adventure," *Look,* Jan. 21, 1969, p. 70.

170 *became Tomsky's deputy:* Perlman, "Scott Carpenter and the Great Sealab Adventure," p. 70; "Record of Proceedings," pp. 122–122-B, 281; Ex. 22: "Sealab On-Scene Organizational and Operations Manual," May 1968, pp. 13, 16–18; Ex. 73: "Commander Tomsky's Understanding as of March 10, 1969, of responsibilities"; "Press Handbook," p. 3-1.

170 *number three man was:* "Record of Proceedings," pp. 122–122-B; Ex. 73: "Commander Tomsky's Understanding as of March 10, 1969, of responsibilities"; "Press Handbook," p. 3-2.

170 *no personal experience with saturation:* Tomsky, interview, May 17, 2002; Bond, *Papa Topside,* p. 153.

170 *diving operations officer:* Mazzone, interview, Jan. 2, 2002; Tomsky, interview, Aug. 5, 2002; "Record of Proceedings," Ex. 23: On-Scene Commander to Distribution List, "Organizational Changes for Sealab III," Feb. 4, 1969; Ex.

22: "Sealab On-Scene Organizational and Operations Manual," May 1968, pp. 26–27.

170   *Deep Submergence Systems Project Technical Office:* Mazzone, e-mail to author, June 4, 2003.

170   *never seemed to be enough:* "Record of Proceedings," pp. 42, 143, 149–50, 213; Carpenter, interview, Jan. 25, 2002.

170   *caused postponements:* "Record of Proceedings," p. 197; Paul Corcoran, Copley News Service, "Delays Plague Sealab Plans but Panic Button Not Hit," *San Diego Union,* Jan. 19, 1969, p. B15; Bunton, *Death of an Aquanaut,* pp. 34–35; Bond, *Papa Topside,* p. 151; "Sealab III Chronicle," pp. 15–16; Bond to family, Oct. 26, 1968 (in author's possession).

170   *flood-out of a PTC:* Bond, *Papa Topside,* pp. 162–63; "Record of Proceedings," pp. 100-I, 200, 279; Corcoran, "Delays Plague Sealab Plans," *San Diego Union.*

170   *initial target depth:* Bond, *Papa Topside,* p. 152; Bond, "Sealab III: Next Step Toward the Depths," *Astronautics & Aeronautics* (a publication of the American Institute of Aeronautics and Astronautics), July 1967, p. 83.

170   *increase of five atmospheres:* Jack Tomsky, taped interview, Escondido, Calif., Aug. 5, 2002.

171   *excursion dives were to be made:* "Record of Proceedings," Ex. 84.

171   *morale booster:* "Record of Proceedings," pp. 141, 245–46.

171   *his team felt the same way:* Ibid., p. 217.

171   *first of five teams scheduled:* Barth, *Sea Dwellers,* pp. 121, 125; "Press Handbook," pp. 2-1, 5-1.

171   *rain-swept Saturday:* Bond, *Papa Topside,* p. 167.

171   *Two symmetrical sections:* "Press Handbook," p. 4-1; *100 Fathoms Deep.*

171   *novel anchoring system:* "Press Handbook," p. 4-3.

171   *pumped in to compensate:* Capt. Nicholson press conference, Feb. 17, 1969.

172   *deal with the problem in situ:* "Record of Proceedings," pp. 25–26, 284, 287; Ex. 47 (see handwritten notes included with typed procedures); Bond, *Papa Topside,* p. 168.

172   *in his more strictly defined role:* "Record of Proceedings," pp. 98, 122; Ex. 73: "Commander Tomsky's Understanding as of March 10, 1969, of responsibilities as of Feb. 14–16, 1969"; Bond, *Papa Topside,* pp. 151–53. While acknowledging his less central role, Bond indicates on p. 153 that he selected Jack Tomsky; he makes a similar claim in his "Sealab III Chronicle," the unedited journal on which the chapters covering Sealab III in *Papa Topside* are based. According to Tomsky and others, it was not actually Bond who picked Tomsky, though Bond may have supported Tomsky's selection.

172   *expedited compression schedule:* "Record of Proceedings," p. 26.

172   *additional soreness:* Ibid., pp. 49, 55.

172   *mini-sub called* Deep Star: "Record of Proceedings," p. 13; Bond, *Papa Topside,* p. 168; *100 Fathoms Deep.*

172 *Carpenter on its three-man crew:* "Record of Proceedings," pp. 25, 41; Capt. Nicholson press conference, Feb. 17, 1969.

172 *bubbles rising from the habitat:* "Record of Proceedings," p. 41; Ex. 47 (see handwritten notes included); *100 Fathoms Deep.*

172 *electrical stuffing tubes:* "Record of Proceedings," pp. 25, 41–42; Capt. Nicholson press conference, Feb. 17, 1969.

172 *On February 16:* "Record of Proceedings," p. 286; Ex. 16: "Sealab III Data Book," p. 27.

172 *a steel pod seven feet:* "Press Handbook," pp. 7-3, 7-6.

172 *ones chosen to get inside:* Bond, *Papa Topside,* pp. 168–69; "Record of Proceedings," p. 27.

173 *had spent six years:* Blackburn, interviews, June 6, 2002, and July 15, 2005.

173 *last-minute replacement:* Blackburn, interview, Sept. 19, 2006.

173 *principal trainer of Tuffy:* "Record of Proceedings," p. 43; "Press Handbook," p. 11-15; Pauli and Clapper, "Project Sealab Report," pp. 408, 410.

173 *lowering took almost an hour:* "Record of Proceedings," pp. 26–27, 286; Ex. 16: "Sealab III Data Book," pp. 27–28.

173 *heater was out of order:* Barth, *Sea Dwellers,* p. 132; Bond, "Sealab III Chronicle," p. 74.

173 *deflector was also missing:* Bond, *Papa Topside,* p. 170; "Record of Proceedings," p. 46; Barth and Blackburn, interviews.

173 *never been so cold:* "Record of Proceedings," pp. 55–56.

173 *called the Mark IX:* Ibid., p. 42; Ex. 44: "Equipment Test Program: LCDR John V. Harter, USN," p. 3.

173 *less than fifty pounds:* "Record of Proceedings," pp. 42-B, 92-A.

174 *cage hung a few feet:* Barth and Blackburn, interviews; "Press Handbook," p. 7-4; Navy photograph, Oct. 4, 1978, no. KN-15824 (in author's possession); *100 Fathoms Deep.*

174 *in the mid-forties:* "Record of Proceedings," Ex. 75: Statement of Robert J. Fyfe, Feb. 20, 1969; Barth, *Sea Dwellers,* p. 134.

174 *sixty-foot length:* "Press Handbook," p. 4-9.

174 *"unbuttoning" procedures:* "Record of Proceedings," pp. 206–8; Barth, *Sea Dwellers,* pp. 133, 135.

174 *"blow the skirt":* Barth, *Sea Dwellers,* p. 133; "Record of Proceedings," pp. 102-I, 207.

174 *a couple of steps up a ladder:* *100 Fathoms Deep.*

174 *yanked a few levers:* Barth, interview, Dec. 16, 2001; Capt. Nicholson press conference, Feb. 17, 1969.

174 *about twelve minutes:* "Record of Proceedings," p. 12; Ex. 16: "Sealab III Data Book," pp. 28–29.

174 *left the hatch and swam:* "Record of Proceedings," pp. 207–8.

174 *beginning to feel uneasy:* Ibid.

174 *he might black out:* Ibid., p. 57.

174 *dazed look:* Ibid., p. 50-A.

174 *"Where's Barth?":* Ibid.; Blackburn, interviews.

174 *Minutes seemed to pass:* "Record of Proceedings," pp. 27, 46, 50-A.

174 *breach of buddy diving etiquette:* Ibid., pp. 205–6, 223–24.

174 *quickly strapped on:* Ibid., pp. 46, 50-A.

175 *Blackburn couldn't understand:* Ibid.

175 *more than 90 percent helium:* "Press Handbook," p. 4-2.

175 *black-and-white image:* "Record of Proceedings," pp. 27, 102-M; *100 Fathoms Deep.*

175 *voice over the intercom told:* "Record of Proceedings," p. 50-A.

175 *left the bottom at 7:25:* "Record of Proceedings," p. 288; Ex. 16: "Sealab III Data Book," pp. 29, 31.

175 *trailer on the main deck:* "Record of Proceedings," p. 102-M; interviews with Sealab III personnel; official Navy photographs, nos. KN-15833 & 15845 (in author's possession.); *100 Fathoms Deep.*

175 *their decompression chamber:* "Press Handbook," p. 7-1; *100 Fathoms Deep.*

175 *process of raising the pod:* Interviews with Sealab III personnel; Navy photographs (in author's possession); *100 Fathoms Deep.*

175 *shaking badly from the cold:* "Record of Proceedings," pp. 50-A, 55, 58; Ex. 16: "Sealab III Data Book," p. 30; Barth, *Sea Dwellers,* p. 133.

175 *almost nine o'clock:* "Record of Proceedings," p. 288; Ex. 16: "Sealab III Data Book," p. 31; Capt. Nicholson press conference, Feb. 17, 1969.

175 *into a blast furnace:* "Record of Proceedings," pp. 50-A, 58; Blackburn, interview, Sept. 19, 2006.

175 *Project Tektite I:* D. C. Pauli and H. A. Cole, eds., "Project Tektite I, a Multi-agency 60-Day Saturated Dive Conducted by the United States Navy, the National Aeronautics and Space Administration, the Department of the Interior and the General Electric Company," ONR Report DR 153, Office of Naval Research, Washington, D.C., Jan. 16, 1970.

176 *reports barely differentiated:* William K. Stevens, "Aquanauts Aim for Depth and Duration Marks," *New York Times,* Feb. 14, 1969, p. 41; Richard D. Lyons, "4 Aquanauts Begin 60-Day Submersion," *New York Times,* Feb. 16, 1969, p. 41; Lyons, "4 Surface After 2 Months on Bottom of the Sea," *New York Times,* April 15, p. 29; Lyons, "President Hails Aquanauts' Feat," *New York Times,* April 16, p. 8; Lyons, "4 Tektite Aquanauts Fit After 60-Day Mission," *New York Times,* April 19, 1969, p. 35.

176 *unlikely to kill:* Pauli and Cole, "Project Tektite I," pp. 14, 24.

176 *no ultralight helium:* Ibid., pp. 37–38.

176 *take less than a day:* Ibid., p. 8.

176 *interconnected vertical tanks:* Ibid., pp. 30–36.

176 *south side of St. John:* Ibid., pp. 8–9.

176 *predict human behavior:* Ibid., pp. 2–3.

176 *topside commanders gathered:* "Record of Proceedings," pp. 25, 37, 41; Mazzone, interview at a reunion of World War II submarine veterans, Buffalo,
N.Y., Aug. 29, 2002; Robert Bornholdt, then a lieutenant, training officer and
team leader, interview, March 4, 2002; Tomsky, interview, Aug. 5, 2002; Bond,
"Sealab III Chronicle," p. 75; *100 Fathoms Deep.*

176 *hashing over a variety:* "Record of Proceedings," p. 25; Cliff Smith, "Sealab
Chief Tells of Role in Tragedy," *San Diego Union,* March 7, 1969, p. B1.

176 *ordered more gas be pumped:* Interviews with Sealab III personnel; "Record
of Proceedings," Ex. 16: "Sealab III Data Book," p. 20; the first entry on this
page, made shortly before 9 A.M. on Feb. 16, 1969, notes that the habitat is
leaking and that the internal pressure is being kept at 6 psi over the outside
water pressure; similar pressure readings appear throughout subsequent log
entries, including those corresponding to the times Barth attempted to open
the hatch.

176 *tricky from six hundred:* Cyril Lafferty, interview, June 9, 2006; Lafferty, an
experienced Royal Navy diver and diving officer who was to lead the fourth
of the five Sealab III teams, was among those in the command van during the
second dive; D. Martin Harrell, a DSSP engineer and aquanaut-to-be, interview, Oct. 27, 2006.

176 *proved to be a bear:* "Record of Proceedings," pp. 102-J, 109; Bornholdt,
interview, March 4, 2002.

177 *as much as eight pounds:* "Record of Proceedings," Ex. 16: "Sealab III Data
Book," pp. 28–29; Bunton, *Death of an Aquanaut,* p. 41.

177 *give a thumbs-up:* Bornholdt, interview, June 7, 2002.

177 *the rest of Team 1:* Ibid.; Bond, *Papa Topside,* p. 171.

177 *his best guess:* Bond, *Papa Topside,* p. 171; "Record of Proceedings," p. 102-B.

177 *very uncomfortable:* "Record of Proceedings," p. 29.

177 *prefer that Barth take Blackburn:* Ibid., p. 28.

177 *Mazzone pointed out:* Ibid., p. 27.

177 *needed time to rest:* Ibid., p. 147.

177 *at two in the morning:* Ibid., p. 148.

177 *Others argued:* Bornholdt, interview, March 4, 2002; Lafferty, interview,
June 9, 2006; Bond, *Papa Topside,* p. 171.

177 *sidestepped the issue:* "Record of Proceedings," pp. 29, 41.

177 *had never been colder:* Ibid., p. 48.

178 *wasn't much conversation:* Ibid., pp. 114–15.

178 *agreed that cold was:* Ibid.

178 *their own nerves:* Ibid.

178 *never in deep water:* Ibid., pp. 113, 197, 283; Ex. 22: "Sealab III On-Scene
Organization and Operations Manual," May 1968, p. 35; interviews with
Sealab III personnel, including Tomsky, Aug. 5, 2002; Bornholdt, June 7,

2002; Blackburn, Sept. 19, 2006; Dr. Paul Linaweaver, who was a commander, senior medical officer in the DSSP Technical Office and an aquanaut-to-be, interview, July 26, 2002.

178   *said he was sorry:* "Record of Proceedings," p. 205.

178   *no need to apologize:* Ibid.

178   *observations but didn't dwell:* Ibid., p. 115.

178   *When Blackburn asked Cannon:* Ibid., p. 50-A.

178   *had to work the system:* Blackburn, interview, June 6, 2002.

178   *to make them sweat:* "Record of Proceedings," p. 100A.

178   *there was a lot to do:* Ibid., pp. 37, 58, Ex. 47: "Procedures that were attached to clipboard in PTC 2 on Feb. 17, 1969."

178   *an hour of sleep:* Ibid., pp. 53, 288.

178   *only Blackburn:* Ibid., pp. 28, 109, 288.

178   *checked over the two Mark IX rigs:* Ibid., pp. 50-D, 109, 129, 289–90.

179   *on the Navy team responsible:* Blackburn, interview, Jan. 27, 2005; "Mark VI Semi-Closed Circuit Breathing Apparatus—Field Evaluation, December 1962–March 1963," Evaluation Report 1-63, conducted by Explosive Ordnance Disposal Unit Two, submitted by J. C. Bladh, project officer, Appendix A, p. viii.

179   *Blackburn scrawled a note:* "Record of Proceedings," pp. 92-C, 129, 180, 290, Ex. 79: Note to P. A. Wells.

179   *put the two rigs back together:* Ibid., pp. 48, 109.

179   *arrived with a backup supply:* Ibid., p. 28.

179   *flooding alarm:* Ibid., pp. 28, 101-G, 289; Ex. 16: "Sealab III Data Book," p. 37.

179   *the two decompression chambers:* Interviews with Sealab III personnel; *100 Fathoms Deep.*

179   *discussed the situation with Barth:* "Record of Proceedings," pp. 29, 58.

179   *no reason to change horses:* Ibid., p. 58.

179   *make the opening dive again:* Ibid., pp. 28, 58.

179   *"Don't be a hero":* Ibid., p. 29.

179   *known as the Wiswell suit:* Ibid., p. 20.

180   *decided to jury-rig an umbilical:* Ibid., p. 50-D.

180   *feel the temperature falling:* Ibid., p. 51.

180   *shortly after three o'clock:* Ibid., p. 290; Capt. Nicholson press conference, Feb. 17, 1969.

180   *twenty hours straight:* "Record of Proceedings," pp. 37, 51.

180   *amphetamine tablets:* Ibid., pp. 40, 100-D, 110, 117, 194, 288.

180   *blowers again chilled them:* Ibid., pp. 45, 58-A.

180   *shared a blanket:* Ibid., pp. 51, 59, 290.

180   *colder than they had on the first ride:* Ibid., p. 47; Barth and Blackburn, interviews.

180 *more than an hour:* "Record of Proceedings," p. 290, Ex. 16: "Sealab III Data Book," pp. 38, 40.

180 *glimpses of the bubbling:* Ibid., p. 45, Ex. 16: "Sealab III Data Book," p. 27; *100 Fathoms Deep.*

180 *only a cool trickle:* Barth, *Sea Dwellers,* p. 134.

181 *no hood or gloves:* "Record of Proceedings," p. 51.

181 *sensation puzzled him:* Barth, *Sea Dwellers,* p. 134; "Record of Proceedings," p. 56.

181 *desperate gasping sound:* "Record of Proceedings," pp. 47, 51, 289, 291.

181 *now shortly after five:* Ibid., pp. 290–91.

181 *tap on his shoulder:* Barth, *Sea Dwellers,* p. 134.

181 *the two paused:* Ibid.

181 *Cannon split off from Barth:* Ibid., p. 135; "Record of Proceedings," pp. 223–24; *100 Fathoms Deep.*

181 *Watching on a TV monitor:* "Record of Proceedings," pp. 59, 101-D, 102-H, 291; *100 Fathoms Deep.*

181 *broken the seal:* "Record of Proceedings," pp. 101-G, 289.

181 *had a crowbar installed:* Barth, *Sea Dwellers,* p. 135.

181 *A little plaque:* "Record of Proceedings," p. 30, Ex. 39: photo of plaque.

181 *saw Barth swim off screen: 100 Fathoms Deep.*

181 *subdued tension:* Interviews with Sealab III personnel, including Robert Bornholdt and Andres Pruna, both of whom were among those on the *Elk River* during the second dive, interviews, June 9, 2002, and Nov. 1, 2002, respectively; Bunton, *Death of an Aquanaut,* pp. 53–54; Bond, "Sealab III Chronicle," pp. 77–78; see also "Record of Proceedings," Ex. 16: "Sealab III Data Book," pp. 24–25, which confirms that Tomsky is present and giving orders.

181 *conflicting opinions:* Mazzone, interview, Aug. 29, 2002; apart from implying that there were disagreements, Mazzone repeatedly declined to go into specifics, beginning with a first interview on Jan. 2, 2002, about what he believed was the problem with topside's approach; he instead deferred to Tomsky's explanation. Others interviewed who were close to the action either couldn't remember exactly what was happening at the topside controls, didn't know, or didn't want to comment.

181 *could not understand why:* Jack Tomsky, interview, Dec. 30, 2003.

181 *undo the two wing nuts:* "Record of Proceedings," pp. 59, 224.

181 *a clumsy method:* Ibid., p. 42.

182 *strange grunting sound:* Barth, interviews, including Dec. 16, 2001; *Sea Dwellers,* p. 135.

182 *startled by a yell:* Blackburn, interviews, including March 16, 2002.

182 *like a man jumping rope:* "Record of Proceedings," p. 59.

182 *reappear on their monitors:* Bond, *Papa Topside,* pp. 171–72; *100 Fathoms Deep.*

182 *immediately sensed trouble:* Bond, *Papa Topside,* p. 172.

182 *tried to lift Cannon:* "Record of Proceedings," p. 60; Barth, interview, Dec. 16, 2001.

182 *jabbed it once, twice:* Bond, "Sealab III Chronicle," p. 77; *100 Fathoms Deep.*

182 *swim for the PTC:* "Record of Proceedings," p. 60.

182 *their umbilicals caught on something:* Ibid.

182 *ringing in his ears:* Ibid., p. 61.

182 *little choice but to leave:* Ibid.

182 *wrangled some more:* Ibid.

182 *as if it were a mile away:* Ibid.

182 *order to get another diver:* Ibid., pp. 44, 169, 102-O, 292.

182 *Berry was in trouble:* Ibid., p. 52.

182 *Blackie squeezed past Barth:* Ibid.

182 *unspool his own umbilical:* Ibid.

182 *could see Cannon's mouthpiece:* Ibid.

183 *as if he were napping:* Ibid.

183 *triggering the bypass:* Ibid.

183 *biggest and fittest:* Blackburn, interviews; Bunton, *Death of an Aquanaut*, p. 55; Navy photograph by J. Reaves, "Views of Aquanaut Training," SL-3-414-7-68, July 9, 1968 (in author's possession); Blackburn is readily recognizable and clearly one of the biggest and tallest of the two dozen aquanauts pictured jogging along a beach.

183 *dragging more than two hundred:* Cannon's weight is given as 157 pounds in "Autopsy Report from the Office of the Coroner, County of San Diego, California, File no. 54110, for Berry Louis Cannon, Feb. 17, 1969"; the dry weight of the Mark IX ("Record of Proceedings," pp. 42-B, 92-A), plus the hot-water suit, adds another fifty pounds or so.

183 *began to feel short of breath:* "Record of Proceedings," pp. 52, 108.

183 *one arm around Cannon:* Ibid., p. 52; Blackburn, interview, June 6, 2002.

183 *wondered if the line was strong enough:* Blackburn, interview, June 6, 2002.

183 *with a fury he had never:* Ibid.

183 *wrestled him out of his Mark IX:* "Record of Proceedings," p. 52.

183 *hoist Cannon into the capsule:* Ibid., p. 53.

183 *had never seen a dead diver:* Ibid., pp. 48, 191–92.

183 *no trace of breath:* Ibid., pp. 189, 193.

183 *fed Cannon a fresh flow:* Ibid., p. 53.

184 *good exchange of gas:* Ibid., pp. 53, 186, 188.

184 *In the awkward circular space:* Ibid.

184 *external heart massage:* Ibid.

184 *Questions from topside:* Ibid., p. 102-O.

184 *Mazzone donned a headset:* Mazzone, e-mail to author, Feb. 2, 2009.

184 *"Take us up!":* "Record of Proceedings," p. 102-O.

184 *ordered the capsule raised:* Ibid.

184 *"Faster, faster"*: Ibid.

184 *might be an hour or more:* "Record of Proceedings," pp. 40, 293; Bond, "Sealab III Chronicle," p. 78; Capt. Nicholson press conference, Feb. 17, 1969.

184 *better part of a week:* Capt. Nicholson press conference, Feb. 12, 1969; Bob Corbett, "Navy Plans Probe of Sealab Death," (San Diego) *Evening Tribune*, Feb. 18, 1969, p. A2 (decompression from a saturation dive typically takes up to twenty-four hours for every hundred feet of depth).

184 *shaking from the cold:* "Record of Proceedings," p. 51.

184 *noticed a froth:* Ibid., pp. 48, 193–94.

184 *about fifteen minutes after:* Bond, "Sealab III Chronicle," p. 78; "Record of Proceedings," p. 102-P; Bond's time estimate seems about right, and even in the various logbooks no precise times of a death announcement from the PTC are noted; see "Record of Proceedings," Ex. 16: "Sealab III Data Book," p. 42; Ex. 18: "Port Control (Main Control Console) Log of Feb. 14 through Feb. 17, 1969," pp. 18–21; Ex. 15: "Sealab III Data Book, Medical Van, Vol. 1 (Portion Of)," p. 39.

184 *"Berry Cannon is dead":* Bond, *Papa Topside*, p. 172; Barth, *Sea Dwellers*, p. 136; "Record of Proceedings," pp. 44, 107-A, and Ex. 18: "Port Control (Main Control Console) Log," p. 18; these sources indicate that both Reaves and Barth, at different, unspecified times, spoke words to this effect over the PTC intercom en route to the surface.

CHAPTER 14: AN INVESTIGATION

Page

185 *An effort to revive:* "Record of Proceedings," pp. 102-P, 107-I, 293.

185 *lowered Cannon's body:* Ibid., p. 103; Barth and Blackburn, interviews.

185 *close his friend's eyes:* Barth, *Sea Dwellers*, p. 137.

185 *rendezvous with the aquanauts:* "Record of Proceedings," pp. 102-P, 103.

185 *into a body bag:* Ibid., p. 103; Barth, *Sea Dwellers*, p. 137.

186 *puffing and bloating:* "Record of Proceedings," p. 74.

186 *found dangling:* Ibid., p. 119; William W. Winters, then an engineman first class and aquanaut-to-be who later made master diver, interview, Gretna, La., March 13, 2007.

186 *two Blackburn had hung up:* "Record of Proceedings," pp. 47, 52.

186 *tossed down the hatch:* Ibid., pp. 44, 47.

186 *board held its first hearing:* Ibid., p. 1.

186 *fact-finding mission:* Judge Advocate General, Department of the Navy, "Procedures Applicable to Courts of Inquiry and Administrative Fact-Finding Bodies That Require a Hearing," Nov. 19, 1990, pp. 1–3, www.jag.navy.mil/Instructions/5830_1.pdf; I. J. Galantin, Chief of Naval Material to Captain John D. Chase, USN, "Formal board of investigation to inquire into the circumstances of the death of Berry L. Cannon in the vicinity of San Clemente

Island, California on Feb. 18, 1969," Feb. 21, 1969 (memo included with "Record of Proceedings").

186 *"parties of interest"*: "Record of Proceedings," pp. 2–5.

186 *free to take part:* Ibid., p. 3.

186 *Wells, a senior chief mineman:* Ibid., p. 89, calls Wells a torpedoman, and he is also listed as such in the "Press Handbook," p. 11–19. However, in Pauli and Clapper, "Project Sealab Report," for Sealab II, Wells is listed as a mineman. Friends, including Barth and Blackburn, say mineman is correct.

186 *ought to be able to participate:* Ibid., p. 6.

186 *represented by legal counsel:* Ibid., pp. 6, 9, 63, 66.

186 *even trial by court-martial:* Judge Advocate General, "Procedures Applicable to Courts of Inquiry," p. 2; "Manual for Courts-Martial United States (2000 edition)," pp. II-19, II-34, www.jag.navy.mil/documents/mcm2000.pdf; "Record of Proceedings," pp. 7–9; Tomsky, interview, Aug. 5, 2002.

186 *late into the evening:* "Record of Proceedings," pp. 1, 5, 12-A, 35, 62, 88, 96, 118; Bond, "Sealab III Chronicle," p. 97.

186 *eighty documentary exhibits:* "Record of Proceedings," pp. iv–x.

187 *nearly fifty people who were called:* Ibid., pp. ii–iii.

187 *"It is quite trying":* Ibid., p. 61.

187 *all three television network newscasts:* Vanderbilt Television News Archive, tvnews.vanderbilt.edu; search for "Sealab" and "Berry Cannon."

187 *splashed onto front pages:* William K. Stevens, "Aquanaut Dies in Dive 600 Feet Down; Test Halted," *New York Times*; Marvin Miles, "Veteran Aquanaut Dies During Dive on Sealab Project," *Los Angeles Times*; Thomas O'Toole, "Sealab Test Halted After Diver Dies," *Washington Post*; Cliff Smith, "Navy Calls for Probe in Tragedy," *San Diego Union*; Mike Darley, "Sikes Investigating Death of Aquanaut; Wants Full Probe of Cannon Fatality," *Panama City* (Fla.) *Herald*; newspaper reports like these appeared on Feb. 18, 1969, p. 1.

187 *open to reporters:* "Record of Proceedings," p. 1; Bond, "Sealab III Chronicle," pp. 92, 102.

187 *"I'm not happy with":* Ibid., pp. 117–18.

188 Apollo 1 *caught fire:* NASA History Division, www.hq.nasa.gov/office/pao/History/Apollo204/

188 Apollo 9 *was launched:* John Noble Wilford, "Apollo 9 Poised for Flight Today," *New York Times*, March 3, 1969, p. 1.

188 *laced with sharp criticism:* Susan Stocking, "Fatal Dive Should Never Have Been Made, Sealab Probe Told," *Los Angeles Times*, March 3, 1969, p. 1.

188 *Blackburn's opening statement:* "Record of Proceedings," pp. 48–53.

188 *their training with the rig:* Ibid., pp. 50-C, 197; Blackburn, interview, Sept. 19, 2006.

188 *trained in less than sixty feet:* "Record of Proceedings," pp. 48, 50-C.

188 *"Training looked good on":* Ibid., p. 50-C.

188 *others had sensed this:* Interviews with Sealab III personnel; Bunton, *Death of an Aquanaut*, pp. 63–64.

188 *"I believe that the second":* "Record of Proceedings," pp. 53, 148.

188 *program was underfunded:* Ibid., p. 149; Carpenter and Stoever, *For Spacious Skies*, pp. 329–30; Carpenter, interviews, Jan. 25, 2002, and Oct. 20, 2006; Tomsky, in the interview of Aug. 5, 2002, recalled Carpenter's funding concerns.

188 *people were pushed harder:* "Record of Proceedings," p. 138.

189 *quick to respond:* Ibid., pp. 5, 122-B, 137, 228–29, 232, 271.

189 *all the funding he had asked for:* Ibid., p. 122-B.

189 *only within reason:* Ibid., pp. 137, 162–63.

189 *"May I clear a point?":* "Record of Proceedings," p. 27.

189 *abridged history of Sealab:* Ibid., p. 98-B.

189 *found himself pelted with:* Bond, "Sealab III Chronicle," pp. 102–4.

189 *a touch of emotion:* Cliff Smith, "Doctor Defends His Actions in Sealab Dive," *San Diego Union*, March 11, 1969, p. B1.

189 *considered Berry Cannon a great friend:* Bond, "Sealab III Chronicle," p. 96.

189 *morbid implication:* Ibid., p. 103; Bond to family, on Deep Submergence Systems Project letterhead, March 10, 1969, p. 3 (in author's possession).

189 *Harrell, an engineer:* "Record of Proceedings," p. 71; D. Martin Harrell, interview during a Navy Experimental Diving Unit tour, March 11, 2005, part of the Sealab reunion that year in Panama City, Fla.; and by telephone, Oct. 27, 2006.

189 *drew immediate objections:* "Record of Proceedings," p. 71.

189 *snafus should be looked into:* Interviews with Sealab III personnel; Bond to Robert C. Sheats, May 1, 1975 (in author's possession); Bunton, *Death of an Aquanaut*, p. 67.

190 *narrate a silent black-and-white:* "Record of Proceedings," p. 57.

190 *spoke dispassionately:* Cliff Smith, "Sealab Panel Views Films of Tragedy," *San Diego Union*, March 4, 1969, p. B1.

190 *footage of Barth and Cannon: 100 Fathoms Deep.*

190 *apparent heart attack:* Bunton, *Death of an Aquanaut*, pp. 57–58; front-page articles published Feb. 18, 1969, previously cited in the notes for this chapter, including "Aquanaut Dies in Dive 600 Feet Down; Test Halted," *New York Times*; "Veteran Aquanaut Dies During Dive on Sealab Project," *Los Angeles Times*; "Sealab Test Halted After Diver Dies," *Washington Post*; Capt. Nicholson press conference, Feb. 17, 1969.

190 *must have been a best guess:* UPI, "Poison Air Linked to Sealab Death," *New York Times*, Feb. 25, 1969, p. 28.

190 *canister with no Baralyme:* "Record of Proceedings," pp. 295, 314; Barth, *Sea Dwellers*, p. 138; William K. Stevens, "Rig of Aquanaut Lacked Chemical," *New York Times*, Feb 21, 1969, p. 57; Bob Corbett, "Diving Rig Had Empty

Canister," (San Diego) *Evening Tribune*, Feb. 20, 1969, p. A1; L. Edgar Prina, Copley News Service, "Missing Filter Cited in Sealab Death," *San Diego Union*, May 15, 1969, p. A5.

190 *with Dr. Bond observing:* "Record of Proceedings," p. 107; Bond, *Papa Topside*, p. 173.

190 *as stated in the autopsy report:* "Record of Proceedings," p. 75; Ex. 51: "Record of Autopsy of Berry L. Cannon" was withheld from the Navy records released in response to the Freedom of Information Act request, but was instead obtained from the San Diego County Medical Examiner's Office, which considered the report to be public information ("Autopsy Report from the Office of the Coroner, County of San Diego, California, File no. 54110, for Berry Louis Cannon, Feb. 17, 1969").

190 *namely drowning:* "Record of Proceedings," p. 80.

190 *indicated that asphyxiation:* Ibid., pp. 82, 301.

190 *there were many unknowns:* Ibid., p. 85.

191 *Raymond, who was to have been:* Ibid., p. 85-A; Ex. 64: Lawrence Raymond, LCDR, MC USN, "Thermal Factors in the Death of a Sealab Diver—A Report to the Investigating Board," Feb. 20, 1969.

191 *left little doubt that:* Ibid., p. 3; Susan Stocking, "Medical Testimony Casts Doubt on Cause of Aquanaut's Death," *Los Angeles Times*, March 8, 1968, Part II, p. 1.

191 *were not entirely known:* "Record of Proceedings," Ex. 64: Raymond, "Thermal Factors," pp. 5, 6.

191 *dead within five to ten minutes:* Ibid., p. 87-A.

191 *extended his breathing time:* Ibid., p. 294, Ex. 64: Raymond, "Thermal Factors," p. 2 (handwritten marginal note about frequent bypass use).

191 *Cannon was electrocuted:* Ibid., p. 301; a number of participants interviewed said electrocution seemed a likely and logical cause of death, including Blackburn, interview, March 16, 2002, and e-mails to author, Oct. 10, 2003; Bornholdt, interview, June 7, 2002; and William Winters, a knowledgeable source and aquanaut-to-be who, like Bornholdt, was working on the *Elk River* that night, in the March 13, 2007, interview cited above.

191 *had some grounding problems:* "Record of Proceedings," p. 290; Ex. 16: "Sealab III Data Book," pp. 15, 16, 39, 44, 47, 48; Ex. 75: Statement of Robert J. Fyfe, Feb. 20, 1969, p. 2; Susan Stocking, "Sealab III Inquiry Ends; Mystery Still Unsolved," *Los Angeles Times*, March 13, 1969, p. 30.

191 *nor the grunting sound:* If Barth, Blackburn, or Reaves mentioned grunting, a yell, or scream during the board hearing, as Barth and Blackburn said they heard, in interviews cited earlier, such testimony does not appear in the "Record of Proceedings," a lengthy but abridged transcript.

191 *as were asphyxiation and thermal stress:* "Record of Proceedings," p. 301.

191 *Cannon's prime killer:* Ibid., pp. 300, 314.

191 *Paul A. Wells:* He declined several requests to be interviewed for this book, including one made by the author on March 7, 2003, by telephone from outside the gates of the mobile home park where Wells was living in Panama City, Fla.

192 *universally admired:* Unequivocal confidence in and admiration for Wells was expressed in numerous interviews with Sealab participants and others who knew and worked with him; see examples in "Record of Proceedings," pp. 92-N, 108, 127, 269, 276; Bond, "Sealab III Chronicle," pp. 82, 85, 97.

192 *thoroughly impressed Bob Sheats:* Sheats Daily Log, entry for Oct. 2, see also Sept. 8, 1965; at the end of his log Sheats ranked Wells as the best of the nine divers he led on Sealab II's Team 3.

192 *most of his Navy career:* "Record of Proceedings," p. 89; "Press Handbook," p. 11-19.

192 *asked to recall his every action:* "Record of Proceedings," p. 90.

192 *"sometimes utter chaos":* Ibid., p. 94-B.

192 *Huss had no such memory:* Ibid., pp. 92-N, 94-A.

192 *"a huge man":* Ibid., p. 94.

192 *personally placed loaded canisters:* Ibid., pp. 91-E, 97.

192 *checked all the rigs previously:* Ibid., pp. 91-C, 92, 315.

193 *"a hurry-up affair":* Ibid., pp. 91-C, 98; Susan Stocking, "Inquiry Told of 'Rush' on Day Aquanaut Died," *Los Angeles Times*, March 11, 1969, p. 10.

193 *carried each rig:* "Record of Proceedings," pp. 91-D, 92-K.

193 *just by lifting the rig:* Ibid., pp. 92–92-A; Susan Stocking, "Sealab Diving Rigs Not Checked on Day Man Died, Quiz Told," *Los Angeles Times*, March 9, 1969, p. B3.

193 *Some in the room were skeptical:* Bond, *Papa Topside*, pp. 174–76; Bond, "Sealab III Chronicle," pp. 98–99; "Record of Proceedings," p. 91-E.

193 *on a forty-seven-pound rig:* "Record of Proceedings," p. 92-A; no disagreement about the Mark IX's actual weight appears here; however, earlier expert testimony, on p. 42-B, put the dry weight at thirty-five pounds, with Baralyme adding another seven or eight pounds, thereby constituting about 20 percent of the lower total weight.

193 *brief supportive comment:* Ibid., p. 92-A.

193 *Others privately agreed:* Interviews with Sealab III personnel.

193 *put refilled canisters:* "Record of Proceedings," pp. 109, 289–90.

193 *sent to Washington for analysis:* Ibid., pp. 92-H, 120–21; William Leibold, e-mail to author, Dec. 7, 2008.

193 *watched closely as Wells packed:* "Record of Proceedings," pp. 92-A, 92-H, 120; Bond, *Papa Topside*, p. 174.

193 *Under pointed questioning:* "Record of Proceedings," pp. 92-A and 92-B; Bond, "Sealab III Chronicle," p. 99 (Bond writes that he, too, questioned Wells about the water weight, but the questioning he describes does not appear in the "Record of Proceedings").

193 *rig had some water:* "Record of Proceedings," p. 92-B.

193 *highly decorated war veteran:* Memo for the press, "Biography of Commander William R. Leibold, USN, member of the Sealab III Board of Investigation," undated (in author's possession).

193 *recently completed a tour as:* Leibold, e-mail to author, Dec. 7, 2008.

193 *"You had water":* "Record of Proceedings," p. 92-B.

193 *"Possibly, but I feel":* Ibid.

194 *no further explanation:* Ibid.; Bond, *Papa Topside*, p. 175.

194 *investigators went to work:* "Record of Proceedings," Ex. 53: Officer in Charge, Navy Experimental Diving Unit to Project Manager, Deep Submergence Systems Project, "Report of Inspection of MK IX Semi-Closed Circuit Mixed Gas Breathing Apparatus on Feb. 21 and 22, 1969," Feb. 26, 1969.

194 *No record had been kept:* Ibid., p. 296.

194 *frame by frame:* Ibid., pp. 124, 124-C.

194 *eight expert interpreters:* Ibid., Ex. 74: Department Head, Evaluation Department, NAVRECONTECHSUPPCEN to Lt. William Marsh, USN, "Analysis of Sealab Video Tape and Film," March 11, 1969.

194 *no one had used rig no. 8:* Ibid., pp. 92-G, 92-H.

194 *could not be a 4:* Ibid., Ex. 74: "Analysis of Sealab Video."

194 *he had worn no. 5:* Blackburn, interview, Aug. 23, 2006.

194 *his first and second dives:* "Record of Proceedings," pp. 57, 61, 196.

194 *One or more of these factors:* Ibid., pp. 180, 222.

194 *strongly believed a saboteur:* Jack Tomsky, interview, Aug. 5, 2002; Craven, *The Silent War*, p. 160.

195 *opportunity for tampering:* "Record of Proceedings," pp. 92-K, 119–21, 306 (Finding of Fact no. 33); Bond, "Sealab III Chronicle," p. 86.

195 *found the probability very small:* "Record of Proceedings," pp. 295 (Finding of Fact no. 58), 306 (Opinion no. 33).

195 *raised the specter of sabotage:* Susan Stocking, "Fatal Dive Should Never Have Been Made, Sealab Probe Told," *Los Angeles Times*, March 3, 1969, p. 1; "Sabotage Hinted by Sealab Chief," *New York Times*, March 3, 1969, p. 12.

195 *mysterious case of a valve:* "Record of Proceedings," p. 33.

195 *Tomsky tried but failed:* Ibid.

195 *set up a security watch:* Ibid., p. 34.

195 *seemed like the kind of mistake:* Mazzone, interview, Jan. 2, 2002.

195 *new to saturation diving:* Ibid.; Bunton, *Death of an Aquanaut*, pp. 33–35.

195 *before convening a final session:* "Record of Proceedings," pp. 131–276.

195 *forty-seven unanimous opinions:* Ibid., pp. 300–310A.

195 *sixty-six findings of fact:* Ibid., pp. 277–99.

195 *not certain:* Ibid., p. 309.

195 *nineteen recommendations:* Ibid., p. 311.

195 *should be officially commended:* Ibid., p. 313.

195   *letters of admonition:* Ibid., p. 313; recommendations 17 and 18, which appear to pertain to the admonition, were redacted here with a legal citation for privacy. But even at the time, the letters were no secret, as reported in papers on Sept. 25, 1969, by the Associated Press, including "Reprimands Sent to 2 in Sealab Death," *San Diego Union*, p. A2, and "2 Admonished by Navy in Sealab Dive Fatality," *New York Times*, p. 24. The letters are also alluded to in the memo cited below, from the DSSP project manager, Capt. William Nicholson. Tomsky discussed his letter in an interview, Aug. 5, 2002.

196   *"None of the alleged":* Cmdr. Jackson Maxwell Tomsky, USN (ret.) to the Secretary of the Navy, Oct. 30, 1969 (this memo was among the supporting documents released with the "Record of Proceedings" in response to a Freedom of Information Act request).

196   *not alone in lamenting:* Interviews with Sealab III personnel, including Tomsky, Aug. 5, 2002; Barth, Dec. 16, 2001; Harrell, Oct. 26, 2006; "Record of Proceedings," p. 70; Ex. 37, 38, 43, "Hull Penetrations" (photos); Carpenter and Stoever, *For Spacious Skies*, pp. 329–30; Bond, "Sealab III Chronicle," pp. 45–49; Bond to family, Dec. 2, 1968, Jan. 10 and Jan. 18, 1969 (in author's possession); "Aquanaut Death Film Shown," *San Diego Union*, Nov. 21, 1969, p. B3.

196   *"unbelievable":* Susan Stocking, "Sealab's Difficulties Called 'Unbelievable,'" *Los Angeles Times*, March 6, 1969, p. 3.

196   *Such a trial, he believed:* Tomsky, interview, Aug. 5, 2002; Tomsky to the Secretary of the Navy, Oct. 30, 1969, p. 4.

196   *two-month-long inquiry:* George C. Wilson, "Naval Inquiry on Pueblo Capture Ends," *Washington Post*, March 14, 1969, p. A6.

196   *did not believe Wells was to blame:* Bond, *Papa Topside*, p. 176; Barth, *Sea Dwellers*, p. 138; interviews with Sealab III personnel.

196   *"extreme and unnecessary":* Memorandum from William Nicholson, Project Manager, Deep Submergence Systems Project to Chief of Naval Material, "Investigation into circumstances of death of Berry Cannon," April 10, 1969, p. 1 (in author's possession).

196   *"the probable impact":* Ibid.

197   *self-imposed exile:* Interviews with Sealab III personnel, notably Barth and Blackburn; see also Bob Barth to Sealab Reunion Roster, April 1989 (in author's possession).

197   *lost two excellent men:* Barth, *Sea Dwellers*, p. 138.

197   *called Venture Out:* Visit by author on March 7, 2003, as explained in above note about Paul A. Wells.

197   *few were satisfied:* Bond, "Sealab III Chronicle," pp. 104–5; interviews with Sealab III personnel; Bunton, *Death of an Aquanaut*, p. 62.

197   *remained convinced that sabotage:* Tomsky, interview, Aug. 5, 2002.

197   *Navy offered public assurances:* "Navy Delays Sealab III for 'Months,'" (San Diego) *Evening Tribune*, Feb. 21, 1969, p. A13; Cliff Smith, "Sealab Work Will

Continue, Admiral Says," *San Diego Union*, March 27, 1969, p. B1; "Sealab III to Resume," (San Diego) *Evening Tribune*, June 18, 1969, p. A12.

197 *Cannon would have wanted it:* Interviews with Sealab III personnel; Barth, *Sea Dwellers*, p. 138.

197 *services for Cannon:* "Berry L. Cannon Services Held," *San Diego Union*, Feb. 20, 1969, p. B1.

197 *his wife's nearby hometown:* Mary Lou Cannon, interview, at her parents' home, March 17, 2002.

197 *Navy chaplain appeared:* Mary Lou Cannon, interview, Jan. 23, 2002.

197 *didn't even look like Berry:* Ibid.

197 *clung to this notion:* Ibid.

197 *"There is a mystery":* Ibid.

198 *burial at Wacahoota:* "Williston High Graduate Dies in Sealab Experiment," *Williston* (Fla.) *Sun*, Feb. 20, 1969, p. 1; Mary Cannon, interview, Jan. 23, 2002.

198 *raised by his grandmother:* Mary Lou Cannon, interview, Jan. 23, 2002.

198 *simple gray headstone:* Alachua Genealogical Society's Virtual Cemetery Project, Wacahoota Baptist Church Cemetery, www.usgennet.org/usa/fl/county/alachua/ACGS/WacahootaB/index.html.

198 *dozens of lobsters:* Richard Cooper, interview during three-day Sealab reunion, Panama City, Fla., March 13, 2005; Barth, *Sea Dwellers*, pp. 125, 137.

198 *improvise a way to raise Sealab:* "Record of Proceedings," pp. 31–33; Craven, *The Silent War*, pp. 156–59.

199 *flew off to the Virgin Islands:* Bond, "Sealab III Chronicle," p. 106; *Papa Topside*, p. 177; Bond, Assistant for Medical Effects, Deep Submergence Systems Project, "Trip Report, Tektite Operation," March 26, 1969 (twenty-one pages, in author's possession).

199 *force the postponement:* "Navy Curbs Sealab for Fiscal Year," (San Diego) *Evening Tribune*, Sept. 23, 1969, p. B1; "Sealab III Victim of Navy Economy," *San Diego Union*, Sept. 24, 1969, p. B1; "Sealab 3 Is Shelved," *New York Times*, Sept. 24, 1969, p. 92.

199 *program had been canceled:* Bob Corbett, "Sealab III Quietly 'Buried,'" (San Diego) *Evening Tribune*, Dec. 16, 1970, p. C1; "Submergence Project Office Discontinued," *San Diego Union*, Oct. 6, 1970, p. B2.

199 *died with Berry Cannon:* Barth, *Sea Dwellers*, p. 138.

199 *Barth was dismayed:* Ibid.

199 *with scant explanation:* Ibid.

199 *Death was always a possibility:* Ibid.

CHAPTER 15: THE OIL PATCH

In addition to books and other documentary sources about oil field diving, some of which are cited below, personal interviews added immeasurably to the material in this chapter, including those that took place during the sixteenth annual divers' reunion at Dick Ransome's roadside bar in Bush, La., a day-long event on March 11, 2007, attended by several dozen old-time oil field divers. Key interviews are cited below, but a number of others whose names are for the most part not specifically cited were also interviewed at various times, including: Harry Connelly, Charles Coggeshall, Harry Lee Coon, Jack Horning, Jack Reedy, Don Risk, John Roat, R. J. Steckel, Bob Tallant, Geoff Thielst, and William Winters. The insights and experiences they shared were invaluable, and information largely attributable to their collective input is cited below as "oil field diver interviews." Harry Connelly shared not just his personal experience but his prized collection of *Undercurrents*, a short-lived and now hard-to-find magazine that provided a further window on the working lives of divers in the offshore industry, circa 1970. Some of the above-named divers worked as supervisors and superintendents who oversaw diving operations. Many, but not all, worked primarily for Taylor Diving & Salvage, a central focus of this chapter. A decidedly non-Taylor perspective came from others, most notably R. Lad Handelman, an industry icon, who could speak with equal authority as a onetime scrappy California abalone diver and as a cofounder and top executive of several influential diving companies, including one of the biggest, Oceaneering International. Author interviews with Handelman include those taped at his home in Santa Barbara, Calif., on Oct. 6, 1998, and Sept. 28, 2005. As in the earlier chapters, documentary sources are cited whenever possible, even when similar information came from interviews, to direct the interested reader or researcher to the most readily accessible source. Especially noteworthy is Christopher Swann's recently published book, cited below. Its 850 pages chronicle the evolution of the offshore industry like no other.

Page

200 *ex-Navy diver and medic:* Alan "Doc" Helvey, who is singled out in this chapter, related his experiences by telephone beginning with a pair of taped interviews on April 9 and May 29, 2007. He answered another round of questions during several taped follow-up interviews on Feb. 4, Feb. 12, and April 9, 2009. Information drawn from these interviews is cited below as "Helvey, interviews."

200 *growing legion of commercial divers:* Joseph A. Pratt, Tyler Priest, and Christopher J. Castaneda, *Offshore Pioneers: Brown & Root and the History of Offshore Oil and Gas* (Houston, Tex.: Gulf Publishing, 1997), p. 153; D. H. Elliott and P. B. Bennett, "Underwater Accidents," in *The Physiology and Medicine of Diving*, Peter B. Bennett and David H. Elliott, eds., 4th ed. (London: W. B. Saunders, 1993), p. 239.

200 *could earn in a day:* Oil field diver interviews; Christopher Swann, *The History of Oilfield Diving: An Industrial Adventure* (Santa Barbara, Calif.: Oceanaut Press, 2007), pp. 257, 576.

201 *oil field called the Forties:* Pratt et al., *Offshore Pioneers*, p. 239.

201 *a first for the United Kingdom sector:* Ibid.

201 *yielded big-money contracts:* Ibid., 226–27, 248–49, 263; Swann, *The History of Oilfield Diving*, pp. 363, 368, 412.

201 *humble origins:* Pratt et al., *Offshore Pioneers*, p. 139.

201 *one of the largest and most influential:* Ibid., p. 155; Swann, *The History of Oilfield Diving*, p. 566.

201 *after the North Sea discoveries:* Hans Veldman and George Lagers, *50 Years Offshore* (Delft, Netherlands: Foundation for Offshore Studies, 1997), pp. 96, 113, 126, 136.

201 *major player in the construction field:* Pratt et al., *Offshore Pioneers*, pp. 45, 53, 65, 66.

201 *laying pipeline at sea:* Ibid., pp. 137, 140, 258.

201 *looked like car engines perched:* Veldman and Lagers, *50 Years Offshore*, pp. 128, 134–35, 164.

201 *crisscrossed with pipelines:* Ibid., p. 178; Oilfield diver interviews.

201 *an estimated fifteen thousand miles:* *Offshore Platforms and Pipelining* (Tulsa, Okla.: Petroleum Publishing Company, 1976), p. 99; Pratt et al., *Offshore Pioneers*, p. 65.

202 *his first experience as a saturation diver:* Helvey, interviews.

202 *age of thirty-five:* Gil T. Webre, "Divers Are Constructors for Offshore Oil Industry" (New Orleans) *Times-Picayune*, Jan. 23, 1976.

202 *study and chase girls:* Helvey, interviews.

202 *their next round of sea floor tasks:* Ibid.

202 *pipe was still attached:* Ibid.

202 *a depth of 320 feet:* Helvey, interviews.

202 *oil patch diver's brand of useful work:* Oil field diver interviews; Nicholas B. Zinkowski, *Commercial Oil-Field Diving* (Cambridge, Md.: Cornell Maritime Press, 1971), pp. 198, 207, 226–41.

203 *With sledgehammers and axes:* Helvey, interviews; Zinkowski, *Commercial Oil-Field Diving*, pp. 152, 156–60, 164.

203 *coordinating the procedures with topside:* Zinkowski, *Commercial Oil-Field Diving*, pp. 232–38; Oil field diver interviews.

203 *looked like the space helmets:* Swann, *The History of Oilfield Diving*, pp. 229–34.

204 *to and from his helmet:* Helvey, interviews; Helvey, e-mails to author, Feb. 18 and July 9, 2009.

204 *cry for help over the intercom:* Helvey, interviews; "North Sea rescue effort fails 320 ft. down," *The Times* (London), Aug. 29, 1973, p. 2.

204 *strained against the combined weight:* Helvey, interviews.

204 *even after rigorous lab tests:* Wallace, interview, Biloxi, Miss., March 12, 2007.

204 *wondering whether he should ever:* Helvey, interviews.

204 *two reported in the North Sea:* Elliott and Bennett, "Underwater Accidents," in *The Physiology and Medicine of Diving*, 4th ed., p. 239; David Blundy,

"Pressures That Kill the Men Who Dive for Britain's Oil," *The Sunday Times* (London), June 23, 1974, p. 3.

204 *coal miners or construction workers:* Elliott and Bennett, "Underwater Accidents," in *The Physiology and Medicine of Diving*, 4th ed., p. 239; M. E. Bradley, M.D., "Commercial Diving Fatalities," *Aviation, Space, and Environmental Medicine*, August 1984, p. 722.

205 *he would have shaken his head:* Helvey, interviews.

205 *nature of the thousand-foot:* Ibid.; D. Michael Hughes, "Many Factors Affect Deepwater Dives," *Offshore*, August 1973, p. 58.

205 *tapped on land in 1859:* Daniel Yergin, *The Prize: The Epic Quest for Oil, Money, and Power* (New York: Simon & Schuster, 1991), p. 27.

205 *shorelines of Southern California:* Veldman and Lagers, *50 Years Offshore*, pp. 13–16; Pratt et al., *Offshore Pioneers*, p. 159.

205 *stand-alone platforms began to appear:* Swann, *The History of Oilfield Diving*, pp. 8–11.

205 *push the conventional depth limits:* Ibid., p. 61.

205 *The use of helium:* Lad Handelman, "Where Did the Major Diving Companies of Today Originate?," *UnderWater*, May/June 2000, pp. 68–69.

205 *revamped decompression tables:* Ibid.

205 *A few companies continued to push:* Donald M. Taylor, "Bounce Diving in 450-600-ft Water Depths and Deeper," *Ocean Industry*, March 1974, pp. 35–37.

206 *case to be made for this new method:* Hughes, "Many Factors Affect Deepwater Dives," p. 64.

206 *first commercial saturation dives:* James W. Miller and Ian G. Koblick, *Living and Working in the Sea*, 2nd ed. (Plymouth, Vt.: Five Corners, 1995), p. 72.

206 *only a few close calls:* Swann, *The History of Oilfield Diving*, p. 241.

206 *two years and cost a lot more:* Miller and Koblick, *Living and Working in the Sea*, p. 72.

206 *Westinghouse's initial enthusiasm:* "In the Absence of Federal Aid, Industry Takes the Initiative," *Marine Engineering/Log*, July 1967, p. 36.

206 *first commercial saturation dives at sea:* Miller and Koblick, *Living and Working in the Sea*, pp. 74–75.

206 *clear out the tangle of debris:* Ibid.

206 *forty-five miles off Grand Isle:* "Prolonged Submergence Opens Door to Deeper Dives," *Marine Engineering/Log*, July 1967, pp. 38–39.

207 *a new company, Ocean Systems Inc.:* Swann, *The History of Oilfield Diving*, p. 169.

207 *transformed his team:* M. C. Link, *Windows in the Sea*, pp. 82, 84; Van Hoek with M. C. Link, *From Sky to Sea*, pp. 283, 291.

207 *former Submersible Portable Inflatable Dwelling participants:* M. C. Link, *Windows in the Sea*, p. 84.

207 *moved to a vacant Linde building:* Swann, *The History of Oilfield Diving*, p. 170.

207 *Sténuit serving as one of the subjects:* Ibid.

207 *his old friend E. C. Stephan:* M. C. Link, *Windows in the Sea*, p. 84.

207 *refused to take a regular desk job:* Ibid., p. 82.

208 *he had quit school:* Wallace, interview, June 1, 2003.

208 *the most significant of his life:* Ibid.

208 *passed along findings about methods:* Swann, *The History of Oilfield Diving*, p. 245.

208 *consultants to Taylor Diving:* Ibid., pp. 245, 252; Robert D. Workman, M.D., "Deep Water Diving Calls for Medical and Physiological Solutions to Problems," *Offshore*, August 1973, p. 48; Wallace, interviews, Nov. 15, 2006, and May 21, 2007.

208 *master diver who had cofounded Taylor:* Pratt et al., *Offshore Pioneers*, pp. 138–39; Swann, *The History of Oilfield Diving*, p. 30.

209 *president and mainstay:* Pratt et al., *Offshore Pioneers*, pp. 146, 155.

209 *could plainly see its potential:* Wallace, interview, March 12, 2007; Pratt et al., *Offshore Pioneers*, pp. 142, 145–47; Swann, *The History of Oilfield Diving*, pp. 245–49.

209 *Brown & Root turned to Taylor:* Pratt et al., *Offshore Pioneers*, pp. 137, 140, 155.

209 *playing piano at Lafitte's:* Ibid., p. 140.

209 *popping up along the Gulf Coast:* Swann, *The History of Oilfield Diving*, p. 26.

209 *a hundred of these smaller companies:* Ken Wallace, "Commercial Diving—Offshore Made It; It's Making Offshore," *Oil and Gas Journal*, March 4, 1974, p. 71.

209 *on Lake Pontchartrain:* Pratt et al., *Offshore Pioneers*, p. 139.

209 *installed a pressure chamber:* Swann, *The History of Oilfield Diving*, p. 34.

209 *sued Westinghouse:* Ibid., p. 244; Wallace, interview, May 21, 2007.

209 *words like "proprietary":* Swann, *The History of Oilfield Diving*, 246.

210 *Gulf of Mexico in the summer of 1967:* Ibid.; Pratt et al., *Offshore Pioneers*, p. 144.

210 *multimillion-dollar pressure complex:* "Industrial Hydrospace Research Center," *Undercurrents*, February 1970, pp. 14–16; Pratt et al., *Offshore Pioneers*, p. 147.

210 *took over as full-time head:* Pratt et al., *Offshore Pioneers*, p. 146; Workman résumé (copy in author's possession).

210 *take the reins from Banjavich:* Robert D. Workman, M.D. to Wallace, March 16, 1976; James E. Fitzmorris Jr., Lt. Governor, State of Louisiana to Wallace, April 11, 1975 (copies in author's possession).

210 *trained in civilian diving schools:* Wallace, "Commercial Diving—Offshore Made It," p. 77.

210 *Clannish tensions:* Oil field diver interviews; Ken Wallace, "Commercial Diving: There Is a Difference," *UnderWater*, Summer 1997, p. 37.

210 *biggest providers of underwater manpower:* Stan Luxenberg, "Deep Beneath the Sea, a New Industry Grows," *New York Times*, Feb. 27, 1977, Section 3, p. 4; Wallace, "Commercial Diving—Offshore Made It," p. 71; Larry L. Booda, "Diving Becoming Big Business," *UnderSea Technology*, Sept. 1968, p. 32.

210 *seized an opportunity to offer:* Henri Germain Delauze, interviews, Marseille, France, Nov. 2 and 3, 2004.

211 *personality conflicts developed:* Ibid.

211 *stuck by their famed boss:* Interviews in France with three longtime members of the Cousteau team: Jean Alinat, Nice, Oct. 24, 2004; André Laban, Juan-les-Pins, Oct. 27 and 31, 2004; Claude Wesly, Marseille, Nov. 2, 2004.

211 *left behind at the office:* Delauze, interviews, Nov. 2 and 3, 2004.

211 *construction of a highway tunnel:* Ibid.; Alain Dunoyer de Segonzac, *Un conquérant sous la mer* (Paris: Editions Buchet/Chastel, 1992), pp. 44–45.

211 *Fulbright scholarship:* Delauze, interviews, Nov. 2 & 3, 2004; Dunoyer de Segonzac, *Un conquérant sous la mer*, p. 47.

211 *met again with Cousteau:* Delauze, interviews, Nov. 2 & 3, 2004; Dunoyer de Segonzac, *Un conquérant sous la mer*, p. 58.

211 *renting a couple of small offices:* Delauze, interviews, Nov. 2 and 3, 2004. Dunoyer de Segonzac, *Un conquérant sous la mer*, p. 57.

211 *headquarters for Delauze's company:* Yves Baix, ed., "Comex—The Conquest of the Ocean Depths," *Océans* magazine (Marseille), special issue commemorating the twentieth anniversary of Comex, trans. John Kingsford (1981?).

211 *Delauze would take diving to depths:* Ibid., p. 61.

211 *notably Dr. Xavier Fructus:* Delauze and Bernard Gardette, scientific director, who began his career at Comex working under Fructus, at Comex, Marseille, Oct. 29, 2004; Dunoyer de Segonzac, *Un conquérant sous la mer*, pp. 83–85.

212 *had its own test chamber:* "COMEX Hyperbaric Experimental Centre, 1965–2000, 36 Years of Deep Diving Development, from Helium to Hydrogen," an in-house pamphlet that catalogues the highlights of company research, April 12, 2001.

212 *Navy made its first thousand-foot:* J. K. Summit and J. W. Kulig, "Saturation Dives, with Excursions, for the Development of a Decompression Schedule for Use During Sealab III," U.S. Navy Experimental Diving Unit Research Report, no. 9-70, 1970.

212 *Navy with scientists at Duke:* P. B. Bennett, "The High Pressure Nervous Syndrome: Man," in *The Physiology and Medicine of Diving and Compressed Air Work*, 2nd ed., p. 249.

212 *"helium tremors":* Ibid., p. 248; Bennett, taped interview, June 8, 2009.

212 *starting point for studies:* Bennett, "The High Pressure Nervous Syndrome: Man," in *The Physiology and Medicine of Diving and Compressed Air Work*, 2nd ed., p. 249.

212 *hyperbaric researcher Ralph Brauer:* Baix, ed., "Comex," p. 34; R. W. Brauer,

"High Pressure Nervous Syndrome: Animals," in *The Physiology and Medicine of Diving and Compressed Air Work*, 2nd ed., p. 231.

212 *Brauer began to experience:* Dr. X. Fructus and Dr. P. Fructus, "Technical Memorandum—Six very deep experimental dives," a collaboration of the Centre Experimental Hyperbare at Marseille with the Wrightsville Marine Bio-Medical Laboratory, Wilmington, N.C., n.d., pp. 19–20; Delauze, interview, Nov. 3, 2004.

213 *sleepiness, dizziness, nausea:* P. B. Bennett and J. C. Rostain, "The High Pressure Nervous Syndrome," in *The Physiology and Medicine of Diving*, 4th ed., p. 194.

213 *causing a hyperexcitability:* Bennett, interview, June 8, 2009.

213 *Delauze did not exhibit:* Delauze, interview, Nov. 3, 2004.

213 *Scientists debated what to call:* Bennett, interview, June 8, 2009.

213 *largely unpredictable:* Ibid.

213 *tried out various forms:* Bennett and Rostain, "The High Pressure Nervous Syndrome," in *The Physiology and Medicine of Diving*, 4th ed., pp. 212–18.

213 *Cannon would not yet have been exposed:* Bennett, interview, June 8, 2009.

213 *Slowing the rate of pressurization:* Bennett and Rostain, "The High Pressure Nervous Syndrome," in *The Physiology and Medicine of Diving*, 4th ed., pp. 212–18.

214 *to 610 meters:* Ibid., pp. 198, 211.

214 *Rolex advertisements: UnderSea Technology*, December 1975, back cover.

214 *depth of sixteen hundred feet:* Bennett and Rostain, "The High Pressure Nervous Syndrome," in *The Physiology and Medicine of Diving*, 4th ed., p. 198.

214 *showed normal arterial gases:* Ibid., p. 218.

214 *another mysterious neural effect:* Bennett, taped interview, April 29, 2009.

214 *experiments like this one continued:* Ibid.; Bennett and Rostain, "The High Pressure Nervous Syndrome," in *The Physiology and Medicine of Diving*, 4th ed., pp. 199–208.

215 *produced and tested the first one:* Swann, *The History of Oilfield Diving*, pp. 261–62.

215 *Taylor Diving was right behind:* Robert Steven, "Hyperbaric Welding Comes of Age," *Offshore*, August 1979, p. 84.

215 *underwater welding habitats:* Ibid.; Oil field diver interviews; Pratt et al., *Offshore Pioneers*, pp. 148–50, 155; Swann, "The Development of Hyperbaric Pipeline Welding," in *The History of Oilfield Diving*, p. 261; Zinkowski, *Commercial Oil-Field Diving*, pp. 341–47; Baix, ed., "Comex," p. 55; "Taylor Diving Hyperbaric Welding," brochure, n.d. (in author's possession).

215 *divers who had trained as welders:* Swann, *The History of Oilfield Diving*, p. 376; Wallace, interview, March 12, 2007.

215 *improved since World War II:* Davis, *Deep Diving and Submarine Operations*, p. 229.

215 *assembly-line style, on pipe-laying barges:* Zinkowski, *Commercial Oil-Field Diving*, p. 198; Oil field diver interviews.

216 *welds of equally high quality:* Swann, *The History of Oilfield Diving,* p. 261.

216 *crucial underwater pipe connection:* Zinkowski, *Commercial Oil-Field Diving,* pp. 225–32.

216 *connected out at sea:* Pratt et al., *Offshore Pioneers,* p. 278; Oil field diver interviews.

216 *lining up two ends of pipe:* Zinkowski, *Commercial Oil-Field Diving,* pp. 236–40; Oil field diver interviews.

216 *greater risk of leaking:* Pratt et al., *Offshore Pioneers,* p. 148.

216 *dominant provider of hyperbaric welding:* Swann, *The History of Oilfield Diving,* p. 566.

217 *Submersible Pipe Alignment Rig:* Ibid., pp. 268, 270; Pratt et al., *Offshore Pioneers,* pp. 151–52.

217 *the two of them worked together:* Oil field diver interviews.

217 *first thousand-foot contract:* Swann, *The History of Oilfield Diving,* p. 391.

217 *Delauze was dismayed:* Ibid., p. 393.

217 *Fast-growing Oceaneering:* Ibid., p. 360; Handelman, "Where Did the Major Diving Companies of Today Originate?," p. 72.

217 *broke with Ocean Systems after:* Handelman, "Where Did the Major Diving Companies of Today Originate?," p. 72.

217 *get a saturation diving system built:* Swann, *The History of Oilfield Diving,* p. 392; "Oceaneering's 1000-ft. Saturation Diving Complex," *UnderSea Technology,* April 1973.

217 *never exceeded about seven hundred feet:* Swann, *The History of Oilfield Diving,* p. 393.

217 *hired by BP Canada:* Ibid.

218 *several dives and spent four hours:* Ibid., p. 394; *Les plongeurs du froid,* a Comex film by Alain Tocco (DVD copy in author's possession); Delauze, e-mail to author, May 14, 2009.

218 *deepest offshore platform ever built:* Ben C. Gerwick, *Construction of Offshore Structures* (New York: John Wiley & Sons, 1986), p. 228; Swann, *The History of Oilfield Diving,* p. 541.

218 *called Cognac:* A. O. P. Casbarian and G. E. Cundiff, "Cognac: Unique Diving Skills Aid Project," *Offshore,* August 1979, p. 51.

218 *backup brigade for system failures:* Ibid., p. 52.

218 *major competitor in marine construction:* Pratt et al., *Offshore Pioneers,* p. 71.

218 *an unusual alliance:* "'Project Cognac' Involves Taylor in World's Deepest Offshore Diving Job," *Taylor Diver* 3, no. 2 (1977): 8–10; Ken Wallace, interview, May 21, 2007.

218 *designed in three sections:* Pratt et al., *Offshore Pioneers,* p. 81.

218 *industrial version of the Eiffel Tower:* Gerwick, *Construction of Offshore Structures,* pp. 230–31; Eiffel Tower information online at www.tour-eiffel.fr.

219 *largest saturation spread ever assembled:* Brownbuilder, Spring 1978, special

issue of Brown & Root's bimonthly company magazine, p. 38 (in author's possession).

219 *process worthy of NASA:* Casbarian and Cundiff, "Diving Skills Aid Project," p. 61; Helvey, interviews.

219 *seriously considered quitting:* Helvey, interviews.

219 *job might just be too risky:* Ibid.

219 *a thousand dollars a day:* Ibid.

219 *more like a* mission: Ibid.

219 *medic for his six-man Cognac team:* Ibid.

220 *best way to avoid HPNS:* Bennett, "The High Pressure Nervous Syndrome: Man," in *The Physiology and Medicine of Diving and Compressed Air Work,* 2nd ed., pp. 249, 255.

220 *over the course of a full day:* Helvey, interviews.

220 *Schwary, an old friend:* Ibid.

220 *sank to his knees:* Ibid.

221 *"bailout bottle":* Ibid.

221 *like a starry night sky:* Ibid.

221 *sheer size of the grouper:* Ibid.

221 *five feet deeper than anticipated:* Ibid.

221 *rest of the Cognac Six:* "World's Deepest Offshore Diving Job," p. 9; Helvey, e-mail to author, Feb. 18, 2009.

222 *industrial dress rehearsal:* Helvey, interviews; Swann, *The History of Oilfield Diving,* pp. 542, 544.

222 *bouts of dyspnea:* Ibid.

222 *massive square base:* Veldman and Lagers, *50 Years Offshore,* p. 172.

222 *nearly twenty stories tall:* "World's Deepest Offshore Diving Job," p. 8, gives the base height as 175 feet.

222 *two dozen steel piles:* Casbarian and Cundiff, "Diving Skills Aid Project," pp. 52, 55.

222 *six hundred feet long:* Gerwick, *Construction of Offshore Structures,* p. 230.

222 *hook up the steel lines:* Helvey, interviews.

222 *"flying eyeball":* Ibid., R. J. Steckel, Cognac project manager for Taylor, interview, March 22, 2007.

222 *disk like a manhole cover:* Helvey, interviews.

223 *could see his bell partner:* Ibid.

223 *During their off-hours:* Ibid.

223 *for thirty-one days:* Ibid.

223 *120 days:* Ibid.; Casbarian and Cundiff, "Diving Skills Aid Project," p. 62.

223 *won $147:* Helvey, interviews.

223 *beginning in the spring of 1978:* Ibid.

224 *docking two spacecraft:* "Construction Completed, Production Begins on Shell Oil's 'Project Cognac,'" *Taylor Diver* 4, no. 2 (1978): 9.

224   *Apollo-Soyuz test:* NASA History Division online, history.nasa.gov/30thastp/.

224   *more plumbing and grouting:* Helvey, interviews.

224   *ahead of schedule:* Ibid.; Swann, *The History of Oilfield Diving,* p. 546.

224   *sixty planned wells:* "Cognac Drilling Begins; Gulf Gas Project Start," *Oil & Gas Journal,* Nov. 6, 1978, p. 110; "Construction Completed, Production Begins on Shell Oil's 'Project Cognac,'" *Taylor Diver,* p. 8.

224   *2,700 offshore wells drilled:* Harry Whitehead, *An A to Z of Offshore Oil and Gas,* 2nd ed. (Houston, Tex.: Gulf Publishing, 1983), p. 321.

225   *industry interest in ROVs:* Eric Bender, "Remotely Operated Vehicles Continue Rapid Growth Offshore," *Sea Technology,* December 1979, p. 14; Swann, *The History of Oilfield Diving,* p. 721.

225   *added an ROV test tank:* Swann, *The History of Oilfield Diving,* p. 686.

225   *global slump in oil prices:* Yergin, *The Prize,* p. 750; U.S. Congress, Office of Technology Assessment, *U.S. Oil Production: The Effect of Low Oil Prices— Special Report,* OTA-E-348 (Washington, D.C.: U.S. Government Printing Office, September 1987), pp. 1, 25, 95–100.

225   *gone looking for other frontiers:* Swann, *The History of Oilfield Diving,* pp. 666, 681.

225   *going out of business or merging:* Ibid., pp. 666, 668; Michael Mulcahy, "Times Hard for Diving Companies, but Not All Signs Are Bad," *Sea Technology,* December 1979, p. 20.

225   *sold off and disbanded:* Wallace, interview, March 12, 2007; Swann, *The History of Oilfield Diving,* pp. 560–61.

225   *swallowed by Oceaneering International:* Swann, *The History of Oilfield Diving,* pp. 598, 668.

225   *because of its ROV business:* Ibid., p. 613.

CHAPTER 16: THE RIVALS PRESS ON

Page

226   *fifteen members produced:* Larry L. Booda, "New Ocean Agency Proposed," *UnderSea Technology,* February 1969, p. 37.

226   *Dr. Julius Stratton:* "President Emeritus Julius Adams Stratton dies at 93," *MIT Tech Talk* 38, no. 37 (June 29, 1994), web.mit.edu/newsoffice/1994/ stratton-0629.html.

226   *report's key recommendations:* Booda, "New Ocean Agency Proposed," pp. 36–38, 43.

226   *an independent civilian agency:* "Our Nation and the Sea; a Plan for National Action," Report of the Commission on Marine Science, Engineering and Resources (Washington, D.C.: U.S. Government Printing Office, January 1969), pp. 4, 233, www.lib.noaa.gov/noaainfo/heritage/stratton/title.html.

226   *a "wet NASA":* Interviews, including James W. Miller, deputy director of the NOAA Manned Undersea Science and Technology Office—from its inception

in 1971 and through 1980—and coauthor of *Living and Working in the Sea*, Aug. 24, 2007.

226 *"This project is based":* "Our Nation and the Sea," p. 162.

227 *goal of two thousand feet:* Ibid., p. 164; Booda, "New Ocean Agency Proposed," p. 38.

227 *millions be spent on research:* "Our Nation and the Sea," p. 166.

227 *formed from a smorgasbord:* "A History of NOAA," compiled by Eileen L. Shea, NOAA representative to the Department of Commerce Historical Council; edited by Skip Theberge, NOAA Central Library, 1999, www.history.noaa. gov/legacy/noaahistory_1.html, pp. 3, 11.

227 *umbrella of the Department of Commerce:* "A History of NOAA," p. 3; Miller, interview, Aug. 2, 2007.

227 *fresh outlook and clout:* "Our Nation and the Sea," p. 233; Miller, interview, Aug. 2, 2007.

228 *diving clubs in Czechoslovakia:* Miller and Koblick, *Living and Working in the Sea*, pp. 315, 350–52, 379.

228 *Soviet Union produced quite a few:* Ibid., pp. 98, 327, 332, 354.

228 *he would call Argyronète:* André Laban, a core member of the Cousteau team from 1952 until the early 1970s, interviews, Juan-les-Pins, France, Oct. 27 and 31, 2004; Dessemond and Wesly, *Les hommes de Cousteau*, pp. 234–37.

228 *prime sponsors backed out:* Swann, *The History of Oilfield Diving*, p. 535.

228 *called* Deep Diver: M. C. Link, *Windows in the Sea*, pp. 90–92.

228 *research sub* Alvin: Woods Hole Oceanographic Institution, www.whoi.edu/page.do?pid=8422.

229 *Link's mini-sub at seven hundred feet:* M. C. Link, *Windows in the Sea*, pp. 174–79.

229 *used for some contract work:* Clark and Eichelberger, "Edwin A. Link," p. 13; Van Hoek with M. C. Link, *From Sky to Sea*, pp. 300, 304.

229 *marine enthusiast J. Seward Johnson:* Walter H. Waggoner, "J. Seward Johnson, a Longtime Director of Family Company," *New York Times*, May 24, 1983, p. D24; as founder of Harbor Branch, www.fau.edu/hboi/AboutHarborBranch. php; place in Johnson & Johnson history, www.jnj.com/our_company/history/history_section_1.htm.

229 *set up its headquarters for:* Clark and Eichelberger, "Edwin A. Link," p. 13; Van Hoek with M. C. Link, *From Sky to Sea*, pp. 305–6, 309.

229 *helicopter without rotors:* M. C. Link, *Windows in the Sea*, pp. 187–90; Van Hoek with M. C. Link, *From Sky to Sea*, p. 309.

229 *entire craft improved over time:* Van Hoek with M. C. Link, *From Sky to Sea*, p. 307.

229 *reach depths of three thousand feet:* M. C. Link, *Windows in the Sea*, p. 188.

229 *under half a million dollars:* Van Hoek with M. C. Link, *From Sky to Sea*, p. 307.

230 *commissioned to the Smithsonian:* M. C. Link, *Windows in the Sea*, pp. 189–90.

230 *dozens of successful undersea projects:* "Marine Casualty Report, Submersible *Johnson Sea Link*, Entanglement off Key West, Florida, June 17, 1973," National Transportation Safety Board and U.S. Coast Guard, Report No. USCG/NTSB-MAR-75-2, Jan. 15, 1975, p. 22.

230 *disturb Ed Link's peace of mind:* Van Hoek with M. C. Link, *From Sky to Sea*, pp. 312, 314.

230 *fifteen miles out at sea:* Ibid.

230 *trying to recover a small fish trap:* Ibid.

230 *At the controls:* Ibid.; "Marine Casualty Report," pp. 20, 22.

230 *veteran submersible pilot:* Van Hoek with M. C. Link, *From Sky to Sea*, p. 315.

230 *Clayton Link:* Ibid., pp. 310–11, 316.

230 *With walkie-talkie in hand:* Ibid., p. 314; Jon Nordheimer, "Submarine Lifted, 2 Trapped Inside 30 Hours Rescued," *New York Times*, June 19, 1973, pp. 1, 78.

230 *The sea was calm:* "Marine Casualty Report," p. 20.

230 *Ed Link was calm:* Van Hoek with M. C. Link, *From Sky to Sea*, p. 314.

230 *calculated that the atmosphere:* "Marine Casualty Report," pp. 6, 25.

231 *Link was confident:* Van Hoek with M. C. Link, *From Sky to Sea*, p. 312.

231 *Tringa was delayed:* Ibid., pp. 312–13.

231 *Floodlights illuminated:* Ibid., p. 314.

231 *They had to turn back:* "Marine Casualty Report," p. 26.

231 *twice considered locking out:* Ibid., pp. 24–27.

231 *wore only shorts and T-shirts:* Ibid., p. 22.

231 *escape to the surface:* Van Hoek with M. C. Link, *From Sky to Sea*, p. 313.

231 *crew, and the trapped divers all agreed:* Ibid., "Marine Casualty Report," pp. 24–25.

231 *mid-forties Fahrenheit:* "Marine Casualty Report," p. 26.

231 *better insulated from the chill:* Van Hoek with M. C. Link, *From Sky to Sea*, p. 313.

231 *Menzies kept telling rescuers:* Stuart Auerbach, "Top Rescue Man Trapped in Sub," *Washington Post*, June 19, 1973, p. A9.

231 *faulty carbon dioxide scrubbers:* "Marine Casualty Report," pp. 25–26.

231 *took off his shirt:* Ibid., p. 25.

232 *Stover and Link increased:* "Marine Casualty Report," pp. 26–27; Jon Nordheimer, "2 Crewmen Dead; Bodies Removed from Submarine," *New York Times*, June 20, 1973, pp. 1, 26.

232 *another dive from the* Tringa: "Marine Casualty Report," p. 27.

232 *Five hours later:* Ibid.

232 *two-man Perry Cubmarine:* Ibid., p. 28; see also www.sfsm.org/submarine.html.

232 *salvage ship, the* A.B. Wood II: Ibid.

232 *they were going to be all right:* Ibid., pp. 28–29; Nordheimer, "2 Crewmen Dead; Bodies Removed from Submarine," p. 26.

232 *Through the portholes:* "Marine Casualty Report," p. 28.

232 *flushed the sealed compartment:* Ibid.; Nordheimer, "2 Crewmen Dead; Bodies Removed from Submarine," p. 26.

232 *doctors on board* Sea Diver *concluded:* "Marine Casualty Report," p. 29.

232 *holding tightly to one another:* Van Hoek with M. C. Link, *From Sky to Sea,* p. 315.

232 *Link left behind a young son:* Ibid., p. 310.

232 *Stover had seven sons:* Ibid.

232 *he might have done differently:* Ibid., p. 314.

233 *borrow Link's first submersible:* Ibid., p. 303.

233 *investigators blamed pilot error and:* "Marine Casualty Report," pp. 2, 31.

233 *"displayed an incredible casualness":* Ibid.

233 *"feeble attempts to rescue":* F. R. Haselton, "After the Sea Link Tragedy: Some Questions About Rescue Equipment," *Washington Post,* June 27, 1973, p. A31.

233 *submersible he called CORD:* Van Hoek with M. C. Link, *From Sky to Sea,* pp. 315, 317.

233 *called Scientist in the Sea:* Bond, *Papa Topside,* pp. 180–81, 184, 189; Advertisement in *UnderSea Technology,* December 1972, p. 26.

233 *watched over a dozen graduate students:* Bond, *Papa Topside,* pp. 194–95.

233 *Hydrolab, one of a handful made:* Miller and Koblick, *Living and Working in the Sea,* pp. 76–84.

234 *most used habitat in the world:* Ibid., p. 84.

234 *featured on* Primus: Brooks and Marsh, *Complete Directory to Prime Time Network and Cable TV Shows,* p. 838; "Shades of Cousteau, James Bond and Jules Verne," *New York Times,* Oct. 24, 1971, p. S16.

234 *the way* Sea Hunt *had:* Brooks and Marsh, *Complete Directory to Prime Time Network and Cable TV Shows,* p. 910.

234 *later refurbished, renamed Tektite II:* Miller and Koblick, *Living and Working in the Sea,* pp. 92–98.

234 *none other than Ed Link:* Ibid., p. 77.

234 *American entrepreneur even took a shot:* Ibid., pp. 119–23.

234 *member of the Stratton commission:* "Our Nation and the Sea," p. iii.

234 *built and tested a habitat:* Jon Pegg, "Five Hundred Sixteen Ft. Five-Day Ocean Saturation Dive Using a Mobile Habitat," *Aerospace Medicine* 42, no. 12 (December 1971): 1257–61.

234 *old friend Charlie Aquadro:* Aquadro, interview, May 16–18, 2003.

234 *$3 billion Skylab:* Roger D. Launius, *Frontiers of Space Exploration* (Westport, Conn.: Greenwood, 1998), p. 48.

234 *moving the historic Experimental:* Lt. Cmdr. Marc Tranchemontagne, "NEDU Celebrates 75 Years," *Faceplate,* April 2003, p. 4.

235 *a "pet" alligator:* Bond, *Papa Topside,* pp. 178, 185; George Bond Jr., interview, Oct. 10, 2003.

235 *Ocean Simulation Facility:* Bond, *Papa Topside,* pp. 184, 187; author visits in December 2001 and March 2003; Navy Experimental Diving Unit pamphlet, Aug. 21, 2001, p. 3 (in author's possession).

235 Pidgeon *and the* Ortolan: Mark V. Lonsdale, *United States Navy Diver: Performance Under Pressure* (Flagstaff, Ariz.: Best Publishing, 2005), p. 295.

235 *But Bond still believed:* Bond, *Papa Topside,* pp. 179–80.

235 *saturation diving school:* Lonsdale, *United States Navy Diver,* pp. 298, 311.

235 *in danger of being scrapped:* Bond, *Papa Topside,* p. 186.

235 *a multinational project:* Ibid., pp. 196, 198, 201.

236 *called Helgoland:* Miller and Koblick, *Living and Working in the Sea,* pp. 110–19.

236 *on a Polish factory ship:* Ibid., p. 116.

236 *several dozen topside participants:* Ibid.

236 *former Sealab aquanauts:* Bond, *Papa Topside,* pp. 200, 205, 220 (Larry Bussey, Dick Cooper, Morgan Wells).

236 *Helgoland's deepest trial:* Miller and Koblick, *Living and Working in the Sea,* p. 116.

236 *rough weather and equipment failures:* Ibid., p. 118; Bond, *Papa Topside,* pp. 198, 202–04, 230.

236 *herring were spawning too far:* Miller and Koblick, *Living and Working in the Sea,* p. 118.

236 *decompress on the bottom:* Ibid., p. 113; Bond, *Papa Topside,* p. 210.

236 *Bond later theorized:* Bond, *Papa Topside,* p. 213.

237 *Wendler died:* Ibid., pp. 208, 212; Miller and Koblick, *Living and Working in the Sea,* p. 264.

237 *as many close calls:* Capt. George F. Bond, MC, USN, "FISSHH Chronicles," Sept. 17, 1975, to Nov. 21, 1975 (unpublished manuscript on which material in *Papa Topside* is based), p. 99 (copy in author's possession).

237 *his habitual writings:* Ibid.

237 *joint salvage venture:* M. J. Lagies, "Treasure Divers Seek Record $5 Billion," (San Diego) *Evening Tribune,* Nov. 19, 1976, p. 1; W. Joe Innis, "Bill Bunton's Team: In Pursuit of the *Awa Maru,*" *The Republican* 14, no. 5, Southern California edition (Nov./Dec. 1976); this issue is largely devoted to Bunton's plans, and includes interviews with George Bond, Scott Carpenter, and Jon Lindbergh; Bunton, interview, San Diego, Calif., Aug. 4, 2002.

237 *often returned to Bat Cave:* Bond, *Papa Topside,* p. 193.

237 *on an old Army cot:* George Foote Bond Jr. and William A. Burch, longtime friend of Bond Sr., interviews at the cabin, Bat Cave, N.C., Oct. 11, 2003.

237 *worked on several books:* Lewis W. Green, "Mountains Keep Calling Him Back," Asheville (N.C.) *Citizen,* Feb. 1, 1970.

CHAPTER 17: THE PROJECTS

Little would be publicly known of the secret saturation diving operations covered in this chapter if not for the publication of *Blind Man's Bluff*, cited in the text and in the notes below. Also cited below are "anonymous sources." These are Navy divers who spoke on condition of anonymity for the reasons explained in the text. Their input in interviews, while often guarded, helped to corroborate and occasionally correct previously published accounts about the diving operations. In some cases their input provided new details. A number of divers involved with the projects declined to be interviewed and would not comment at all. Details of the Navy's rejection of the author's Freedom of Information Act requests are noted below.

Page

238 *were certain aquanauts and support:* Jack Tomsky, taped interview, Escondido, Calif., Dec. 30, 2003; Anonymous sources.

238 Hughes Glomar Explorer: Sherry Sontag and Christopher Drew with Annette Lawrence Drew, *Blind Man's Bluff: The Untold Story of American Submarine Espionage* (New York: HarperPerennial, 2000; page numbers below correspond to this paperback edition; original hardcover edition is by PublicAffairs, New York, 1998), pp. 204–5.

238 *with an international cast:* "Press Handbook," p. 12-1; Barth, *Sea Dwellers,* p. 173.

238 *could have produced real results:* Richard A. Cooper, Ph.D., Sealab III aquanaut in charge of the lobster experiment, interview, Panama City, Fla., March 15, 2005.

238 *provided a further smoke screen:* Tomsky, interview, Dec. 30, 2003; John Piña Craven, *The Silent War: The Cold War Battle Beneath the Sea* (New York: Simon & Schuster, 2001), p. 277.

238 *retrieve the remains:* Anonymous sources, who say that a missile retrieval mission came before the cable tapping, a different chronology than presented in Sontag and Drew, *Blind Man's Bluff*, pp. 183, 186–87.

238 *who knew that "the projects":* Tomsky, interview, Dec. 30, 2003.

239 *called back to duty:* Tomsky, interview, Sept. 1, 2009.

239 *more satisfactory coda:* Ibid.

239 *Inside one of the buildings:* Anonymous sources.

239 *Deep Submergence Rescue Vehicle:* Official U.S. Navy Web site: www.navy.mil/navydata/fact_display.asp?cid=4100&tid=500&ct=4; Lonsdale, *United States Navy Diver,* pp. 284–88; "Navy DSRV launched," *Undercurrents,* March 1970, p. 15.

240 *"U.S. Navy DSRV Simulator":* Navy photograph (in author's possession).

240 Glomar-*style cover story:* Sontag and Drew, *Blind Man's Bluff*, pp. 67, 181;

"Navy Bares Secret Role of M.I. [Mare Island] Sub," Vallejo (Calif.) *Times-Herald*, Sept. 25, 1969, p. 1.

240 *four main compartments:* Anonymous sources.

241 *learn to use this novel system:* Ibid.

241 *known as the "flat fish":* Ibid.

241 *to the desolate Sea of Okhotsk:* Sontag and Drew, *Blind Man's Bluff*, p. 183.

241 *winter months of 1971 and 1972:* Ibid., pp. 181, 191; Anonymous sources.

241 *carefully anchored:* Sontag and Drew, *Blind Man's Bluff*, p. 185.

241 *dropping through the hatch:* Anonymous sources.

241 *range of four hundred feet:* Ibid.; Tomsky, interview, Sept. 1, 2009; Sontag and Drew, *Blind Man's Bluff*, pp. 196, 225.

241 *lowering of a "gondola":* Anonymous sources; Sontag and Drew, *Blind Man's Bluff*, p. 250.

241 *capture of Francis Gary Powers:* James Reston, "U.S. Concedes Flight over Soviet, Defends Search for Intelligence; Russians Hold Downed Pilot as Spy," *New York Times*, May 8, 1960, p. 1.

241 *some lighthearted moments:* Anonymous sources; Sontag and Drew, *Blind Man's Bluff*, p. 193.

242 *Bradley had a hunch:* Sontag and Drew, *Blind Man's Bluff*, pp. 171–74.

242 *device worked by induction:* Ibid., p. 186.

242 *NSA eavesdroppers on board:* James Bamford, *Body of Secrets: Anatomy of the Ultra-Secret National Security Agency from the Cold War Through the Dawn of a New Century* (New York: Doubleday, 2001), pp. 370–71.

242 *record for months or even a year:* Sontag and Drew, *Blind Man's Bluff*, p. 189.

243 *resembled a twenty-foot section of pipe:* Ibid.; also shown in photograph section.

243 *acknowledged that secret saturation dives:* Craven, *The Silent War*, pp. 154, 161, 187–88, 277–79.

243 *published in 2005 with official blessings:* Lonsdale, *United States Navy Diver*, pp. xiii–xv.

243 *Navy rejected two recent requests:* Author's first Freedom of Information Act request filed Aug. 30, 2004. Letter to author from Rear Adm. Joseph A. Walsh, director of Submarine Warfare, dated Nov. 16, 2004, denied the request in its entirety because the requested information "remains classified." Appeal of the decision in author letter of Dec. 7, 2004, was denied in a letter from Capt. A. W. Whitaker IV, deputy assistant Judge Advocate General, dated May 13, 2005. Author's second FOIA request filed May 1, 2008. E-mail from Jenna Watson, Office of Naval Intelligence Freedom of Information/Privacy manager, dated June 9, 2008, denied the request. Appeal of the decision in author letter of July 31, 2008, was denied in a letter from G. E. Lattin, deputy assistant Judge Advocate General, dated Sept. 16, 2008. As of June 2011, the author had received no final response to a third FOIA request, dated Sept. 1, 2010. An official response was due by the following month, according to statute.

243 *bound by their long-standing oaths:* Anonymous sources.

243 *trailblazing* Halibut *missions:* Sontag and Drew, *Blind Man's Bluff*, p. 198.

243 *code-named "Ivy Bells":* Ibid.

243 *first the* Seawolf: Ibid., pp. 223–24; Lonsdale, *United States Navy Diver*, p. 309.

244 *learned of the Okhotsk taps: Blind Man's Bluff*, pp. 251–53.

244 *defector put U.S. officials onto the trail:* Ibid., p. 273.

244 *Pelton was arrested:* Ibid., pp. 274, 276.

244 *convicted of selling intelligence secrets:* Sharon La Franiere, "Sentencing Is Delayed in Pelton Espionage Case," *Washington Post*, Nov. 11, 1986, p. 16.

244 *Russians were able to pluck:* Sontag and Drew, *Blind Man's Bluff*, p. 252.

244 Parche *placed a first tap:* Ibid., pp. 232–33, 237–38; Ed Offley, "Secret Navy Sub Finds New Home at Bangor Base," *Seattle Post-Intelligencer*, Nov. 23, 1994, p. A1.

244 *and others followed:* Sontag and Drew, *Blind Man's Bluff*, pp. 240, 256, 267; Lonsdale, *United States Navy Diver*, p. 310.

244 *divers had to work at greater depths:* Sontag and Drew, *Blind Man's Bluff*, p. 238; Lonsdale, *United States Navy Diver*, p. 310; Tomsky, interview, Sept. 1, 2009.

244 *reading of Soviet military minds:* Sontag and Drew, *Blind Man's Bluff*, p. 197.

244 *not designed for a first nuclear strike:* Ibid., p. 268.

244 *conduit for real-time taps:* Ibid., p. 270.

244 *the* Richard B. Russell: Ibid., pp. 262, 297.

244 *"research and development" equipment:* Lonsdale, *United States Navy Diver*, p. 311.

244 *tap cables and to pick up pieces:* Sontag and Drew, *Blind Man's Bluff*, p. 297.

244 *shifted to other parts of the world:* Ibid.

244 USS Jimmy Carter *came specially equipped:* Ibid., p. 298; Matt Apuzzo, "Last and Largest of the Seawolves, *USS Jimmy Carter* to Be Commissioned," Associated Press State and Local Wire, Feb. 19, 2005.

245 *"multi-mission platform":* Official U.S. Navy Web site, www.navy.mil/list_all. asp?id=17173.

245 *"tactical surveillance":* Journalist 3rd Class Adam Vernon, "*USS Jimmy Carter* Arrives at New Home," Naval Base Kitsap Public Affairs, Department of Defense, U.S. Navy Releases, Nov. 15, 2005.

245 *"intelligence collection":* Master Chief Communications Specialist Gerald McLain and Chief Mass Communication Specialist Terry L. Rhedin, "*USS Jimmy Carter* Gets 'Depermed,'" Northwest Region Fleet Public Affairs, Department of Defense, U.S. Navy Releases, Aug. 17, 2006.

245 *very little doubt among experts:* Norman Polmar, former consultant to senior Navy and Defense Department officials who also worked for the Deep Submergence Systems Project and author of more than thirty books on naval, aviation, and intelligence projects, taped interview, March 17, 2009; official Navy

Web site schematic drawing of USS *Jimmy Carter* showing hull extension with its "innovative ocean interface module for accommodating new capabilities in Naval Special Warfare, tactical surveillance, and mine warfare," online as of June 2009, www.navy.mil/navydata/cno/n87/usw/issue_11/ship_sensors_weapons.html; Bill Sweetman, "Exposing the Spy Sub of the Future—Robot Mini Subs, Navy SEAL Launches, High-Tech Espionage: The Submarine of the 21st Century Has Arrived," *Popular Science*, Aug. 1, 2005, p. 81.

245 *whose namesake approved:* Sontag and Drew, *Blind Man's Bluff*, p. 229.

245 *continue the cable-tapping tradition:* Ibid., p. 298; Apuzzo, "Last and Largest of the Seawolves," Feb. 19, 2005; John J. Lumpkin, "USS *Carter* Will Be Able to Tap Undersea Cables, Experts Say," *AP Worldstream*, Feb. 18, 2005; James Bamford, "NSA Confidential," *Newsweek* Web exclusive, May 19, 2001; Bob Drogin, "NSA Blackout Reveals Downside of Secrecy," *Los Angeles Times*, March 13, 2000, p. 1.

245 *fiber-optic lines too problematic:* Neil King Jr., "As Technology Evolves, Spy Agency Struggles to Preserve Its Hearing; Its Limited Success in Tapping Undersea Cable Illustrates Challenges Facing NSA," *Wall Street Journal*, May 23, 2001, p. A1.

245 *individual lines can be tapped:* Govind P. Agrawal, interviews, Oct. 8 and 9, 2009; e-mail to author, Oct. 22, 2009; Agrawal, author of *Fiber-Optic Communication Systems* and coauthor and series editor of *Undersea Fiber Communication Systems*, is professor of optics at the University of Rochester's Institute of Optics.

245 *usefulness of a mobile workshop:* King, "As Technology Evolves, Spy Agency Struggles," p. A10.

CHAPTER 18: ANSWERS AND QUESTIONS

Page

247 *depth of eighteen hundred feet:* Bennett and Rostain, "The High Pressure Nervous Syndrome," in *The Physiology and Medicine of Diving*, 4th ed. (London: W. B. Saunders, 1993), pp. 203, 213; *U.S. Navy Diving Manual*, revision 4, Jan. 20, 1999, p. 1–24.

247 *get worse over time:* Bennett and Rostain, "The High Pressure Nervous Syndrome," in *The Physiology and Medicine of Diving*, 4th ed., p. 214.

247 *A similar British experiment:* Ibid., pp. 202, 213.

247 *Navy established a diving limit:* Peter Bennett, interview, June 8, 2009.

247 *called Atlantis:* Bennett and Rostain, "The High Pressure Nervous Syndrome," in *The Physiology and Medicine of Diving*, 4th ed., pp. 218–22.

248 *to cover the substantial expenses:* Peter Bennett, interview, May 10, 2007.

248 *Duke's new experimental chambers:* Michael Mulcahy, "Duke's F. G. Hall Laboratory—A Center for Excellence in Experimental Diving," *Sea Technology*, December 1979, p. 25.

248 *shot of Scotch:* Bennett and Rostain, "The High Pressure Nervous Syndrome," in *The Physiology and Medicine of Diving,* 4th ed., p. 218.

248 *trimix seemed to alleviate:* Ibid., pp. 215, 218.

249 *made eating uncomfortable:* Bennett, interview, May 10, 2007.

249 *Send a robot or a machine:* Swann, *The History of Oilfield Diving,* pp. 721, 724, 736.

249 *After the final Atlantis dive:* Bennett and Rostain, "The High Pressure Nervous Syndrome," in *The Physiology and Medicine of Diving,* 4th ed., p. 221.

249 *these Comex dives:* Guy Imbert, Thomas Ciesielski, and X. Fructus, "Safe Deep Diving Using Hydrogen," *Marine Technology Society Journal* 23 (December 1989): 29; *Hydra 8,* an in-house Comex video of the dives.

249 *replicating some common oil field jobs: Hydra 8* video.

249 *method for adding hydrogen:* Imbert et al., "Safe Deep Diving Using Hydrogen," p. 26; Bennett and Rostain, "The High Pressure Nervous Syndrome," in *The Physiology and Medicine of Diving,* 4th ed., p. 226; Claude Gortan, manager, Comex Hyperbaric Experimental Center, interview at Comex, Oct. 29, 2004.

250 *push downward largely ended:* Bennett and Rostain, "The High Pressure Nervous Syndrome," in *The Physiology and Medicine of Diving* (New York: W. B. Saunders, 2003), 5th ed., p. 337.

250 *West German government had poured:* Ibid., p. 339; Bennett, interview, June 11, 2009.

250 *real modern-day marvel:* Bennett, interview, June 11, 2009.

250 *continuing through 1990:* Ibid.; Bennett and Rostain, "The High Pressure Nervous Syndrome," in *The Physiology and Medicine of Diving,* 5th ed., p. 339.

250 *experimental series called Hydra:* "COMEX Hyperbaric Experimental Centre, 1965–2000, 36 Years of Deep Diving Development, from Helium to Hydrogen," an in-house pamphlet that catalogues the highlights of company research, April 12, 2001.

251 *divers were compressed slowly:* B. Gardette, J. Y. Massimelli, M. Comet, C. Gortan, H. G. Delauze, "HYDRA 10: A 701 msw Onshore Record Dive Using 'Hydreliox,'" *Proceedings of the 19th Annual Meeting of the European Underwater and Baromedical Society,* Trondheim, Norway, 1993, p. 32; Bennett and Rostain, "High Pressure Nervous Syndrome," in *The Physiology and Medicine of Diving,* 5th ed., p. 343.

251 *Just one of the three:* Gardette et al., "HYDRA 10," p. 36.

251 *minor signs of HPNS:* Ibid.

251 *held at 2,200 feet:* Ibid., p. 33; Bennett and Rostain, "High Pressure Nervous Syndrome," in *The Physiology and Medicine of Diving,* 5th ed., p. 343.

251 *they were not eliminated:* Gardette et al., "HYDRA 10," p. 36.

251 *forty-two consecutive days:* Ibid.

251 *deepest diver in the world:* "COMEX Hyperbaric Experimental Centre," p. 12; H. G. Delauze, B. Gardette, C. Gortan, "COMEX Hydra Program: 35 Years

of Research on Hydrogen," *MAJ* (?), February 2000, p. 4 (copy in author's possession).

251 *at least as meaningful a record:* Bennett, interview, May 10, 2007.

251 *Bond had optimistically predicted:* George Bond, "Man-in-the-Sea Program" (paper presented to Association of American Medical Colleges, Seventeenth annual session, San Diego, Calif., Sept. 29, 1966, in author's possession), p. 10.

252 *pressure created the ultimate barrier:* Bennett and Rostain, "High Pressure Nervous Syndrome," in *Bennett and Elliott's Physiology and Medicine of Diving*, 5th ed., p. 349; Peter Bennett, interviews, May 10, 2007, and April 29, 2009.

252 *some people, like Theo:* Bennett and Rostain, "The High Pressure Nervous Syndrome," in *Bennett and Elliott's Physiology and Medicine of Diving*, 5th ed., p. 328.

252 *no one knew for sure:* Bennett, interview, April 29, 2009.

252 *breathing a liquid instead:* J. A. Kylstra, "Liquid Breathing and Artificial Gills," in *The Physiology and Medicine of Diving and Compressed Air Work*, 2nd ed., p. 155.

252 *Bond, too, once saw great potential:* Bond, "Man-in-the-Sea Program" (paper presented to AAMC), pp. 10–11; Bond, "Sealab III: Next Step Toward the Depths," *Astronautics & Aeronautics*, July 1967, p. 88.

252 *still couldn't escape the effects of pressure:* Bennett, interview, June 8, 2009.

253 *"carried Cousteau's bags":* Delauze, interview, Comex headquarters, Marseille, Oct. 29, 2004.

253 *spent millions:* Michael Balter, "SAGA Submarine Fulfills Fantasy of Undersea Life," *International Herald Tribune*, Oct. 31, 1990, p. 16; Delauze, e-mail to author, Oct. 19, 2009, says Comex and IFREMER each spent $20 million.

253 *Cousteau asked for royalties:* Swann, *The History of Oilfield Diving*, p. 536.

253 *Sous-marin d'Assistance:* J. Mollard and D. Sauzade, "SAGA, premier sous-marin industriel autonome," *IFREMER, Actes de Colloques*, no. 12, 1991, p. 95, www.ifremer.fr/docelec/doc/1990/acte-1166.pdf; Delauze, e-mail to author, Oct. 20, 2009, confirming proper wording of the sub's French name, which can be found in several forms and is referred to only by acronym in this paper.

253 *hoped to build a larger version:* Dunoyer de Segonzac, *Un conquérant sous la mer*, p. 182.

253 *submerged for up to four weeks:* Delauze, e-mail to author, Oct. 16, 2009.

253 *depths down to nearly fifteen hundred feet:* Ibid.

253 *launch of the first SAGA:* Swann, *The History of Oilfield Diving*, p. 536.

254 *end of August 1989:* Yvon Calvez, "Saga: vingt ans d'histoire sous-marine," *Navires Ports et Chantiers*, October 1989, p. 514.

254 *depth of 1,040 feet:* Mollard and Sauzade, "SAGA, premier sous-marin industriel autonome," p. 101.

254 *relics from a Roman shipwreck:* Capt. Jacques-Yves Cousteau, "Fish Men Discover a 2,200-year-old Greek Ship," *National Geographic*, Jan. 1954.

254   *archaeological missions:* Delauze, interviews, Marseille, Nov. 2 and 3, 2004;
      Sténuit, interview, Nov. 26, 2004.

254   *where Cousteau shot the film footage: Omnibus,* no. 16, Part 1, on CBS, Jan. 17,
      1954, viewed at Museum of Television and Radio, New York City, March 28,
      2003.

254   *with Nicolas Hulot:* Delauze, e-mail to author, April 27 and May 20, 2009.

254   *storage hangar:* Delauze, e-mail to author, April 27, 2009.

254   *solution looking for a problem:* Swann, *The History of Oilfield Diving,* p. 538.

254   *could not foot the bill alone:* Ibid., p. 536.

255   *NOAA got a million and a half dollars:* "A History of NOAA," p. 12.

255   *vessel envisioned was called Oceanlab:* Ibid.

255   *four hundred research proposals:* "Oceanlab Concept Review—Report of a
      Study by a Subcommittee of the Ocean Sciences Board," National Academy
      of Sciences, Washington, D.C., 1980, p. 9.

255   *got a look at the plans:* "A History of NOAA," p. 12.

256   *Russians were planning:* Lee H. Boylan, "Underwater Activities in the Soviet
      Union" (revision of a paper prepared for the Workshop on Medical Aspects
      of Noncombatant Submersible Operations, San Diego, Calif., Nov. 19–20,
      1974), pp. 45, 56–60; Miller and Koblick, *Living and Working in the Sea,*
      p. 406.

256   *press scarcely took note:* United Press International, "$21.5 Million Lab Is
      Being Designed for Oceanic Studies," *New York Times,* July 24, 1977, p. 37.
      This brief wire story, which contains the *Times*'s only mention of Oceanlab,
      appeared below the daily weather reports and forecast. Reports in other
      media are similarly rare.

256   *Ed Link died on Labor Day:* Van Hoek with M. C. Link, *From Sky to Sea,*
      p. 320.

256   *George Bond died:* Author visit to Bat Cave, Oct. 10, 2003; Sara L. Bingham,
      "Dr. George Bond Was a True Pioneer" (Hendersonville, N.C.) *Times-News,*
      Jan. 4, 1983, p. 5.

256   *dedicated to Captain Bond:* Navy Experimental Diving Unit Dedication
      Ceremony program, May 17, 1991 (in author's possession); bronze plaque
      at the facility; Bond, *Papa Topside,* Foreword by Walt Mazzone, p. ix.

256   *two more Conshelf habitats:* Capt. Jacques-Yves Cousteau, "At Home in the
      Sea," *National Geographic,* April 1964, p. 469.

256   *Cousteau died in 1997:* Gerald Jonas, "Jacques Cousteau, Oceans' Impresario,
      Dies," *New York Times,* June 26, 1997, p. 1.

256   *eased out of saturation diving:* Lonsdale, *United States Navy Diver,* p. 298.

257   *eighteenth-century designs:* Davis, *Deep Diving and Submarine Operations,* p.
      583.

257   *Modern one-atmosphere suits:* U.S. Navy Lt. Mike Thornton, Dr. Robert
      Randall, and Kurt Albaugh, P.E., "Then and Now: Atmospheric Diving

Suits," *UnderWater*, March/April 2001, www.underwater.com/archives/arch/marapr01.01.shtml; Swann, *The History of Oilfield Diving*, p. 742.

257 *record depth of two thousand feet:* Mark G. Logico, "Navy Diver Sets Record with 2,000 Foot Dive," www.military.com/features/0,15240,108883.00.html.

257 Pigeon *and the* Ortolan: Lonsdale, *United States Navy Diver*, p. 299.

257 *salvage of the USS* Monitor: Ibid., p. 299; oceanexplorer.noaa.gov/explorations/02monitor/monitor.html; also oceanexplorer.noaa.gov/explorations/monitor01/monitor01.html.

257 *rented a saturation system:* Lonsdale, *United States Navy Diver*, pp. 299, 303.

257 *sunk in gunnery practice:* Ibid., p. 298.

257 *cut up for scrap:* Jim Osborn, former Navy engineer and Sealab III aquanaut, interview, Panama City, Fla., March 11, 2005; Bond, *Papa Topside*, p. 189, apparently errs in the assertion that the habitat was scuttled in the Santa Barbara Channel.

258 *Bob Barth was one of the many:* Barth, interviews, Dec. 3, 2004, and Dec. 13, 2006.

258 *marked "Dinosaur Locker":* Barth, interview at Navy Experimental Diving Unit, March 10, 2003.

258 *the very PTC:* Barth, interview, April 8, 2009.

258 *moved down to the Florida Keys:* Ibid.

258 *Marinelab:* Miller and Koblick, *Living and Working in the Sea*, pp. 409–11.

258 *La Chalupa:* Ibid., pp. 123–34.

258 *Jules' Undersea Lodge:* Ibid., pp. 415–19.

258 *names in the guest book:* Jules' Undersea Lodge Web site, www.jul.com/mediainfo.html.

258 *Ian Koblick, has long been active:* Miller and Koblick, *Living and Working in the Sea*, p. 439.

259 *world's only undersea research station:* Aquarius Web site, uncw.edu/aquarius.

259 *initially dubbed the George F. Bond:* Craig Cooper, Aquarius manager from 1991 to 2009, e-mail to author, Feb. 11, 2011.

259 *run on a minimal budget:* Cooper, interview, April 24, 2009.

259 *affixed to the seabed in 2007:* Ibid.

259 *This surface-based unit:* Cmdr. Scott W. Thomas, "Saturation Fly-Away Diving System (SAT FADS)," *Faceplate*, October 2010, p. 8.

260 *staff had developed way stations:* Cooper, taped interview, Feb. 9, 2011.

260 *in early May 2009 to create a haven:* Ibid.

260 *Smith was a few paces away:* Corey Seymour, Navy hospital corpsman senior chief, interview, Feb. 19, 2011.

260 *Smith made it clear that he was okay:* Ibid.; Cooper, interview, Feb. 9, 2011; Robert Silk, "Aquarius Diver's Death Remains a Question," *The* (Key West, Fla.) *Citizen*, May 9, 2009, keysnews.com/node/13119.

260 *Minutes had passed:* Seymour, interview, Feb. 19, 2011.

261 *hurriedly picked up Smith:* Ibid.

261 *Attempts to resuscitate:* Ibid.

261 *given a clean bill of health:* Autopsy Report for Dewey Dwayne Smith, from the Office of the Medical Examiner, Marathon, Fla., May 6, 2009; e-mailed from E. Hunt Scheuerman, M.D., District 16 Medical Examiner, to author, Feb. 8, 2011.

261 *A triathlete:* Seymour, interview, Feb. 19, 2011.

261 *a meticulous diver:* Autopsy Report; Cooper, interview, Feb. 9, 2011.

261 *worry that the NOAA habitat program:* Cooper, interview, Feb. 9, 2011.

INDEX